Geobotany Studies

Basics, Methods and Case Studies

The series includes outstanding monographs and collections of papers on a range of topics in the following fields: Phytogeography, Phytosociology, Plant Community Ecology, Biocoenology, Vegetation Science, Eco-informatics, Landscape Ecology, Vegetation Mapping, Plant Conservation Biology, and Plant Diversity. Contributions should reflect the latest theoretical and methodological developments or present new applications on large spatial or temporal scales that will reinforce our understanding of ecological processes acting at the phytocoenosis and vegetation landscape level. Case studies based on large data sets are also considered, provided they support habitat classification refinement, plant diversity conservation or vegetation change prediction. **Geobotany Studies: Basics, Methods and Case Studies** is the successor to Braun-Blanquetia, a journal published by the University of Camerino from 1984 to 2011 in cooperation with the Station Internationale de Phytosociologie (Bailleul, France) and the Dipartimento di Botanica ed Ecologia (Università di Camerino, Italy) and under the aegis of the Société Amicale Francophone de Phytosociologie, the Société Française de Phytosociologie, the Rheinold-Tüxen-Gesellschaft and the Eastern Alpine and Dinaric Society for Vegetation Ecology. This series promotes the expansion, evolution, and application of the invaluable scientific legacy of the Braun-Blanquet school.

Gheorghe Coldea • Vasile Cristea

The Vascular Plant Communities of the Retezat National Park (Southern Carpathians)

Springer

Gheorghe Coldea
Department of Ecology
Institute of Biological Research
Cluj-Napoca
Cluj-Napoca, Romania

Vasile Cristea
Department of Taxonomy and Ecology
Babeş-Bolyai University
Cluj-Napoca, Romania

ISSN 2198-2562 ISSN 2198-2570 (electronic)
Geobotany Studies
ISBN 978-3-031-05617-8 ISBN 978-3-031-05618-5 (eBook)
https://doi.org/10.1007/978-3-031-05618-5

This Springer imprint is published by the registered company Springer Nature Switzerland AG
The registered company address is: Gewerbestrasse 11, 6330 Cham, Switzerland

1887–1971

This book is dedicated to the memory of the Academician Prof. Dr. Alexandru Borza, extraordinary botanist and phytocoenologist, the founder of the Retezat National Park and of the Romanian phytocoenology.

Preface

The flora of the Retezat Mountains, which are located in the southwestern part of the Southern Romanian Carpathians, was continuously influenced by species migration from different regions of Europe during the deluge. The floristic richness of the Retezat Mountains captivated the interest of botanists since the beginning of the nineteenth century (Baumgarten 1816; Rochel 1828; Heuffel 1858; Schur 1866), and their interest in this area continued through the twentieth century (Jávorka 1911; Pax 1898, 1919; Zahn 1928; Prodan 1930, 1931; Borza 1934; Csűrös et al. 1956a, b). Based on available literature and his personal research, Nyárády (1958) had published the first global enumeration of vascular plants in Retezat, including 920 species. Subsequent research was conducted mainly in the Calcareous Retezat. Csűrös and others (1962, 1972) updated the floristic list to a total of 1152 species and 104 subspecies.

The dominant floristic element in the flora of Retezat is the Alpine-Middle European (37.8%), while the Arctic-Alpine-Boreal and Arctic-Atlantic element is present in small percentage (4.6%) (*Salix herbacea, Carex atrata, Dryas octopetala, Silene acaulis, Veronica alpina, Cerastium alpinum, C. fontanum, Lloydia serotina, Loiseleuria procumbens, Juncus trifidus, Carex rupestris, Epilobium anagallidifolium,* and *E. alsinifolium*). Compared to these, the Balkan and Caucasian elements are present in higher numbers (12.2%) in Retezat (*Alyssum repens, Bruckenthalia spiculifolia, Campanula abietina, C. transsilvanica, Carduus kerneri, Carex nigra* subsp. *dacica, Pseudorchis frivaldii, Hypericum richeri* subsp. *grisebachii, Lathyrus hallersteinii, Phyteuma confusum, Plantago gentianoides, Poa media, Pulmonaria rubra, Saxifraga pedemontana* subsp. *cymosa, Senecio doronicum* subsp. *transsilvanicum, Symphyandra wanneri, Saxifraga marginata, Veronica bachofenii,* and *V. baumgartenii*). Also, the pan-Carpathian endemic elements and the endemic taxa in the Southeastern Carpathians are well represented (12.5%) in the Retezat Mountains. Pan-Carpathian and Dacian paleoendemics are supposed to originate from Tertiary taxa and have heterogeneous florogenetic origin (Boşcaiu 1971). The high speciation potential of the Retezat area is also revealed by the exuberant variation of the *Hieracium* populations, from which numerous local and regional microendemisms have been described (Zahn 1928; Nyárady 1958).

All these floristic elements have left their phytogeographical markings within the plant communities, particularly from the types described in this massif. Borza (1934), in his first phytocoenological work on Retezat, noted that some of the associations differed floristically from those in the Tatra Mountains and the Alps, which is why he considered them regional variants specific to the massif or to the Southern Romanian Carpathians. Impressed by the floristic richness, Prof. Alexandru Borza managed to establish the Retezat National Park in 1935. He also intended writing a geobotanical monograph of the park that he unfortunately did not achieve, although at his urging, research in the Retezat National Park resumed for a short period of time in 1965. Some data on several types of vegetation in the Calcareous Retezat area were published in specialized journals of the time (Boşcaiu et al. 1972, 1978; Resmeriță 1987). In this work, we analyzed syntaxonomically the multitude of unpublished relevés, which were later completed with new ones for certain plant associations that were less outlined floristically, to elaborate the monograph of the vegetation of Retezat. Based on 545 relevés, representative for the habitat types occurring in Retezat, we distinguished and described 67 plant associations on floristic and ecological grounds. These were classified into 13 vegetation classes, 19 orders, and 28 alliances. Of the 67 plant associations described in the paper, three were described as new syntaxa: *Salici kitaibelianae–Dryadetum octopetalae, Phyteumo confusi–Juncetum trifidi*, and *Aconito taurici–Rumicetum alpini*. Also, the relict association *Vaccinio myrtilli–Pinetum sylvestris* was identified and described from the studied area as a new sub-association: *Jovibarbetosum heuffelii*. This sub-association is individualized by several regional species, like *Jovibarba heuffelii, Moehringia pendula, Dianthus tenuifolius, Symphyandra wanneri*, and *Silene nutans* subsp. *dubia*.

To support the classification of the 67 plant associations within the European syntaxonomic framework, we briefly characterized floristically and ecologically the higher rank syntaxa.

Through such an arrangement and presentation of community floristic composition, we consider that we have revealed more prominently the regional specificity of the distinguished syntaxa and their similarity or differences compared to their counterparts described in the neighboring geographical areas.

Cluj-Napoca, Romania

Gheorghe Coldea
Vasile Cristea

About the Book

Situated in the western part of the Southern Romanian Carpathians, the **Retezat National Park,** founded by Alexandru Borza in 1935, covers 38,300 ha, while in the north-western part of the Retezat Massif the "Gemenele Reserve" (1947 ha) was established, which is strictly protected.

The heterogeneous geological structure, comprising both crystalline and calcareous rocks and the diverse landscape of the Retezat Massif, incessantly influenced the evolution of the flora and vegetation of this area, during the Pleistocene.

The floristic studies carried out during the nineteenth and twentieth centuries in the Retezat Massif identified 1152 plant species and 104 subspecies from the *Cormobionta* sub-regnum. Of these, about 12% are endemic Carpathian and Dacian-Balkan taxa that provide the local specificity of the associations hosting them. The phytocoenological research that we performed resulted in the description of 67 associations that we grouped into 28 alliances, 19 orders, and 13 vegetation classes. These classes are *Asplenietea trichomanis, Thlaspietea rotundifolii, Salicetea herbaceae, Montio-Cardaminetea, Scheuchzerio-Caricetea fuscae, Oxycocco-Sphagnetea, Molinio-Arrhenatheretea, Caricetea curvulae, Loiseleurio-Vaccinietea, Elyno-Seslerietea, Mulgedio-Aconitetea, Carpino-Fagetea*, and *Vaccinio-Piceetea*. We mention that the following plant associations that we are describing are new syntaxa: *Phyteumo confusi-Juncetum trifidi, Salici kitaibelianae-Dryadetum octopetalae*, and *Aconito taurici-Rumicetum alpine*. For the protection of some rare plant species and vulnerable plant associations from the area of the "Limestone Retezat," we propose the delimitation of two new botanical reserves, the first on the peaks Albele-Scorota-Piatra Iorgovanului (3500 ha) and a second one on the Piule Mountain (2500 ha).

Contents

Chapter 1
Physical, Geographical and Geological Description of the Retezat Mountains

1.1 Location and Boundaries of the Retezat Mountains

Located in the northwestern part of the Southern Romanian Carpathians, the Retezat Mountains stand out through altitudes surpassing 2500 m (Peleaga peak 2509 m, Păpuşa peak 2508 m) and occupy a total area of approx. 700 km^2, between 45°15′15″–45°28′52″ N and 22°43′31″–23°22′41″ E.

The Retezat Mountains come into contact to the north with the Haţeg depression, at altitudes of 500–700 m, on the southern alignment of Brazi, Râul de Mori, Nucşoara, Coroieşti, Hobiţa. To the east the valleys of Râul Bărbat and Pilugului separate the Retezat Mountains from Tulişa Massif (1972 m). The southern limit of the unit is the Jiu Românesc valley, which separates it from the Vâlcan-Oslea Mountains. Next, also to the south, the Soarbele and Paltina valleys separate the Retezat from the Godeanu Mountains (2230 m). The western limit of the Retezat is the Râului Mare valley, which separates it from the Ţarcu Mountains (Ţarcu peak 2186 m, Căleanu peak 2196 m) (Mihăilescu 1963).

1.2 Geological Structure

The geological substrate of the Retezat Mountains is mainly composed of crystalline rocks but sedimentary formations are present on the southern and south-eastern sides. The crystalline foundation of the autochthonous Danubian consists of massive eruptive granitoid and epimetamorphic crystalline schists attributed to the Proterozoic. The crystalline schists are poorly metamorphosed and belong to several series with local names. They include chlorite-biotite schists and micaschists (Rof Series), quartzite schists and biolytic filites (Râuşor Series), the Drăgşan Series (with a

G. Coldea, V. Cristea, *The Vascular Plant Communities of the Retezat National Park (Southern Carpathians)*, Geobotany Studies,
https://doi.org/10.1007/978-3-031-05618-5_1

sericite-chlorite and an amphibolytic complex), and also the Tulişa Series, made up of conglomerate complexes and quartzite schists.

Two large eruptive masses are inserted in the mass of crystalline schists from the central part of the Retezat massif, consisting of granodiorites and biotites, and to the south the Buta massif, consisting of granodioritic gneisses of magmatic origin (Pavelescu 1958; Codarcea and Pavelescu 1963).

The sedimentary formations are present on the southern side of the Retezat (Stănuleţii, Piatra Iorgovanului, Albele, Piule) and consist of reef limestones from the Upper Jurassic and Lower Cretaceous, reason for the name Retezatul Calcaros (Calcareous) given by geologists and Retezatul Mic by geomorphologists, because the subunit does not exceed 2100 m altitude.

The Getic nappe is present in the southern part of the Jiului valley basin and to the north of Retezat, in the southern part of the Haţeg basin, between the localities Hobiţa and Nucşoara (Murgoci 1912). It consists of mesometamorphic crystalline schists of sedimentary origin, strongly metamorphosed at great depths.

The biotite pegmatites and granites, biotite micaschists, and biotite amphibolites predominate in the structure of these schists.

The complex geological structure of the Retezat Mountains is reflected both in the terrain's morphology and the floristic diversity of the plant cover.

1.3 Oro-hydrographic Characterization

Through their altitude, massiveness, and variation of the landscape forms, the Retezat Mountains have an outstanding geographical individuality, dominating the neighboring mountains by at least 200 m. Their altitudes range between 500 m, in the area where the Râul Mare leaves the montane surroundings, and 2509 m (Peleaga peak), corresponding roughly to 2000 m of altitudinal zonation of the mountain landscape. The fluvial relief belt, where river modeling predominates, stretches between the mountain base (550–600 m) up to 1500 m, followed by the glacial one, above 1550 m. The massiveness of the Retezat is given by the existence of two parallel ridges with west-east development (Schreiber and Sorocovschi 1993).

The northern ridge, higher in the central part, runs between the peaks Zlata (2142 m), Zănoaga (2269 m), Şesele (2323 m), Judele (2382 m), Bucura (2436 m), Peleaga (2509 m), Păpuşa (2508 m), and Lănciţa (2095 m) on a distance of 18 km, with altitudes above 2000 m. Secondary peaks originating from this ridge develop to the north, towards the Haţeg Depression, with several high peaks such as Retezat (2485 m), Pietrele (2206 m), Valea Rea (2309 m), and Vârful Mare (2455 m). Some shorter ridges run also to the south, the one towards Slăvei peak (2340 m) and Muchia Slăveiului (2115 m) being the most important. The northern ridge, with the aforementioned peaks, regularly exceeding 2100 m altitude and consisting mainly of crystalline rocks, is also known as Retezatul Mare. This subunit is characterized by the maximum extension of glacial and cryo-nival relief (Mihăilescu 1963).

The southern ridge of the Retezat Mountains comprises mostly reddish-white limestones, and it includes in the southwestern sector the peaks Stănuleț (2030 m), Piatra Iorgovanului (2016 m), Albele (2013 m), Scorota (2080 m), Piule (2083 m), Drăgșanu (2080 m) and continues to the northeast with the peaks Custura (2463 m) and Gruniu (2302 m). The lower altitudes in this sector and the steeper slopes resulted here in a less developed glacial landscape; instead, the karst landscape predominates, resulting in the name Retezatul Mic or Retezatul Calcaros (Calcareous) given by specialists (Schreiber and Sorocovschi 1993).

The two main peaks of the massif, positioned SW-NE, open the region to the circulation of Atlantic and Mediterranean air masses, resulting in the highest humidity and runoff in the Southern Romanian Carpathians.

The river networks in the massif follow two main directions. Most of the territory (76.4%) is drained by the rivers from the northern part (Râul Bărbat, Pârâul Alb, Râușor, Șerel, Paroș, Sălaș) and western part—Râul Mare with its tributaries (Lăpușnicu Mare, Zlata, Râușor, Nucșor) that merge in Strei and then pour into Mureș. The southern part of the massif, with a smaller area (23.6%) is drained by the Jiul de Vest, through the tributaries Buta, Valea Lazărului, and Pilugu (Ujvari 1972).

During the last Ice Age, abundant snows that had accumulated in the pre-existing valleys changed into massive glaciers in the Riss and Würm periods and generated a post-glacial cryo-nival relief consisting of cirques, lakes and glacial valleys of Alpine-Pyrenean type. The 58 glacial lakes from the Retezat Mountains (Pișota 1971) were formed around 2000 m altitude and include some of the largest (Bucura 8.8 ha, Zănoaga 6.5 ha, Tăul Negru 4.04 ha) and the deepest (Zănoaga 29 m, Tăul Negru 24.8 m, Galeș 20.5 m and Bucura Mare 15.7 m) lakes in the Southern Romanian Carpathians. The glacial lakes and the originating streams are usually grouped in limnological complexes located in the river basins: Bărbat (Tăul Custura, Tăul Țapului), Nucșoara (Galeș, Pietrele, Stânișoarei), Zlătuia (Gemenele, Tăul Negru, Tăul Secat, Zănoaga Mică), Judele (Zănoaga, Judele, Tăul Ascuns, Tăul Urât, Tăul Răsucit), and Bucura (Bucura, Florica, Viorica, Ana, Lia, Tăul Agățat, Peleaga, and Peleguța).

The lakes and the very rich hydrographic network of the Retezat Mountains have created favorable conditions for the development of hygrophilous and meso-hygrophilous plant communities with distinct particularities compared to other mountains of the Southern Romanian Carpathians.

1.4 Pedo-climatic Characterization

The lack of meteorological stations in the Retezat Mountains area determined the climatologists to use data from the meteorological stations from the neighboring massifs (Țarcu, Parâng, Semenic) for its climatic characterization (Fărcaș and Sorocovschi 1993). The following concise description is based on those data.

The arrangement of the ridges in the massif on two main SW–NE alignments and their different position in relation to the neighboring massifs determine the western and north-western flanks to be exposed to wet Atlantic circulation while the south-western and southern flanks are prone to the Mediterranean cyclonic circulation.

The average annual temperatures are between 6 °C at the base of the massif and −2.3 °C at the top. Cold islands, with average annual temperatures below 0 °C are found on peaks higher than 2100 m. The air temperature drops vertically by 0.3–0.4 °C in winter and by 0.5–0.7 °C in summer. The coldest month is January, the warmest being July (below 1100 m) and August (above 1100 m). The average air temperature drops to −5 °C in January at the mountain base and to −11 °C on the summits. The average monthly temperature oscillates around −7 °C at the upper limit of the forest (1700–1800 m). Cold islands with temperatures below −10 °C appear on the peaks above 2200 m (Peleaga, Păpușa, Retezat, Bucura). The average temperature varies in July between 16 °C at the mountain base and 6 °C on higher (up to 1900 m) peaks.

Frost is an almost permanent phenomenon as on higher peaks it may also occur in summer. The number of frost days in the alpine area (2200–2500 m) is 250–275 per year, decreasing to 175–200 per year in the lower areas (800–1500 m). The summer isotherm of 10 °C, specific to the upper limit of trees, is located at 1900 m altitude on the southern flank of the massif and at approx. 1800 m altitude on the northern one (Geanana 1975).

Atmospheric precipitation is unevenly distributed over the mountain range, depending on the advance of Atlantic and Mediterranean wet air masses. The average annual rainfall ranges between 900 mm at the base of the mountain and 1300 mm at 1800 m altitude. The annual rainfall decreases to 1130 mm at 2500 m altitude. The highest amounts of rainfall occur on the western and northwestern slopes whereas sheltered, eastern and north-eastern slopes receive 200–300 mm less rainfall at the same altitude. The maximum precipitation (120–150 mm) occurs in June as opposed to October-November (50–70 mm). Most precipitation falls as snow in the cold season. The maximum frequency of snowy days occurs between December-February. The average thickness of the snow layer increases from 70 to 90 cm at the base of the mountain to up to 1–3 m above 2000 m altitude. On the northern slopes, in wind-sheltered areas, the snow layer persists until the end of the summer season. In these microthermal areas, chionophilic phytocoenoses with a regional and local specificity develop.

Wind is a dominant element of the mountain climate, determining the distribution of precipitation over the massif's area. The ridges and high plateaus are often influenced by an advective air circulation, being exposed to the currents that cross the region. The average annual wind speed is 6–7 m/s on the peaks, with a maximum in February (14 m/s) and a minimum in July-August (4–5 m/s). The dominant wind directions during the year are from north (19.7%) and south (17.1%), according to the weather records of the Țarcu peak (Boșcaiu 1971). Due to the high wind speeds, at high altitudes snow is scattered leaving soil exposed to strong and long lasting frosts. Such phenomena have favored the development and maintenance of arctic-alpine phytocoenoses of primary grasslands and arctic-alpine dwarf shrubs.

The relative air humidity increases from 74–75% at the base of the massif to 85–87% on the summits. The highest values (90–92%) appear in May-June and the lowest (80%) in October. The maximum moisture deficit occurs in October, when the number of days with humidity below 30% is the highest. Fog distribution and stagnation of in the different sectors of the massif is the key element in vegetation stratification. The annual frequency of fogs increases from 50–75 days at the base of the massif up to 250 days at 1800 m altitude and almost 300 days on the highest peaks. Summit fog has favorable consequences for the development of phytocoenoses since the waves of humid air cause continuous wetting of the substrate due to mist droplets and drizzle that fall on soil and moisten plants.

The pedogenetic processes in the massif were determined, on the one hand, by the bioclimatic factors specific to the altitudinal belts, and on the other hand, by the nature of the lithological substrate underlying the soil. The predominance of acid, crystalline and granitic substrate in the Retezat Mountains favored the development of oligobasic pedogenetic series as a normal altitudinal succession in the formation of zonal automorphic soils. In relation to these, soils formed on eubasic lithological substrate have an intrazonal character. The altitudinal succession of soils coincides, in general, with the altitudinal zonation of vegetation in the Retezat Mountains. In the lower mountain belt of the Retezat Mountains, eutricambosols and shallow lithosols are associated with mixed forests (sessile oak, hornbeam, linden and sessile oak, pine and beech) developed on acidic rocks with a coarse texture. They are located in the basin of the Râul Mare valley, at altitudes between 650 and 800 m, on the eastern and south-western slopes. In the middle mountain belt, districambosols (acid brown and argilluvial brown soils) and eutricambosols (eu- and mesobasic brown soils with mull humus type) are associated with woody vegetation dominated by beech and beech mixed with fir and spruce. They are present between 800 and 1300 m and evolved both on crystalline schists and granites in the western part of the Retezat and on limestone in its southern part. The largest areas consist of districambosols with mull or moder humus types, moderately deep, weakly skeletal, with high trophicity and low-moderate acidity (pH = 5.5–6.5). Both soil trophicity and climatic conditions in these areas offer favorable conditions for the development of pure and mixed beech forests. In the upper mountain belt of the massif (1300–1800 m) districambosols (acidic brown soils) and cryptopodzols (brown acidic cryptospodic soil) alike are associated with dominating spruce forests, whereas lithosols have a patchy distribution, on steep slopes. They are superficial or medium-deep soils, semi-skeletal, with moder type humus, small-medium edaphic volume, and high acidity (pH = 4.5–5.5). The pedo-climatic specificity of these forests in the Romanian Carpathians was presented in detail in a synthesis paper (Chiriță et al. 1967). In the subalpine belt (1800–2200 m), podzolic soils (typical podzol, feriiluvial podzol, and lithic podzol) are associated with mountain pine (*Pinus mugo*) and alpine juniper (*Juniperus communis* subsp. *alpina*) shrubs. These soils have simple morphology of genetic horizons (Au-R), with humiferous horizon (A umbric) no thicker than 15–20 cm and strong acidic reaction (pH = 4.2–4.5). The secondary grasslands with *Festuca nigrescens* and *Nardus stricta* installed after scrub removal facilitate soil erosion and drive the pedological process towards lithosols. In the alpine belt (2200–2500 m) of the Retezat Mountains, lithosols (alpine humico-silicatic soils) are associated with primary grasslands

Photo 1.1 Retezat Peak (2485 m) and Ştevia lake (photo M. Ciobanu)

of *Carex curvula*, *Festuca supina,* and *Juncus trifidus* and the dwarf shrubs of *Loiseleuria procumbens* and *Vaccinium gaultherioides*. They are present on acidic substrates, have shallow profile (10–15 cm), usually with an abundant skeletal content. Rendzinas are present in the lower alpine belt (2050–2200 m) on calcareous substrate. They have an intrazonal distribution and a neutral or basic reaction (pH = 6.6–7.3). Plant communities developed on these soils are richer in species as compared to those evolved on acidic substrates (Photos 1.1–1.6).

Photo 1.2 Pietrele Peak (2270 m) and Pietricelele lakes (photo M. Ciobanu)

Photo 1.3 Pietrele Peak (2270 m) and Pietrele Lake (photo M. Ciobanu)

Photo 1.4 Peleaga Peak (2509 m) and Bucura, Florica and Viorica lakes (photo M. Ciobanu)

Photo 1.5 Bucura Ridge (2206 m) and Bucura Lake (photo M. Ciobanu)

Photo 1.6 Judele Peak (2398 m) and glacial Judele Valley (photo M. Ciobanu)

Chapter 2
History of Floristic and Vegetation Research

The geomorphological specificity of the Retezat Mountains, characterized by imposing peaks that exceed 2450 m altitude (Peleaga 2509 m, Păpușa 2508 m, Retezat 2485 m) and intramontane valleys with numerous lakes and glacial cirques (Gemenele, Zănoaga, Bucura) indicating the location and extent of glaciers of the glacial period (Würm 3) in the area, captivated the attention of botanists since the early nineteenth century. The first plant records from the Retezat Mountains are provided in the J.C.G. Baumgarten's floristic monographic work (Baumgarten 1816) titled "Enumeratio Stirpium Magno Transsilvanicae Principatui". Among the many mountain species cited, we mention some with phytogeographical and phytocoenological interest, such as *Anthemis carpatica, Viola declinata, Senecio abrotanifolius* subsp. *carpathicus, Saxifraga pedomontana* subsp. *cymosa, Saxifraga bryoides, Leucanthemopsis alpina, Phyteuma confusum, Draba siliquosa, Senecio doronicum* subsp. *transylvanicus (S. glaberrimus), Juncus trifidus, Luzula alpinopilosa* subsp. *obscura, Helianthemum canum, Thlaspi dacicum (T. alpestre), Heracleum palmatum, Campanula transsilvanica, Symphyandra wanneri, Ranunculus crenatus, Oxytropis halleri, Laserpitium krapfii, Biscutella laevigata, Plantago holosteum, Carex pyrenaica, Pedicularis hacqueti, Carex aterrima, Agrostis rupestris, Trisetum alpestre.*

A few years later (Rochel 1828) the botanist A. Rochel published in his paper on Banat four rare taxa, unknown in the region—*Plantago atrata, Campanula abietina, Pimpinella saxifraga* subsp. *alpestris* and *Achillea distans* subsp. *alpina.* Later on, the botanists from Banat and Transylvania, who continued their research in the Retezat Mountains, retrieved the species mentioned by the two explorers and identified new taxa in this mountain massif. Thus, Heuffel (1858), in his comprehensive work on the Banat flora, with more than 2000 taxa, described several species from the Retezat Mountains such as *Anthemis macrantha, Centaurea kotschyana, Carex nigra* subsp. *dacica, Hieracium transsilvanicum, Hieracium kotschyanum*, and *Draba dorneri*, the latter being locally endemic to Retezat. He also mentioned for the first time the following species from the massif: *Cardamine resedifolia,*

G. Coldea, V. Cristea, *The Vascular Plant Communities of the Retezat National Park (Southern Carpathians)*, Geobotany Studies,
https://doi.org/10.1007/978-3-031-05618-5_2

Allium victoriale, Poa laxa, Spiraea chamaedryfolia, Streptopus amplexifolius, and *Aconitum variegatum* subsp. *paniculatum.*

Schur (1866) mentioned in his Transylvanian plant list: *Anthoxanthum alpinum, Anthemis cretica* subsp. *kitaibeliana, Cerastium alpinum, Carex curvula, Calamagrostis pseudophragmites, Silene petraea, Festuca supina, Poa alpina, Poa cenisia* subsp. *contracta,* and *Sedum fabaria* from the Retezat Mountains.

Csató (1868) carried out research in different areas of the Retezat massif and reported 82 plant species. Some of these species are rare in the massif (*Campanula cochleariifolia, Narcissus poeticus* subsp. *radiiflorus*), while others are phytocoenologically important (*Cerastium cerastioides, Cerastium alpinum* subsp. *lanatum, Primula minima, Plantago gentianoides, Scorzonera rosea, Sesleria rigida, Soldanella major, Soldanella pusilla, Helleborus purpurascens, Veronica baumgartenii*) because they define their ecological and phytocoenological specificity.

In his work "Enumeratio Florae Transsilvanicae vasculosae critica" (Simonkiai 1886), the famous botanist L. Simonkai published the results of his intensive floristic research in the Apuseni Mountains and the western part of the Southern Romanian Carpathians (Retezat, Godeanu, Parâng Mountains), less explored botanically, as well as other regions in Transylvania. The author reported 128 vascular plants from the Retezat Mountains, among them several montane-subalpine species with phytogeographic significance such as *Achillea schurii, Achillea lingulata, Alyssum repens, Bupleurum falcatum* subsp. *cernuum (B. diversifolium), Campanula serrata, Campanula kladniana, Carduus kerneri, Leucanthemum rotundifolium, Dianthus giganteus, Edraianthus graminifolius, Euphorbia carniolica, Festuca picturata, Pseudorchis frivaldii, Hepatica transsilvanica, Heracleum palmatum, Juncus thomasii, Laserpitium archangelica, Lathyrus hallersteinii, Moehringia pendula, Pedicularis haquetii, Poa laxa, Poa media, Potentilla ternata, Pulmonaria rubra, Saxifraga carpathica, Silene nutans* subsp. *dubia, Silene lerchenfeldiana, Senecio incanus* subsp. *carniolicus,* and *Thymus comosus.* In addition to these, he mentioned some rare species such as *Agrimonia repens, Euphrasia minima* subsp. *tatrae, Minuartia setacea* subsp. *banatica,* and *Tozzia alpina* subsp. *carpathica.* These rare and regional plant species revealed the great floristic diversity of the Retezat Mountains and stimulated the interest of botanists in studying the region.

Jávorka (1911) carried out floristic research in the calcareous area of the Retezat (Stănulete, Piatra Iorgovanului, Stânişoara, Custura) and reported some interesting plants, new for the massif, such as *Festuca carpatica, Salix alpina, Rumex scutatus, Oxyria digyna, Dianthus petraeus, Dianthus spiculifolius, Kernera saxatilis, Papaver alpinum* subsp. *corona-sancti-stephani, Erysimum comatum, Erysimum witmannii, Hesperis matronalis* subsp. *obtusa, Oxytropis carpatica, Onobrychis montana* subsp. *transsilvanica, Hedysarum hedysaroides, Helianthemum alpestre, Leontodon montanus* subsp. *pseudotaraxaci, Athamanta turbith* subsp. *hungarica, Hieracium villosum, Hieracium bifidum, Hieracium krásánii,* and *Hieracium chloribracteum.* He also described two taxa new to science (*Hieracium filarszkyi* Jáv. & Zahn and *Hieracium paltinae* Jáv. & Zahn). These new taxa identified in the eastern part of the massif complete the group of montane and subalpine species of the Retezat Mountains.

In the same period (1911–1912), F. Pax accompanied by the young botanist A. Borza visited Retezat during several floristic trips to gather botanical information necessary to elaborate the work "Pflanzengeographie von Rumänien" (Pax 1898, 1919).

After the First World War, the Retezat Mountains were studied floristically by Haret (1922), who reported 106 species of phanerogams. Among the rarest species in the massif, they mentioned *Luzula forsteri, Centaurea nigrescens, Koeleria macrantha* subsp. *transsilvanica, Festuca porcii, Poa laxa, Poa minor, Leucanthemopsis alpina, Cardamine resedifolia, Plantago gentianoides,* and *Leontodon croceus.*

A few years later Zahn (1928) reported in his work entitled "Hieracia transsilvanica ..." 64 species of *Hieracium,* of which 58 taxa occurred also in the Retezat Mountains.

Later on, in his monographic works "Centaureele României" and "Achileele României", Prodan (1930, 1931) mentioned rare or specific plant species of the Retezat Mountains such as *Centaurea phrygia* subsp. *retezatensis, Centaurea nervosa, Centaurea kotschyana, Achillea lingulata,* and *Achillea crithmifolia.*

Likewise, Borza (1934) published the first phytocoenological work on the crystalline massif of Retezat, based on research done in 1933. The author described 24 syntaxa and provided chorological data for the 300 taxa included in the plant list that he identified there. Among the regional, endemic and Carpathian-Balkan plant species presented in the coenological tables, we mention *Alopecurus laguriformis, Avenula planiculme, Campanula abietina, Campanula serrata, Carex nigra* subsp. *dacica, Carex pyrenaica, Centaurea nervosa, Chrysosplenium alpinum, Crocus heuffelianus, Dianthus compactus, Festuca supina, Pseudorchis frivaldskii, Heracleum palmatum, Hieracium fritzei, Hieracium paltinae, Laserpitium krapfii, Luzula alpinopilosa* subsp. *obscura, Phyteuma vagneri, Plantago gentianoides, Poa media, Potentilla ternata, Rhododendron myrtifolium, Scorzonera rosea, Silene nutans* subsp. *dubia, Soldanella major, Thlaspi dacicum, Thymus pulcherrimus, Veronica baumgartenii,* and *Viola declinata.* The author considered some of these species characteristic of the plant associations described and employed others for the more comprehensive characterization of higher-ranking syntaxa (alliances or orders).

An interesting floristic and ecological work was published by B. Pawlowski (1939), as a result of the long trip he made in 1937 in the Southern Romanian Carpathians. The author reported 56 species from the Retezat Mountains, among them some rare species with phytogeographic significance, such as *Cerastium transsilvanicum, Cerastium lerchenfeldianum, Dianthus petraeus, Dianthus glacialis* subsp. *gelidus, Gypsophila petraea, Papaver alpinum* subsp. *corona-sancti-stephani, Draba lasiocarpa, Oxytropis pyrenaica, Androsace villosa* subsp. *arachnoides, Pedicularis baumgartenii, Campanula transsilvanica, Edraianthus graminifolius, Jasione orbiculata, Leontodon montanus* subsp. *pseudotaraxaci,* and *Crepis viscidula.* The paper also includes a genuine altitudinal distribution scheme of 56 alpine species he found on the Peleaga, Custura, and Bucura peaks in Retezat.

A decade later (1948–1962) Şt. Csürös carried out floristic and vegetation research in the Retezat Mountains, mostly in the limestone area, and co-authored several scientific papers (Csűrös et al. 1956a, b, 1962, 1964). Among the rare and phytogeographically important species reported in these works, he mentioned *Saxifraga corymbosa, Saxifraga adscendens, Saxifraga androsacea, Saxifraga marginata, Linum uninerve, Acinos alpinus* subsp. *majoranifolius, Jurinea glycacantha, Leontodon croceus, Saussurea discolor, Bromus riparius, Festuca saxatilis, Festuca versicolor, Carex capillaris, Carex rupestris, Lomatogonium carinthiacum, Onobrychis montana* subsp. *transsilvanica, Eritrichium nanum, Scrophularia laciniata, Swertia perennis* subsp. *punctata, Carduus kerneri* subsp. *lobulatiformis, Trisetum fuscum, Agrostis alpina, Iris reichenbachii, Lilium jankae,* and *Nigritella nigra*. We note that 426 plant species were listed in the floristic work on the Calcareous Retezat (Soltés et al. 2001), whereas 13 types of herbaceous and woody plant communities were described in the vegetation studies, mainly based on species dominance (Csűrös et al. 1956a, b, 1964).

During the same period, Nyárady (1958) compiled all the floristic data previously published by the botanists from the Retezat Mountains, to which he added his personal data collected during years in the region. He finally published a list of 920 species of cormophytes, mentioning also the locations from where each species was observed. The work is noteworthy because it reported a high number of *Hieracium* species (71) identified in the massif and enumerated plant species associated with certain landscape forms (peaks, glacial lakes) and with herbaceous and woody vegetation types (beech, spruce and pine forests).

The floristic, ecological, phytocoenological, and palynological research in the Retezat National Park multiplied after 1960 and continued intermittently until 2012. The algological studies were carried out thoroughly by Péterfi (1993), those of lichenology by Ciurchea and Crişan (1993), those of bryology by Plămadă (1993), and those on cormophytes by Coldea et al. (1972). Several results of phytocoenological research on particular biotopes in Retezat were published over time by Boşcaiu et al. (1972, 1978), Resmeriţă (1987) and Täuber (1987). A synthetic overview of the plant associations identified in the perimeter of the Retezat National Park was published at the end of the last century by Coldea (1993). The history of vegetation in Retezat was documented in several works by Pop et al. (1970) and Fărcaş et al. (1999).

Chapter 3
Areal-Geographical Characterization of the Retezat Flora and the Endangered Species

The botanical studies carried out in the Retezat National Park area throughout the time documented the presence of 1152 plant species and 104 subspecies belonging to the phyla *Pteridophyta* and *Spermatophyta,* as well as 52 hybrid species of the genus *Hieracium* described by E.I. Nyárády in the Romanian Flora, Vol. X.

The 1152 species belong to 45 orders, 79 families, and 378 genera. Regarding the distribution of taxa by altitude, we specify that only 360 can be considered of high altitude (alpine) in the broad sense, and of these only 110 are actual alpine species occurring primarily in the alpine grasslands. The remaining 250 species are considered subalpine species typical to dwarf and juniper shrublands and secondary grasslands distributed in the subalpine belt (1700–2200 m). Approximately 420 species are present in the mountain spruce forests and mixed (spruce, beech, and fir) forests (1250–1700 m). The remaining 372 species occur in the secondary mesophilic grasslands, in beech-mountain maple (*Acer pseudoplatanus*) forests and pure beech forests, and in the phytocoenoses of sessile oak mixed with linden and pine (600–1150 m).

The areal-geographical analysis of the Retezat flora reveals that the European elements are dominant (24.5%), followed by the Eurasian (23.2%), European alpine (13.3%), and circumpolar (8.3%). The Carpathian endemic element (12.5%) and the Dacian-Balkan one (12.2%) are well-represented in the flora of the massif. The other phytogeographical elements (Pontic, Pontic-Mediterranean, continental and cosmopolitan) account together for only 6% of all the species inventoried. The regional, Carpathian, and Dacian-Balkan endemic elements are of major importance in the study of plant communities, differentiating floristically the syntaxa of the region from those described in other neighboring regions. Therefore, they will next be presented in detail.

The Carpathian endemic species identified over time in the Retezat Mountains are *Alopecurus laguriformis*, *Anthemis cretica* subsp. *kitaibelii*, *Aconitum lycoctonum* subsp. *moldavicum*, *Athamantha turbith* subsp. *hungarica*, *Achillea oxyloba* subsp. *schurii*, *Barbarea vulgaris* subsp. *lepuznica*, *Cardamine glanduligera*, *Campanula*

G. Coldea, V. Cristea, *The Vascular Plant Communities of the Retezat National Park (Southern Carpathians)*, Geobotany Studies,
https://doi.org/10.1007/978-3-031-05618-5_3

carpatica, *Campanula rotundifolia* subsp. *polymorpha*, *Campanula rotundifolia* subsp. *kladniana, Campanula serrata*, *Carduus kerneri* subsp. *lobulatiformis*, *Centaurea phrygia* subsp. *retezatensis*, *Centaurea triniifolia*, *Cerastium arvensis* subsp. *lerchenfeldianum*, *Cerastium transsilvanicum*, *Chrysosplenium alpinum*, *Dianthus henteri*, *Dianthus glacialis* subsp. *gelidus*, *Dianthus spiculifolius*, *Dianthus tenuifolius*, *Doronicum carpaticum*, *Draba dorneri*, *Draba simonkaiana*, *Erysimum witmanni* subsp. *witmanni*, *Erysimum witmanni* subsp. *transsilvanicum*, *Eritrichium nanum* subsp. *jankae*, *Festuca carpatica*, *Festuca stricta* subsp. *saxatilis*, *Festuca versicolor*, *Festuca stricta* subsp. *rumelica*, *Festuca porcii*, *Gypsophila petraea*, *Genista tinctoria* subsp. *oligosperma*, *Galium kitaibelianum*, *Koeleria macrantha* subsp. *transsilvanica*, *Hepatica transsilvanica*, *Linum uninerve*, *Leontodon montanum* subsp. *pseudotaraxaci*, *Leontodon repens*, *Luzula alpinopilosa* subsp. *obscura*, *Leucanthemum rotundifolium*, *Papaver alpinum* subsp. *corona-sancti-stephani*, *Pedicularis baumgarteni*, *Poa granitica* subsp. *disparilis*, *Pulmonaria filarszkyana*, *Phyteuma tetramerum*, *Phyteuma vagneri*, *Silene nutans* subsp. *dubia*, *Salix kitaibeliana*, *Scabiosa lucida* subsp. *barbata*, *Soldanella major*, *Thymus pulcherrimus*, *Thymus comosus*, *Trisetum fuscum*, *Symphytum cordatum*, *Viola declinata,* and *Onobrychis montana* subsp. *transsilvanica*.

The Dacian and Dacian-Balkan species present in the Retezat flora are *Achillea lingulata*, *Achillea crithmifolia*, *Alyssum repens*, *Alyssum repens* subsp. *transsilvanicum*, *Anthemis macrantha*, *Bruckenthalia spiculifolia*, *Campanula abietina*, *Campanula transsilvanica*, *Carduus kerneri* subsp. *kerneri*, *Carex nigra* subsp. *dacica*, *Centaurea atropurpurea*, *Centaurea kotschyana*, *Crepis viscidula*, *Citisus ciliatus* subsp. *alpestris*, *Dianthus compactus*, *Dianthus giganteus*, *Echinops commutatus*, *Pseudorchis frivaldii*, *Helleborus purpurascens*, *Hesperis matronalis* subsp. *obtusa*, *Hieracium sparsum* subsp. *sparsum*, *Hypericum richeri* subsp. *grisebachii*, *Jasione orbiculata*, *Juncus thomasii*, *Lathyrus hallersteinii*, *Minuartia hirsuta* subsp. *frutescens*, *Minuartia recurva*, *Moehringia pendula*, *Dactylorhiza cordiger*, *Phyteuma confusum*, *Plantago gentianoides*, *Poa media*, *Poa cenisia* subsp. *contracta*, *Potentilla ternata*, *Plantago holosteum*, *Pulmonaria rubra*, *Rhododendron myrtifolium*, *Saxifraga pedemontana* subsp. *cymosa*, *Scleranthus uncinatus*, *Jovibarba heuffelii*, *Sempervivum marmoreum*, *Senecio abrotanifolius* subsp. *carpathicus*, *Senecio doronicum* subsp. *transylvanicus*, *Senecio papposus*, *Sesleria bielzii*, *Silene heuffelii*, *Silene saxifraga*, *Sesleria heufleriana*, *Symphyandra wanneri*, *Telekia speciosa*, *Thymus balcanus*, *Verbascum lanatum*, *Veronica bachofenii*, *Viola baumgartenii*, *Viola dacica*, *Saxifraga corymbosa*, *Laserpitium archangelica*, *Scrophularia heterophylla* subsp. *laciniata*, and *Bupleurum falcatum* subsp. *cernuum*.

Regional *Hieracium* species such as *Hieracium transsilvanicum*, *H. oreophilum*, *H. apiculatum*, *H. sparsum* subsp. *magocsyanum*, *H. sparsum* subsp. *kotschyanum*, *H. sparsum* subsp. *tomiasae*, *H. paltinae*, *H. pisaturense*, *H. prenanthoides* subsp. *lanceolatum*, *H. prenanthoides* subsp. *perfoliatum*, *H. lumbricicaule*, *H. paxianum*, *H. nigrilacus*, *H. pocuticum*, *H. pseudobifidum*, *H. pseudocaesium*, *H. retyezatense*, *H. tubulare*, and *H. pelagae* complete the two types of phytogeographic elements.

All together they attest the regional specificity of the flora and vegetation of the Retezat Mountains.

Based on the areal-geographical analysis of the Retezat Mountains, we can confirm the affiliation of this territory to the Central European, Eastern Carpathian province, the Southern Romanian Carpathians circumscription, and the Retezat-Godeanu Mountains subdivision, according to the assertion of Borza and Boşcaiu (1965).

Regarding the rare and endangered plants in the Retezat Mountains, according to the "Red Book of Vascular Plants in Romania" by Dihoru and Negrean (2009), the following species are threatened by extinction at different degrees in the immediate or more distant future: *Barbarea vulgaris* subsp. *lepuznica* (CR), *Centaurea phrygia* subsp. *retezatensis* (VU), *Draba dorneri* (CR), *Draba simonkaiana* (CR), *Lilium jankae* (VU), *Oxytropis pyrenaica* (CR), *Pseudorchis frivaldii* (CR), *Pulsatilla vulgaris* subsp. *grandis* (CR), *Salix alpina* (CR), *Scutellaria alpina* (CR), and *Sorbus chamaemespilus* (EN).

The following plants, considered natural monuments and protected by law, are present in the national park area: *Angelica archangelica*, *Cypripedium calceolus*, *Gentiana lutea*, *Leontopodium alpinum*, *Nigritella nigra,* and *Trollius europaeus*. The areas where these species occur must be declared as strictly protected and the management plan must include measures for their conservation within the national park.

Chapter 4
Mapping Rare, Endangered Angiosperm Species of Phytogeographical Interest from the Retezat National Park

Out of the 1229 cormophyte plant species and subspecies published by various botanists from the nineteenth century till present from the Retezat Mountains, some have a limited distribution area restricted to a single locality, such as *Draba dorneri* and *D. simonkaiana*, others occur in two or more localities (Fundort) in the massif, e.g. *Asperula capitata* and *Campanula transsilvanica*. We selected for mapping 60 species of angiosperms in relation to species distribution in the massif and their cenotic and phytogeographical importance. The chorological maps drawn up for these species show both their distribution in the massif and their conservation status in the perimeter of the National Park. Of the total park area (38,300 ha), only 1950 ha are included in the Gemenele Scientific Reserve, which offers strict protection to species and habitats. The other surface of the park, according to the management plan, is included in the integral protection area, where grazing and ecotourism occur freely. The rare plant species and the habitats in which they coexist are continuously subject to zoo-anthropic impact, and pedo-ecological conditions make them unsuitable for vulnerable species. Through the chorological maps, we want to draw the attention and support the park administration in the actions of protection and conservation of endangered species of the park's integral protection area. In elaborating the chorological maps of the plant species we were guided by two recently published works: The Red Book of Vascular Plants in Romania (Dihoru and Negrean 2009) and Phytosociology (Cristea et al. 2015). The chorological data for the mapped species were collected from the floristic lists published by the nineteenth and twentieth century botanists, those from the herbarium of University of Cluj, as well as plant associations we described or identified in the Retezat National Park. We transposed all these localities (Fundort) as dots on the park area, both in the strict protection area (Gemenele Reserve) and in the integral protection area. Of the 60 mapped species, only 20 are present in the Gemenele Scientific Reserve, the other 40 species occur only in the integral protection zone.

We propose the scientific council of the park the establishment of two botanical reserves in the Calcareous Retezat (Retezatul Calcaros) to protect the rare and

G. Coldea, V. Cristea, *The Vascular Plant Communities of the Retezat National Park (Southern Carpathians)*, Geobotany Studies,
https://doi.org/10.1007/978-3-031-05618-5_4

vulnerable species in this area of the park. Below we list alphabetically the 60 plants species for which we elaborated chorological maps: *Achillea lingulata, Achillea oxyloba* subsp. *schurii, Aconitum lycoctonum* subsp. *moldavicum, Aconitum napellum* subsp. *tauricum, Allium victorialis, Alopecurus laguriformis, Androsace willosa* subsp. *arachnoidea, Alyssum repens, Anthemis carpatica, Asperula capitata, Barbarea vulgaris* subsp. *lepuznica, Campanula carpatica, Campanula transsilvanica, Carex limosa, Carex pyrenaica, Carduus kerneri* subsp. *lobulatiformis, Centaurea kotschyana, Cerastium lerchenfeldianum, Cerastium transsilvanicum, Chrysosplenium alpinum, Crepis viscidula, Cypripedium calceolus, Dianthus barbatus* subsp. *compactus, Dianthus glacialis* subsp. *gelidus, Dianthus petraeus, Dianthus spiculifolius, Draba dorneri, Draba simonkaiana, Festuca carpatica, Festuca porcii, Festuca stricta* subsp. *rumelica, Festuca xanthina, Gentiana frigida, Gentiana lutea, Gypsophila petraea, Hepatica transsilvanica, Hieracium paltinae, Hieracium sparsum* subsp. *kotschyanum, Jasione orbiculata, Laserpitium krapfii, Leontodon croceus* subsp. *rilaensis, Leontodon montanus* subsp. *pseudotaraxaci, Leontopodium alpinum, Leucanthemopsis alpina, Lilium jankae, Nigritella nigra, Oxytropis pyrenaica, Papaver alpinum* subsp. *corona-sancti-stephani, Pedicularis baumgarteni, Phyteuma vagneri, Plantago atrata* subsp. *carpatica, Poa cenisia* subsp. *contracta, Poa granitica* subsp. *disparilis, Pseudorchis frivaldii, Pulsatilla vulgaris* subsp. *grandis, Saxifraga corymbosa, Saxifraga marginata, Senecio abrotanifolius* subsp. *carpathicus, Thalspi dacicum, Tozzia alpina* subsp. *carpatica.* The list above includes species of different origins, but the Carpathian and Carpathian-Balkan ones predominate, which, together, give the plant communities described from the Retezat National Park a local and regional specificity (Figs. 4.1–4.60).

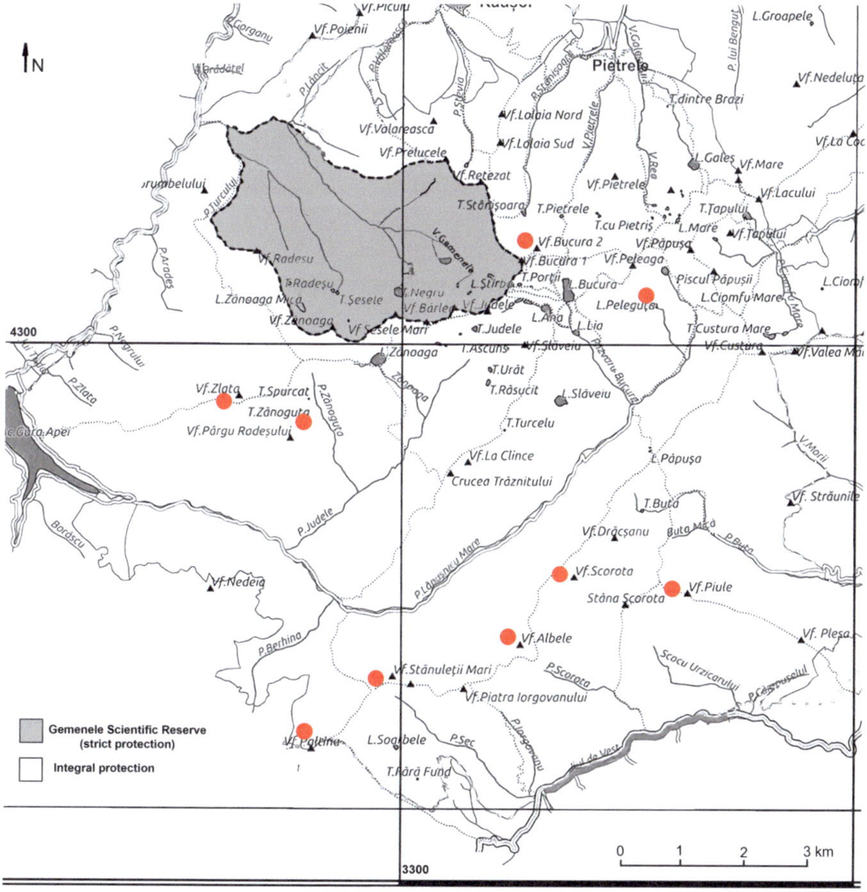

Fig. 4.1 Distribution of *Achillea lingulata* in the Retezat National Park

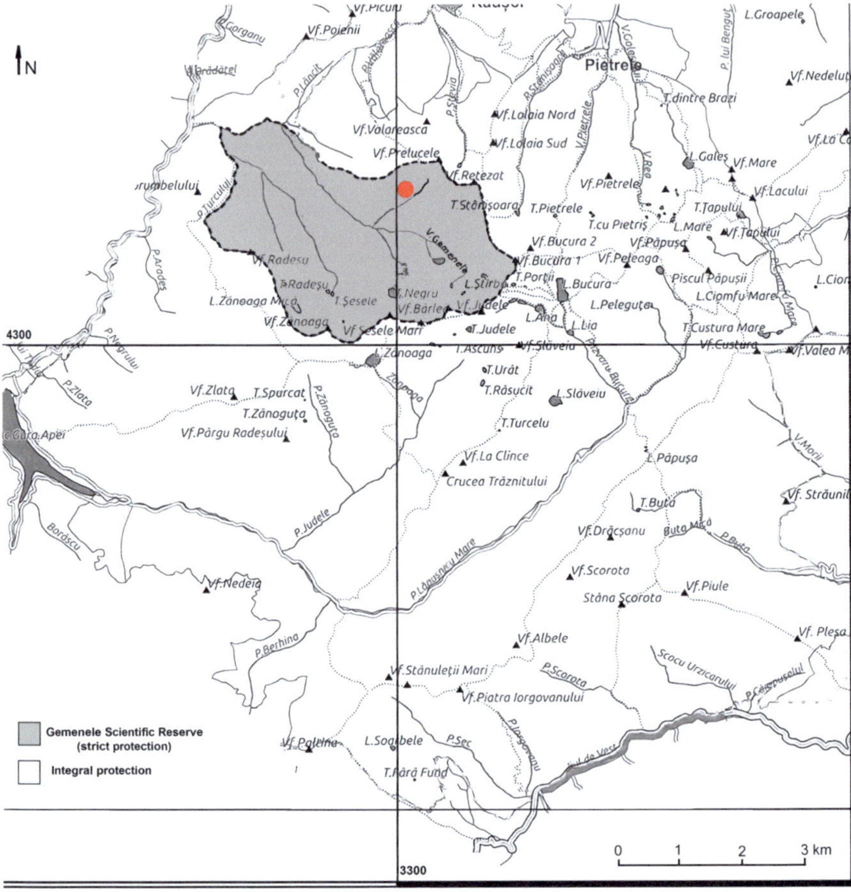

Fig. 4.2 Distribution of *Achillea oxyloba* subsp. *schurii* in the Retezat National Park

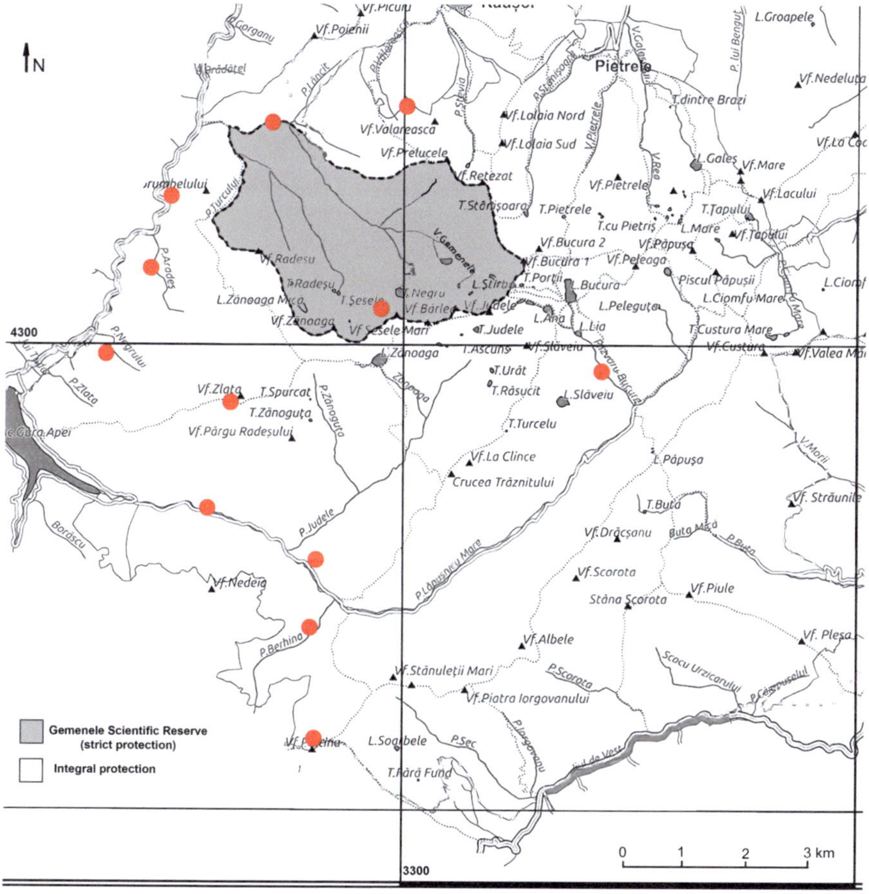

Fig. 4.3 Distribution of *Aconitum lycoctonum* subsp. *moldavicum* in the Retezat National Park

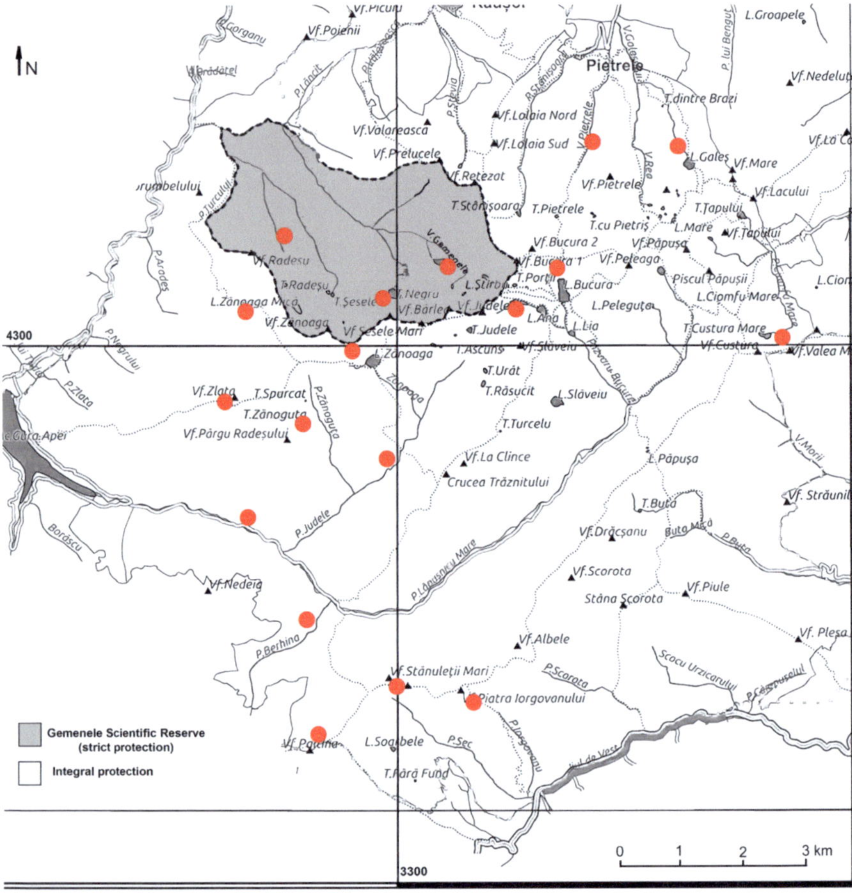

Fig. 4.4 Distribution of *Aconitum napellus* subsp. *tauricum* in the Retezat National Park

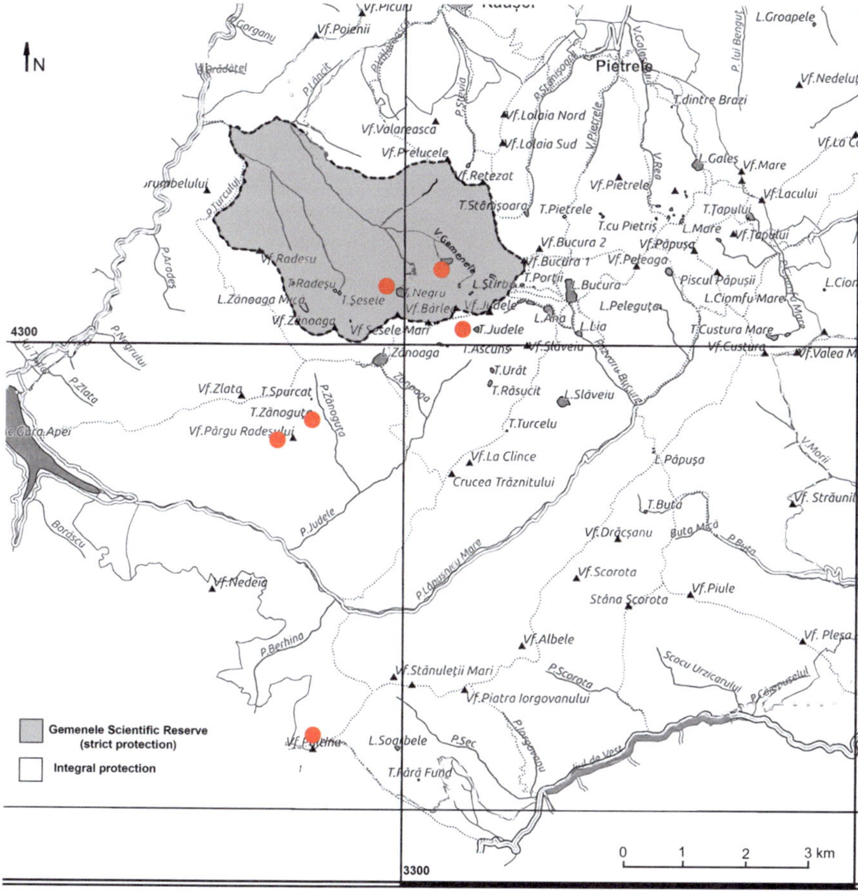

Fig. 4.5 Distribution of *Allium victorialis* in the Retezat National Park

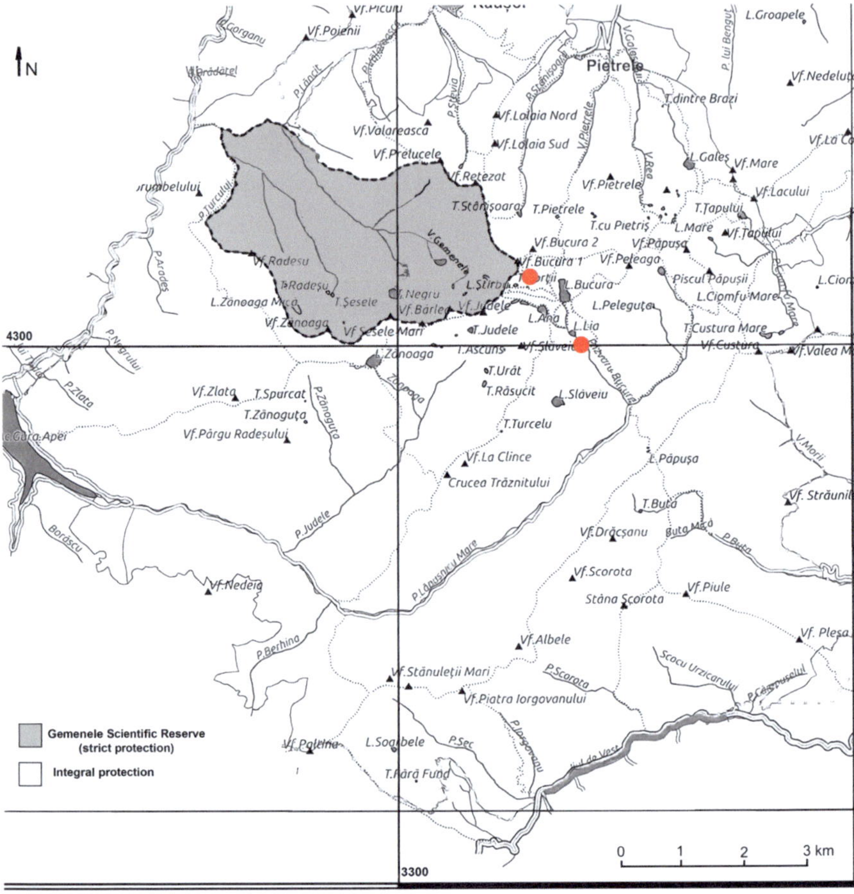

Fig. 4.6 Distribution of *Alopecurus laguriformis* in the Retezat National Park

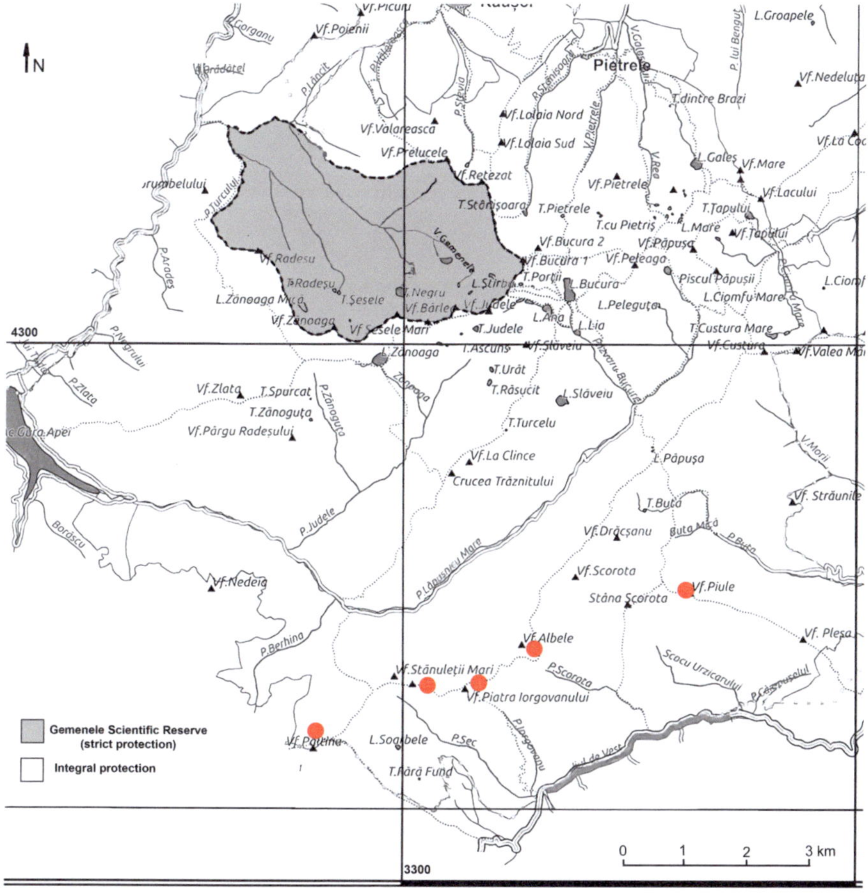

Fig. 4.7 Distribution of *Alyssum repens* in the Retezat National Park

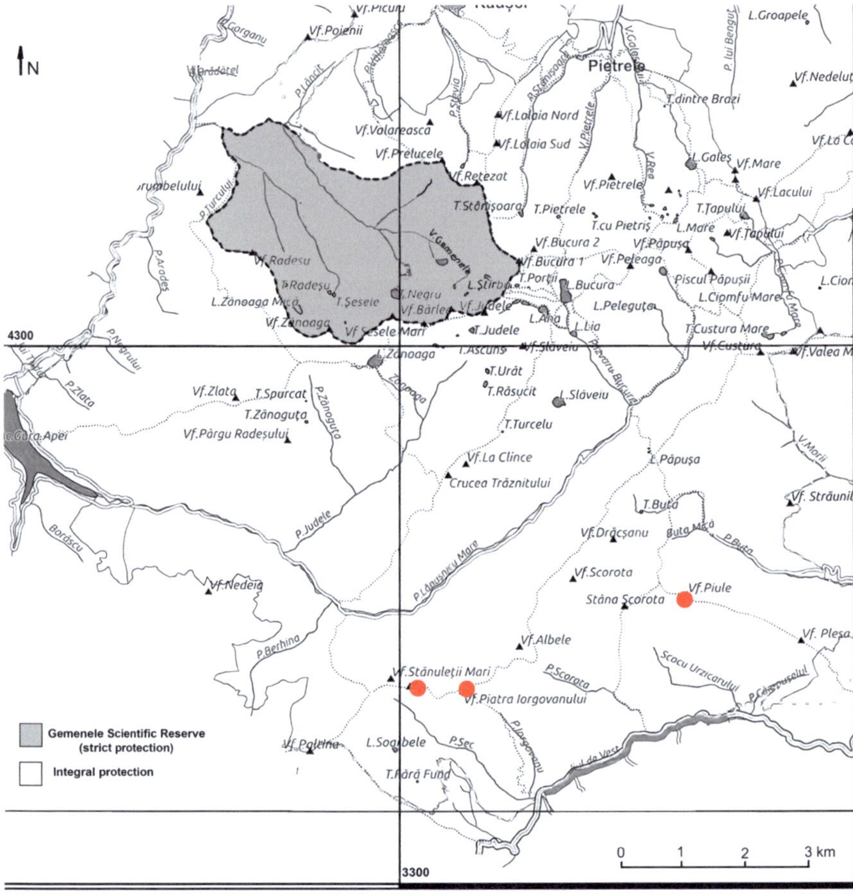

Fig. 4.8 Distribution of *Androsace villosa* subsp. *aracnoidea* in the Retezat National Park

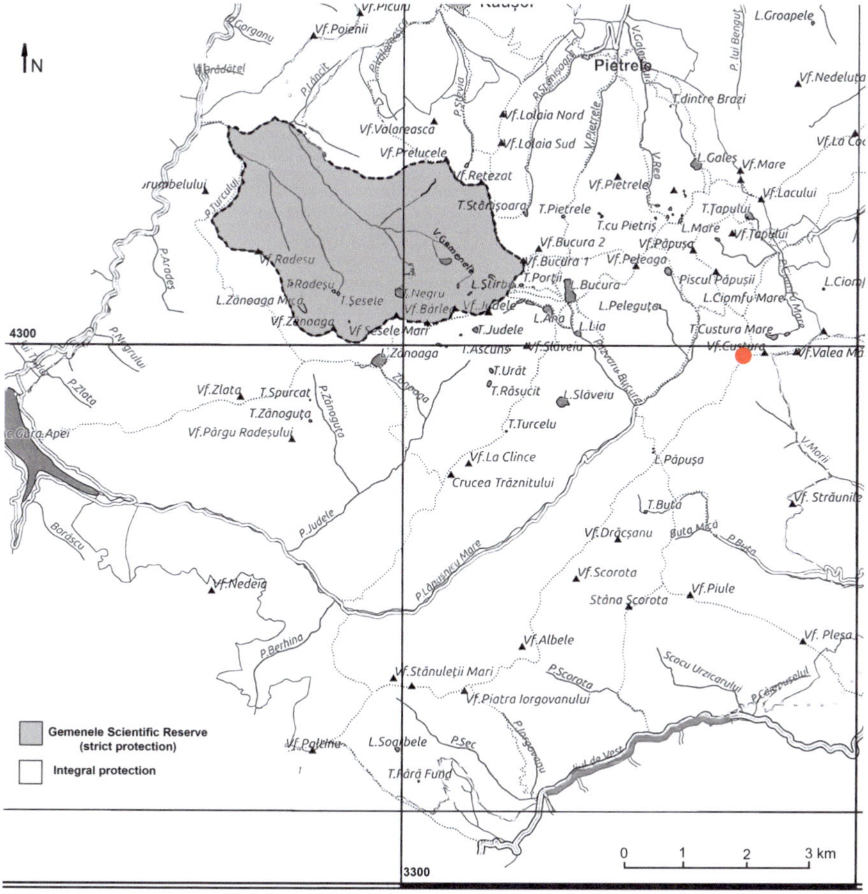

Fig. 4.9 Distribution of *Anthemis carpatica* in the Retezat National Park

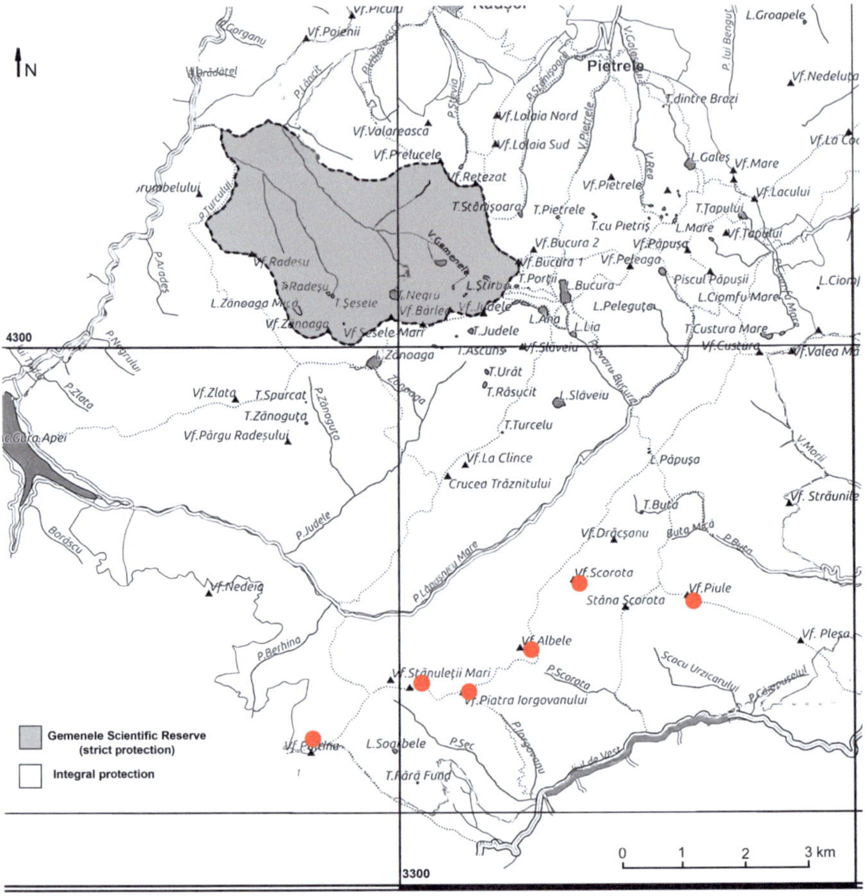

Fig. 4.10 Distribution of *Asperula capitata* in the Retezat National Park

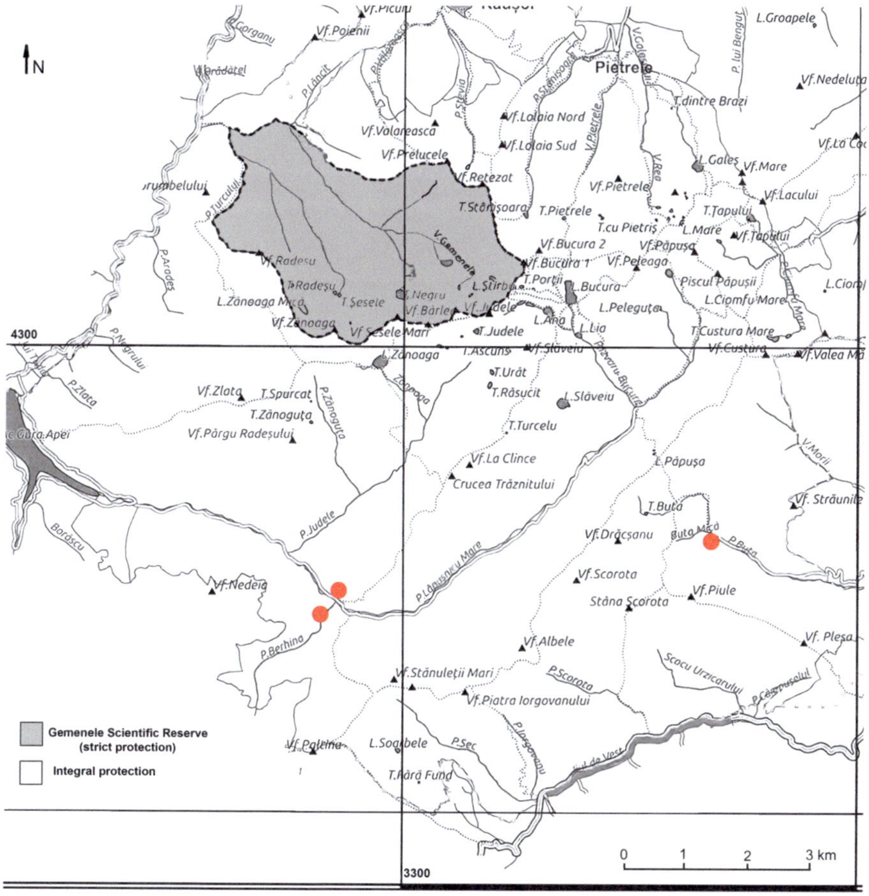

Fig. 4.11 Distribution of *Barbarea vulgaris* subsp. *lepuznica* in the Retezat National Park

Fig. 4.12 Distribution of *Campanula carpatica* in the Retezat National Park

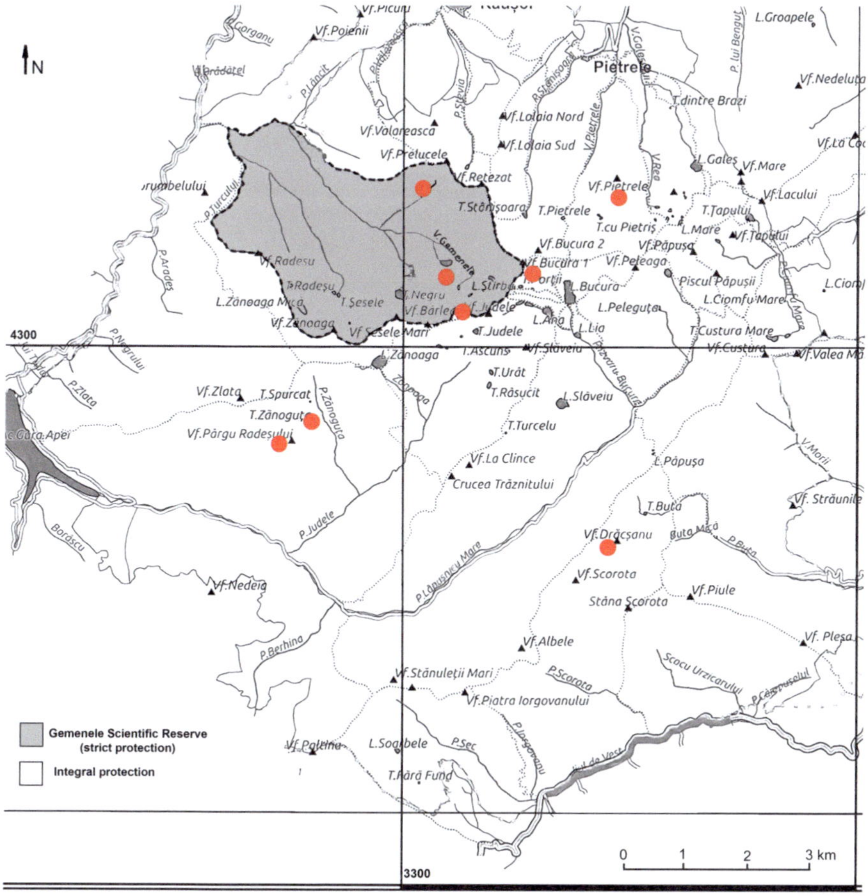

Fig. 4.13 Distribution of *Campanula transsilvanica* in the Retezat National Park

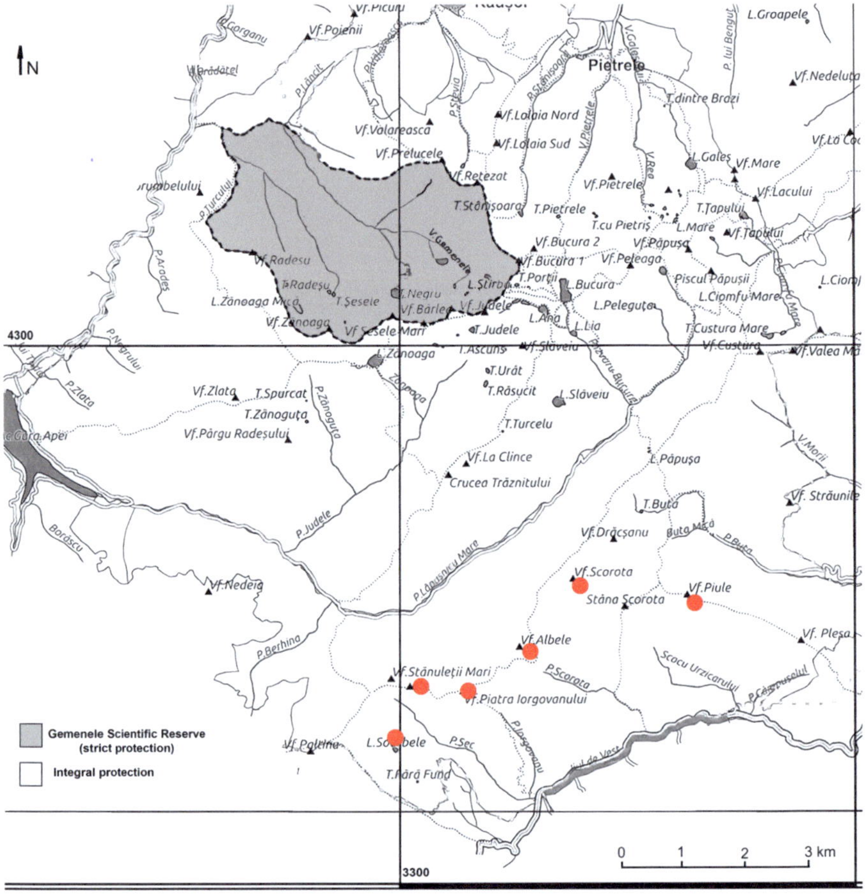

Fig. 4.14 Distribution of *Carduus kerneri* subsp. *lobulatiformis* in the Retezat National Park

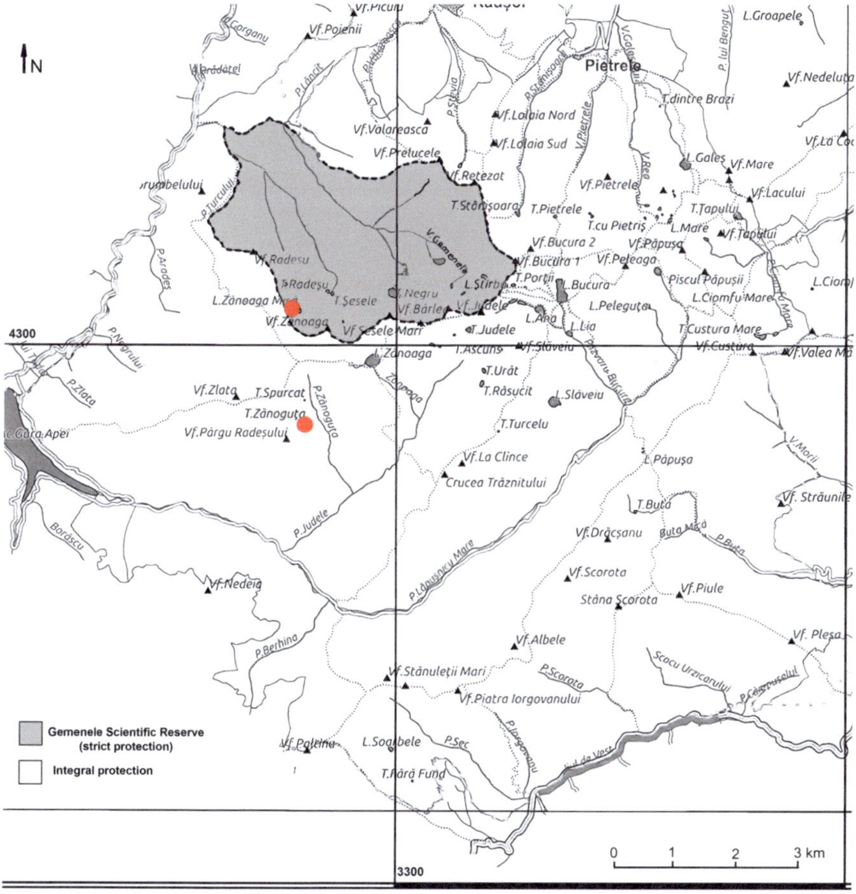

Fig. 4.15 Distribution of *Campanula limosa* in the Retezat National Park

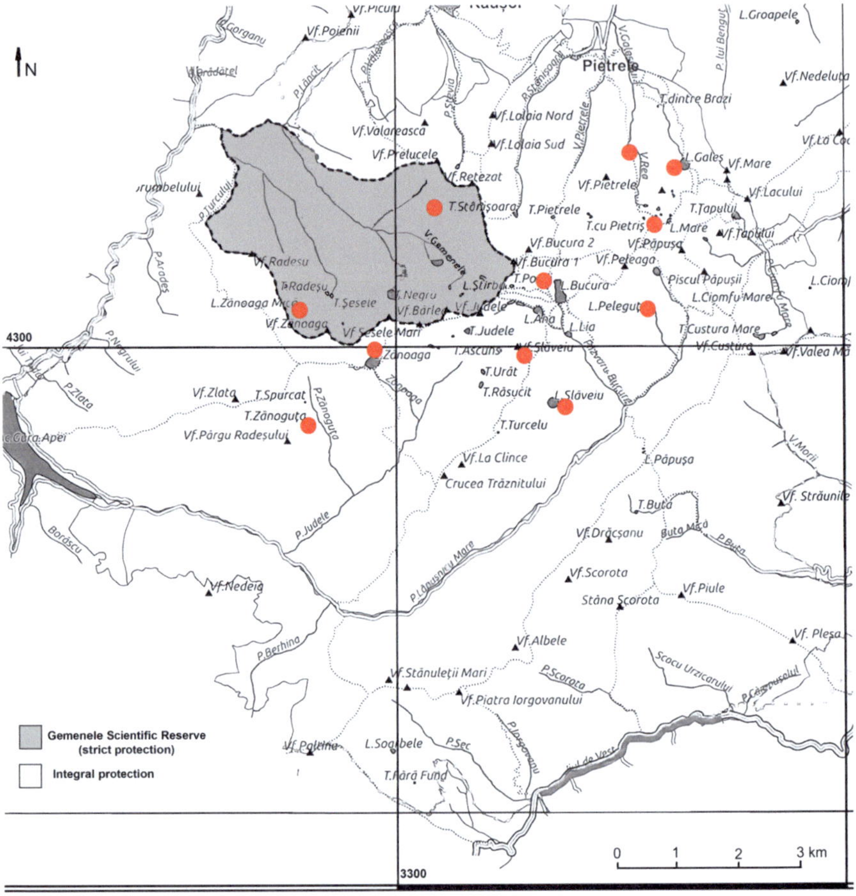

Fig. 4.16 Distribution of *Carex pyrenaica* in the Retezat National Park

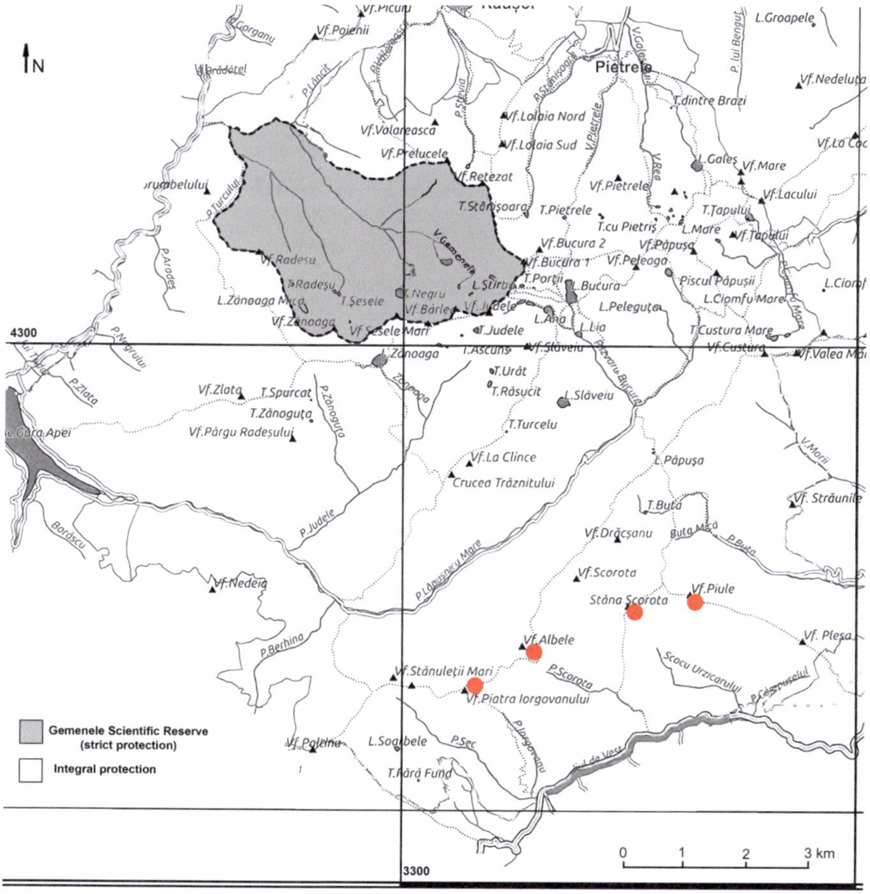

Fig. 4.17 Distribution of *Centaurea kotschyana* in the Retezat National Park

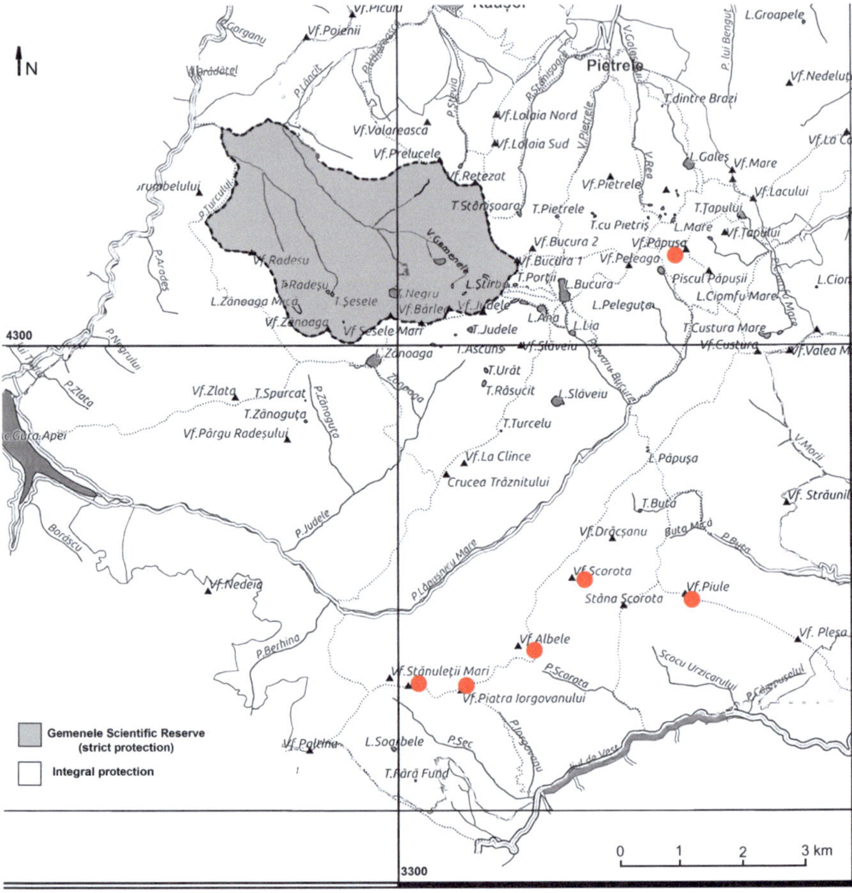

Fig. 4.18 Distribution of *Cerastium lerchenfeldianum* in the Retezat National Park

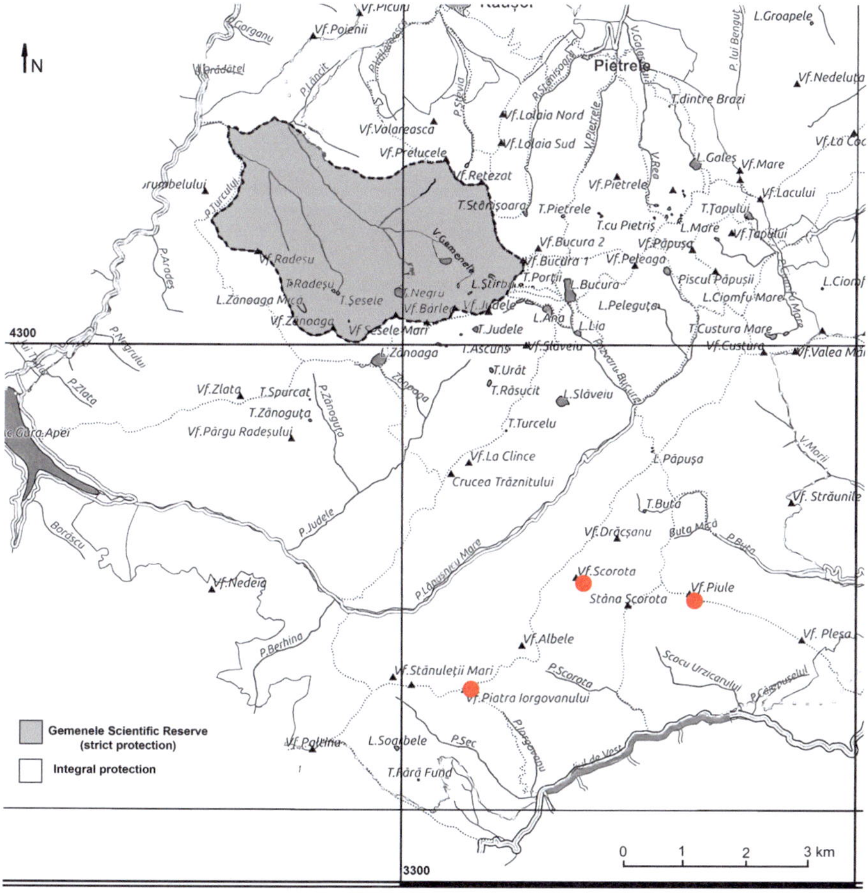

Fig. 4.19 Distribution of *Cerastium transsilvanicum* in the Retezat National Park

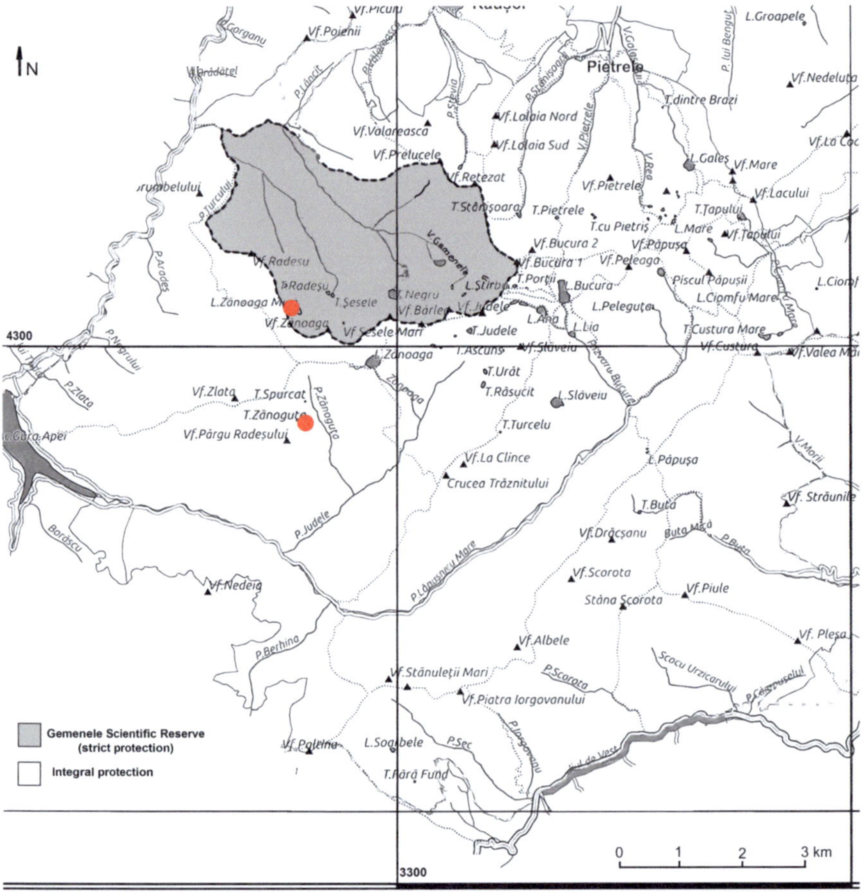

Fig. 4.20 Distribution of *Chrysosplenium alpinum* in the Retezat National Park

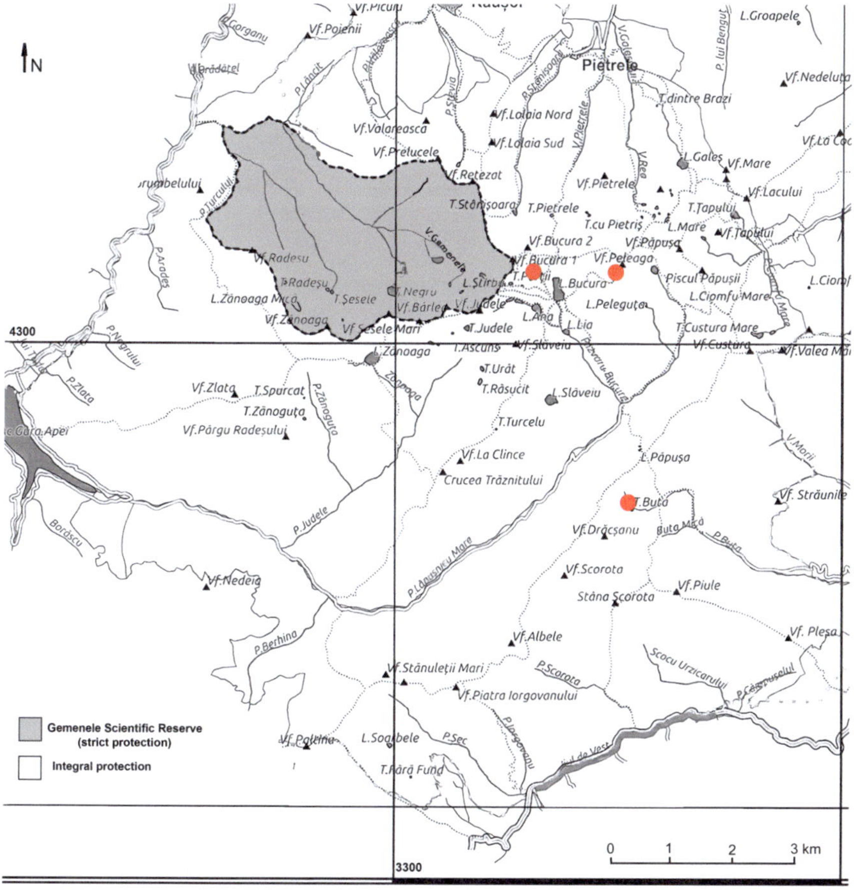

Fig. 4.21 Distribution of *Crepis viscidula* in the Retezat National Park

Fig. 4.22 Distribution of *Cypripedium calceolus* in the Retezat National Park

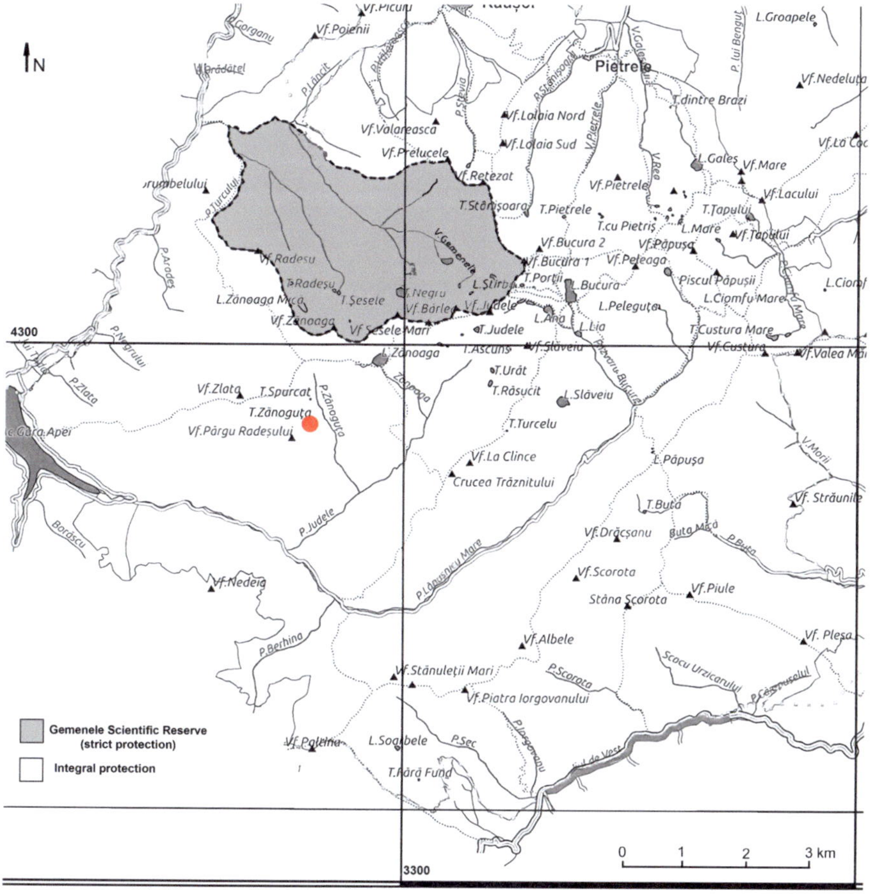

Fig. 4.23 Distribution of *Dianthus barbatus* subsp. *compactus* in the Retezat National Park

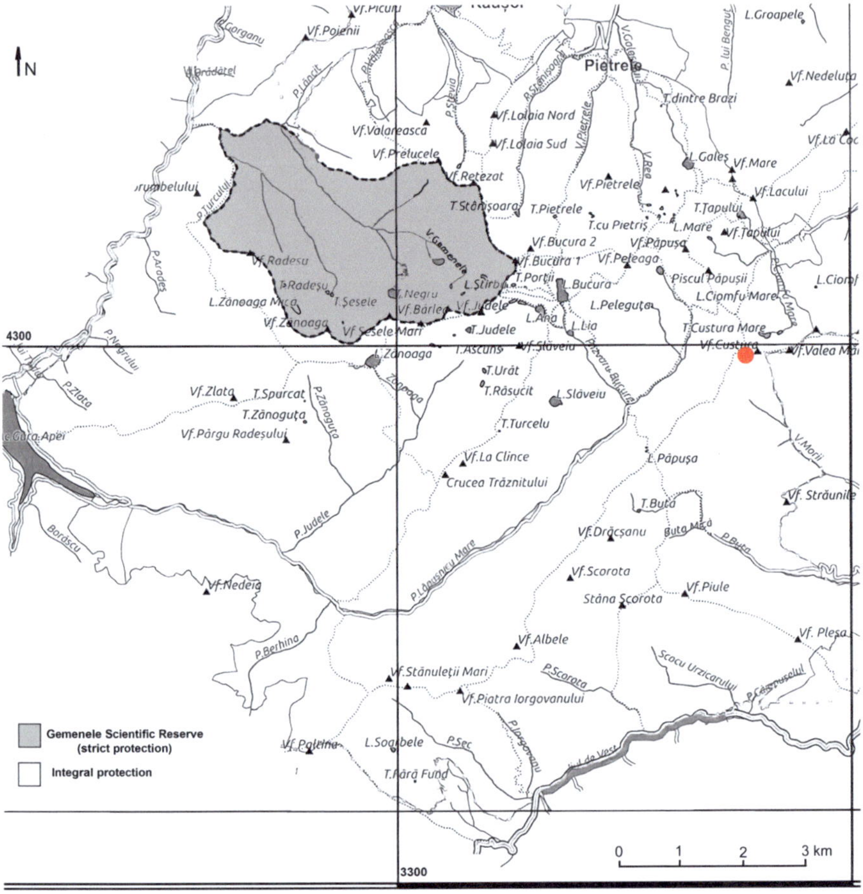

Fig. 4.24 Distribution of *Dianthus glacialis* subsp. *gelidus* in the Retezat National Park

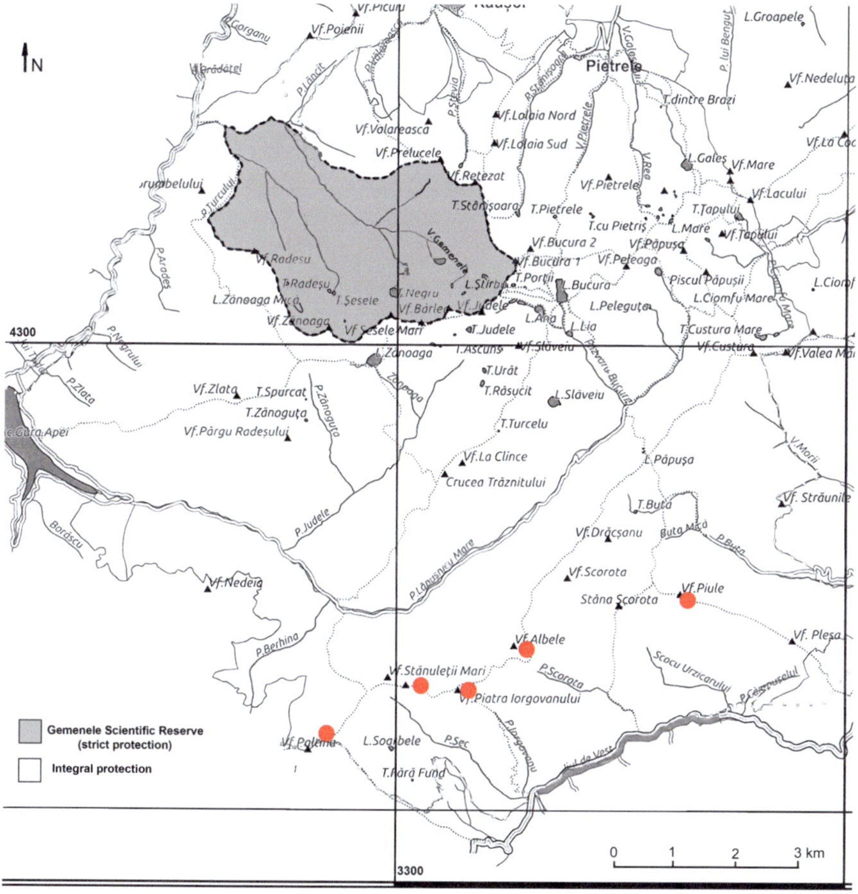

Fig. 4.25 Distribution of *Dianthus petraeus* Tutin in the Retezat National Park

Fig. 4.26 Distribution of *Dianthus spiculifolius* in the Retezat National Park

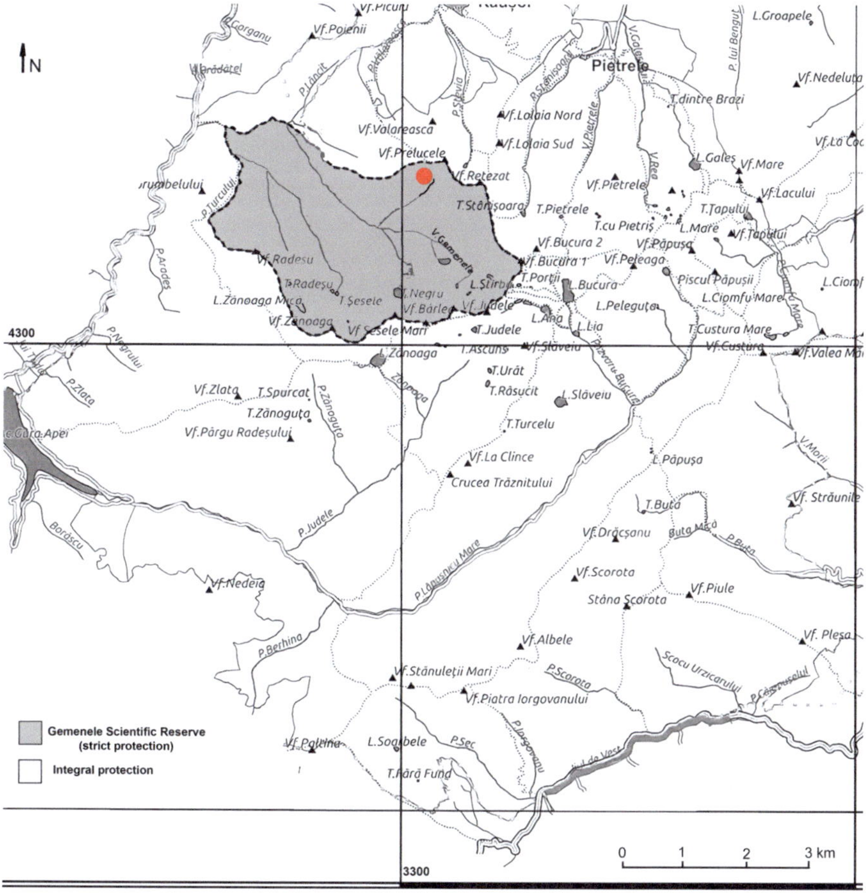

Fig. 4.27 Distribution of *Draba dorneri* in the Retezat National Park

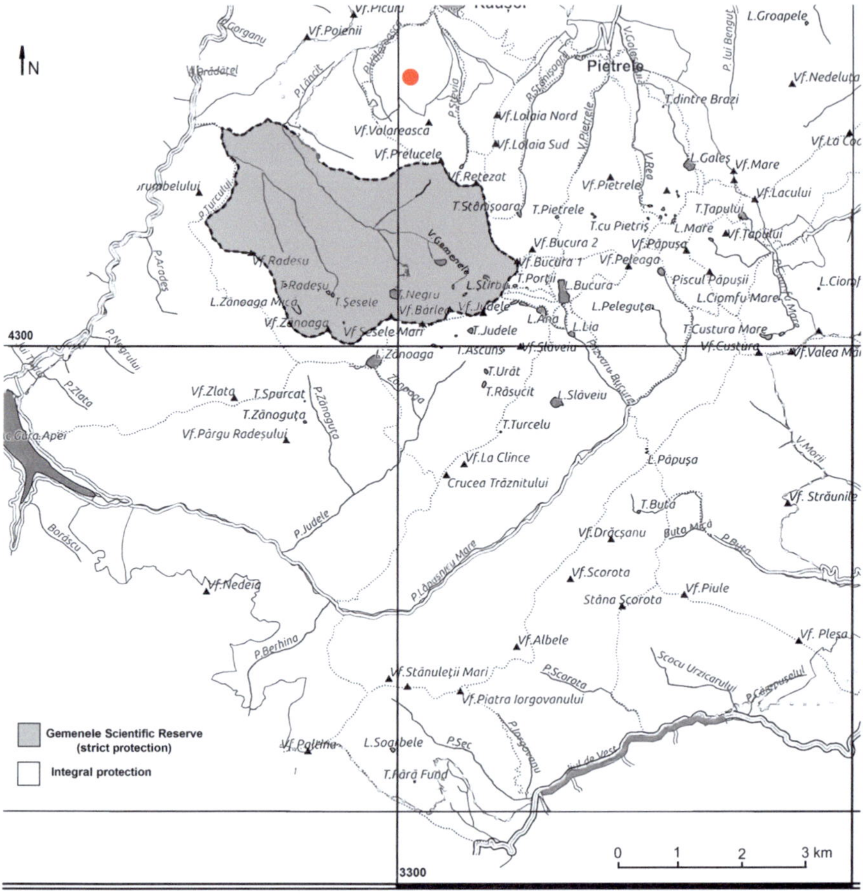

Fig. 4.28 Distribution of *Draba simonkaiana* in the Retezat National Park

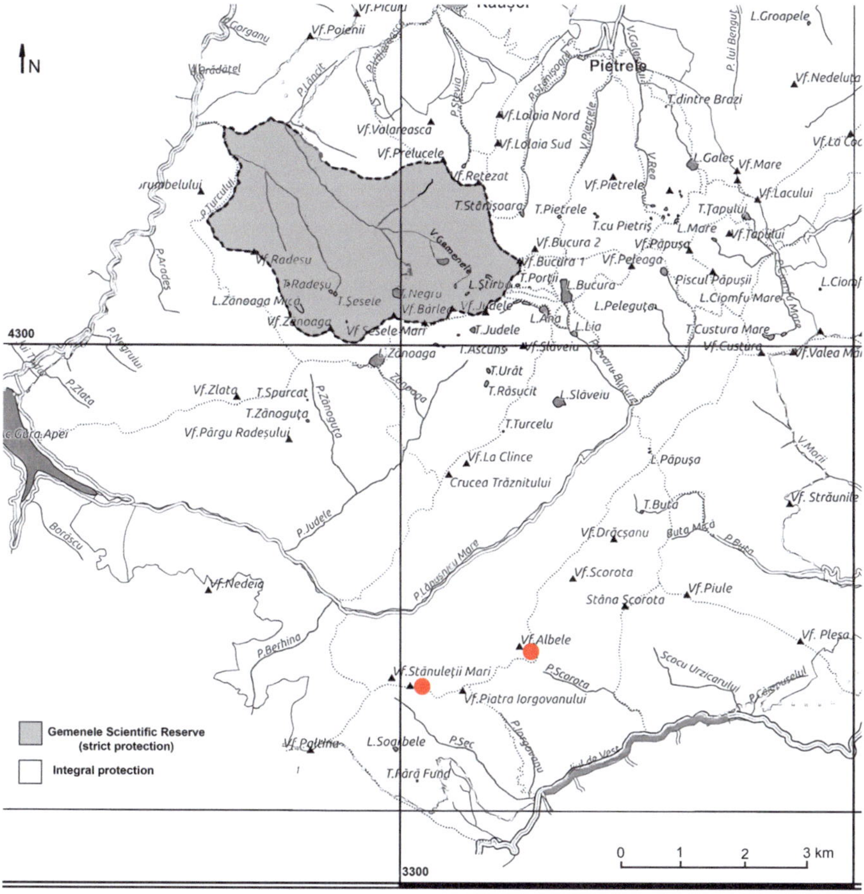

Fig. 4.29 Distribution of *Festuca carpatica* in the Retezat National Park

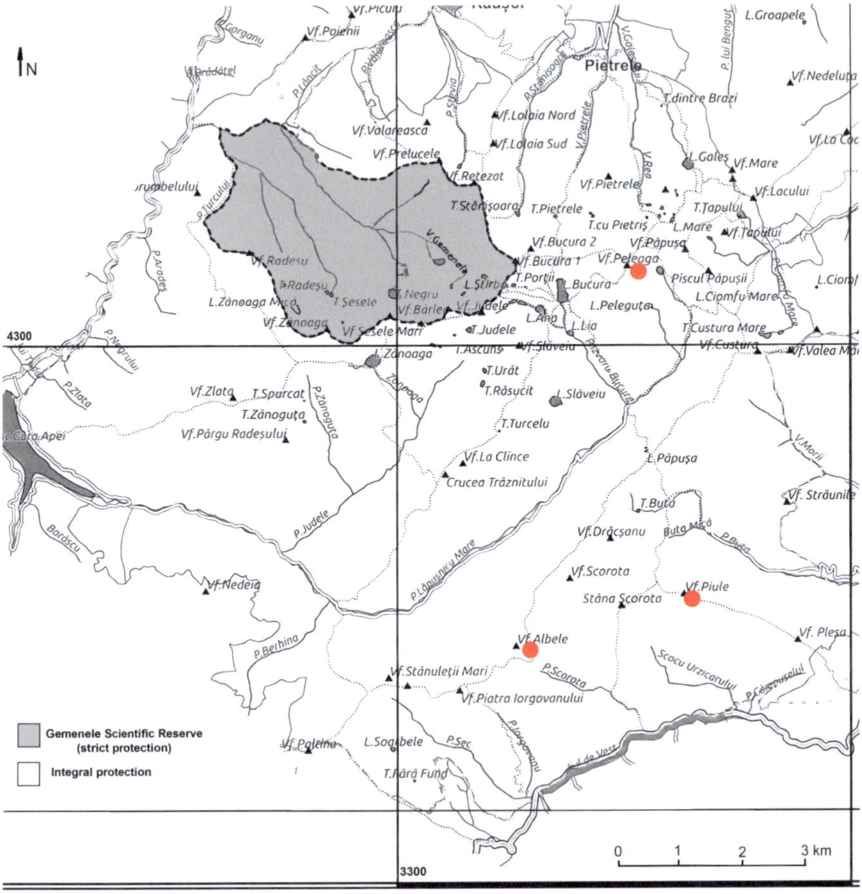

Fig. 4.30 Distribution of *Festuca porcii* in the Retezat National Park

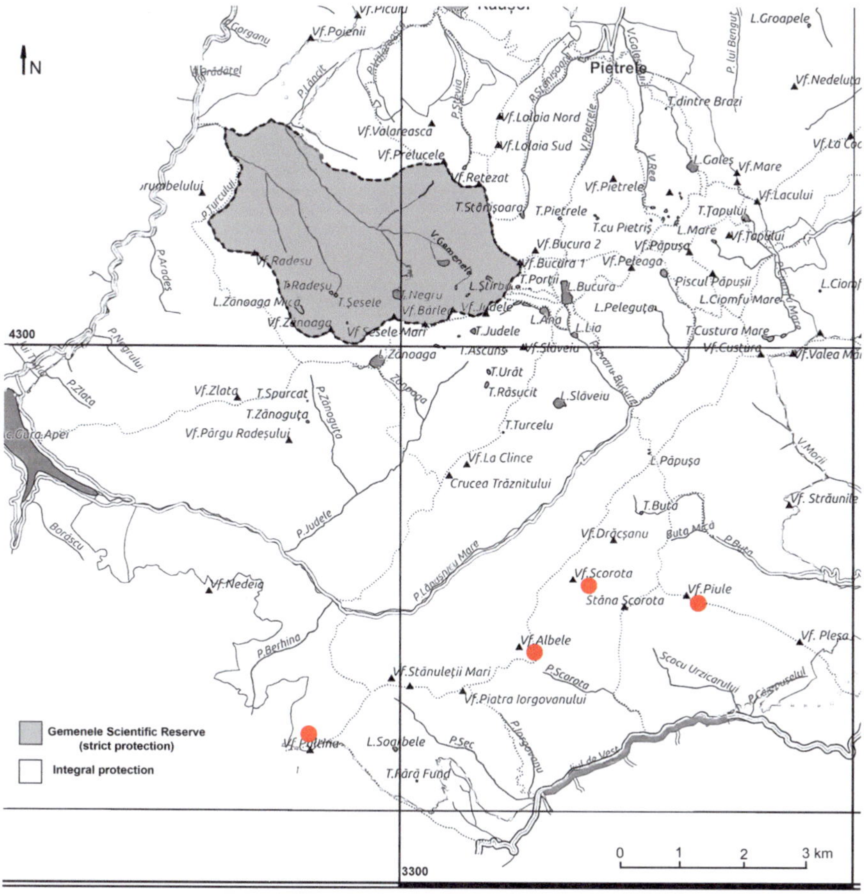

Fig. 4.31 Distribution of *Festuca stricta* subsp. *rumelica* in the Retezat National Park

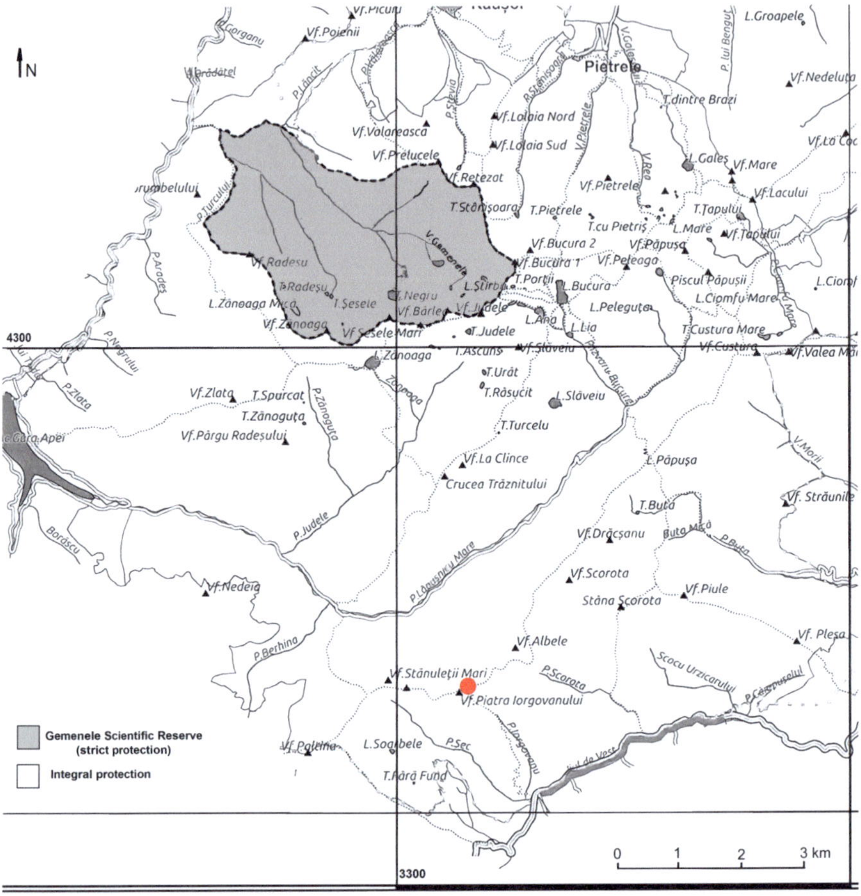

Fig. 4.32 Distribution of *Festuca xanthiana* in the Retezat National Park

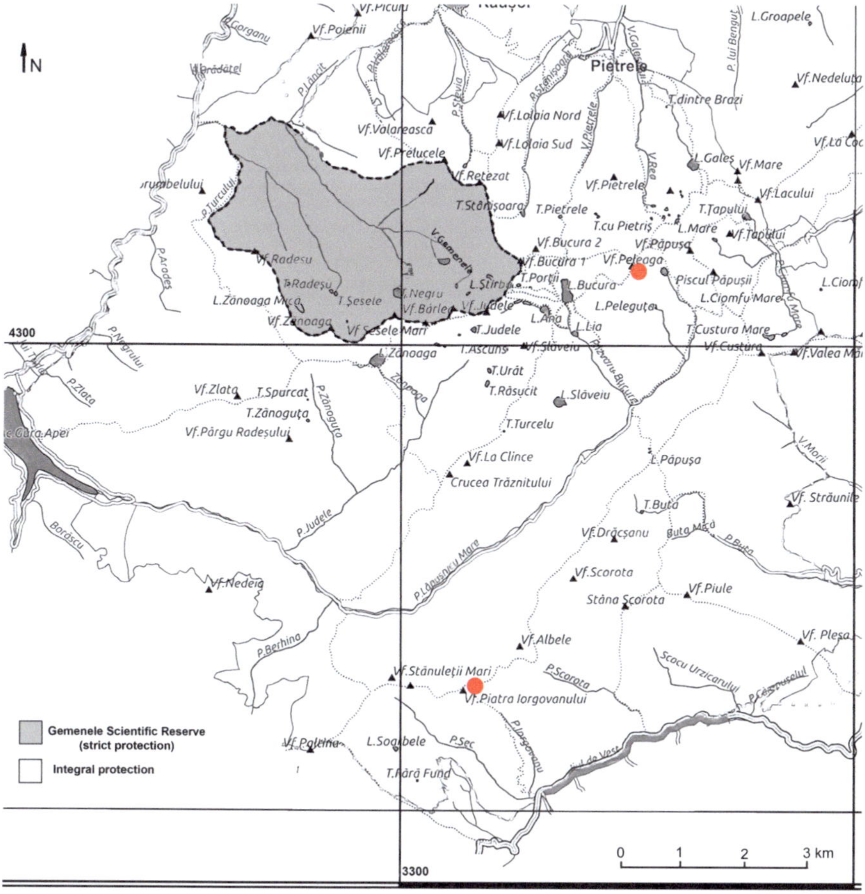

Fig. 4.33 Distribution of *Gentiana frigida* in the Retezat National Park

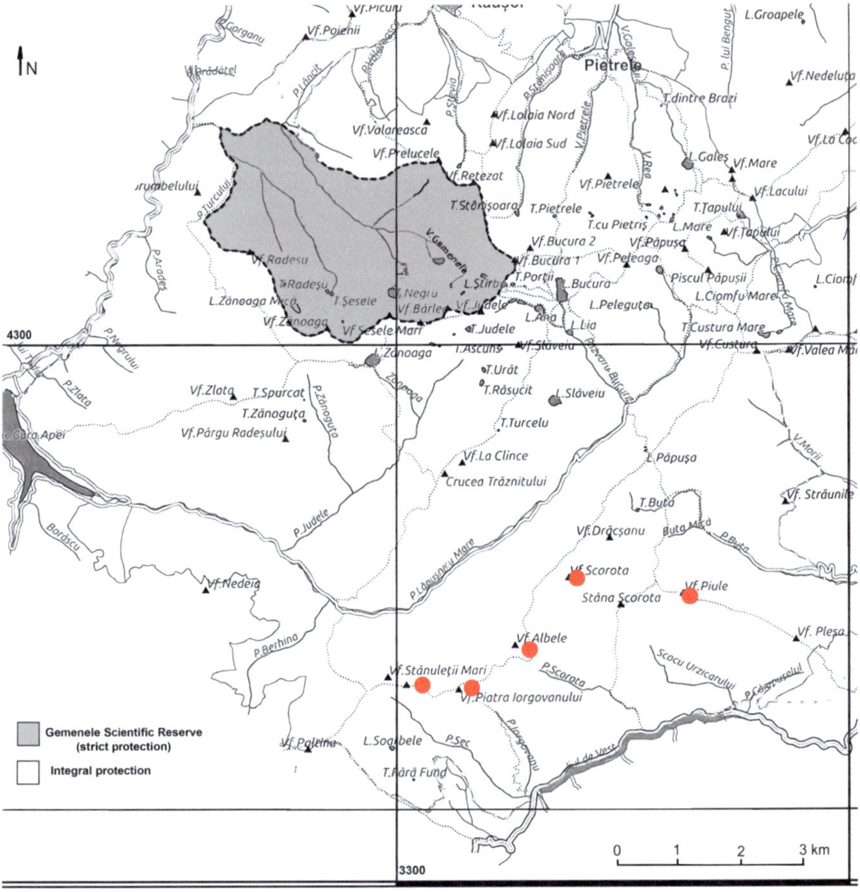

Fig. 4.34 Distribution of *Gentiana lutea* in the Retezat National Park

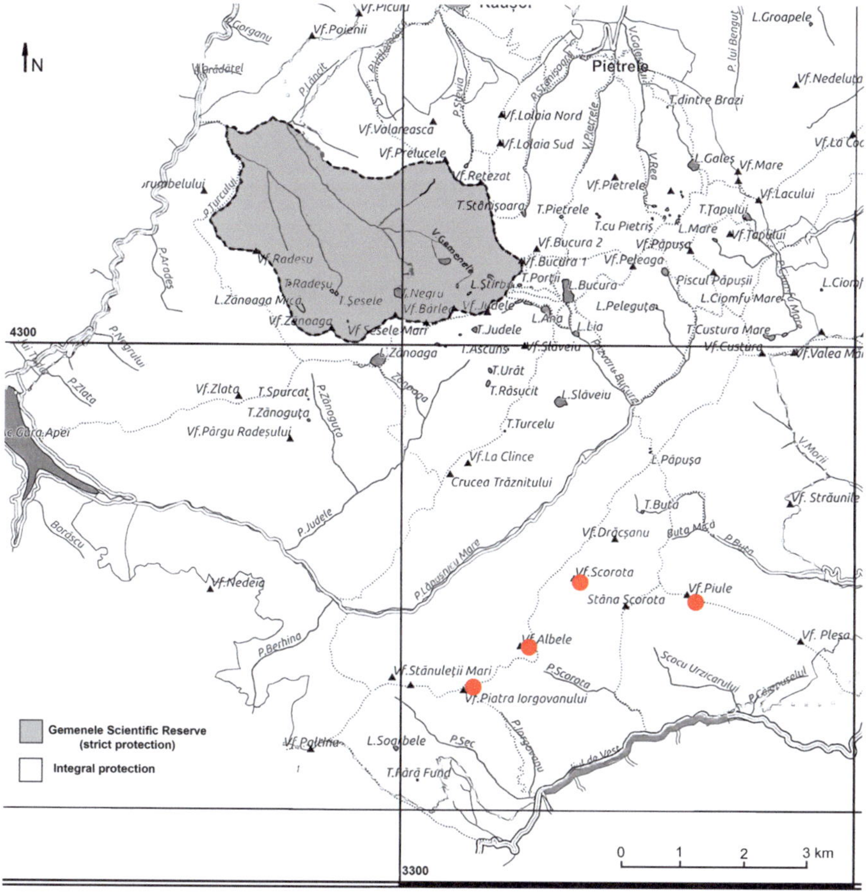

Fig. 4.35 Distribution of *Gypsophila petraea* in the Retezat National Park

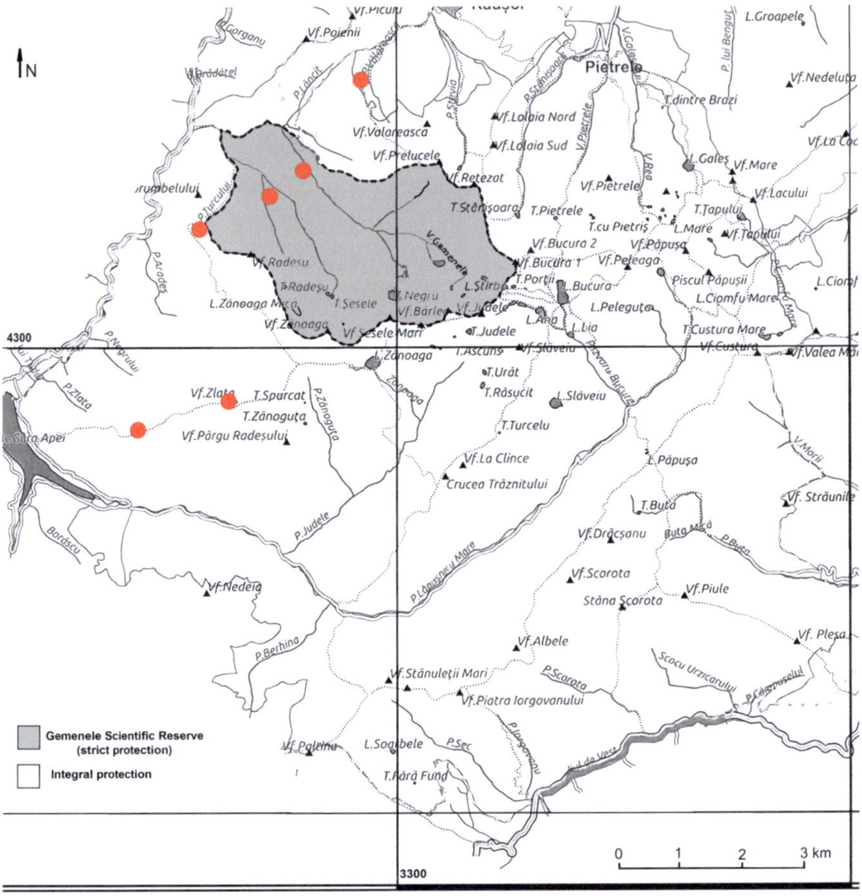

Fig. 4.36 Distribution of *Hepatica transsilvanica* in the Retezat National Park

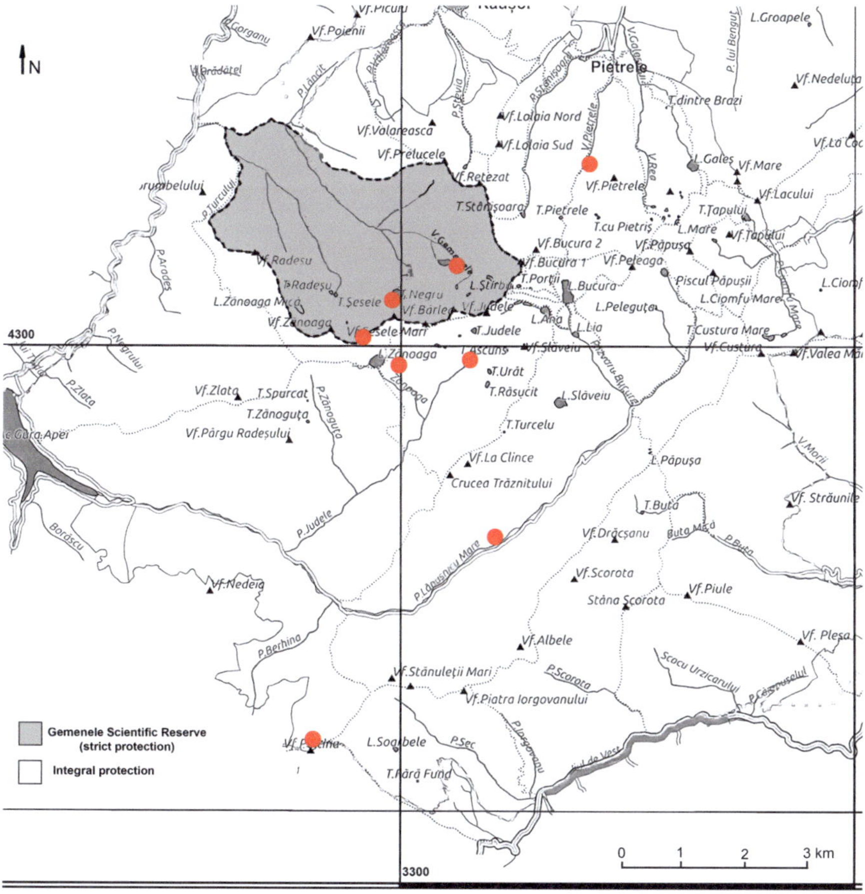

Fig. 4.37 Distribution of *Hieracium paltinae* in the Retezat National Park

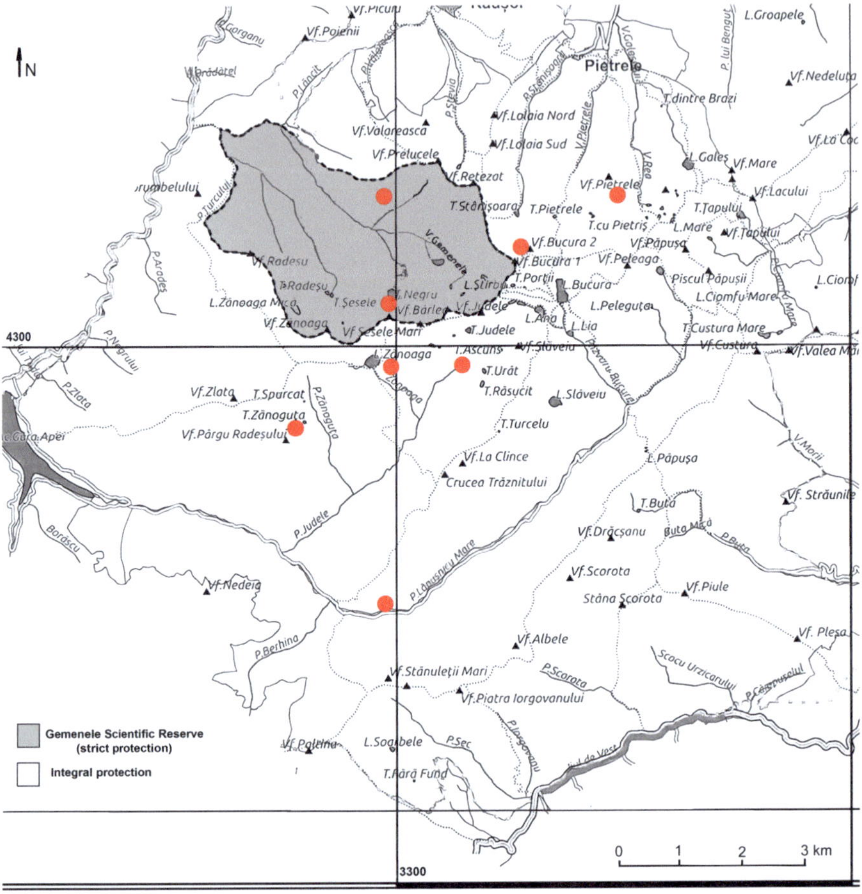

Fig. 4.38 Distribution of *Hieracium sparsum* subsp. *kotschyanum* in the Retezat National Park

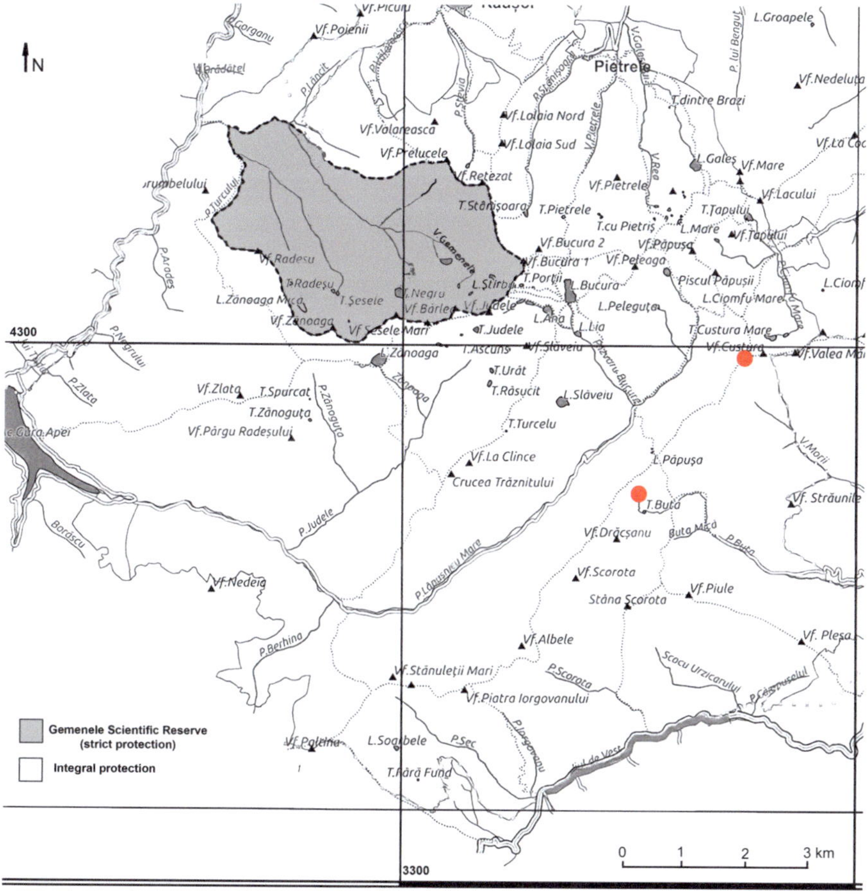

Fig. 4.39 Distribution of *Jasione orbiculata* in the Retezat National Park

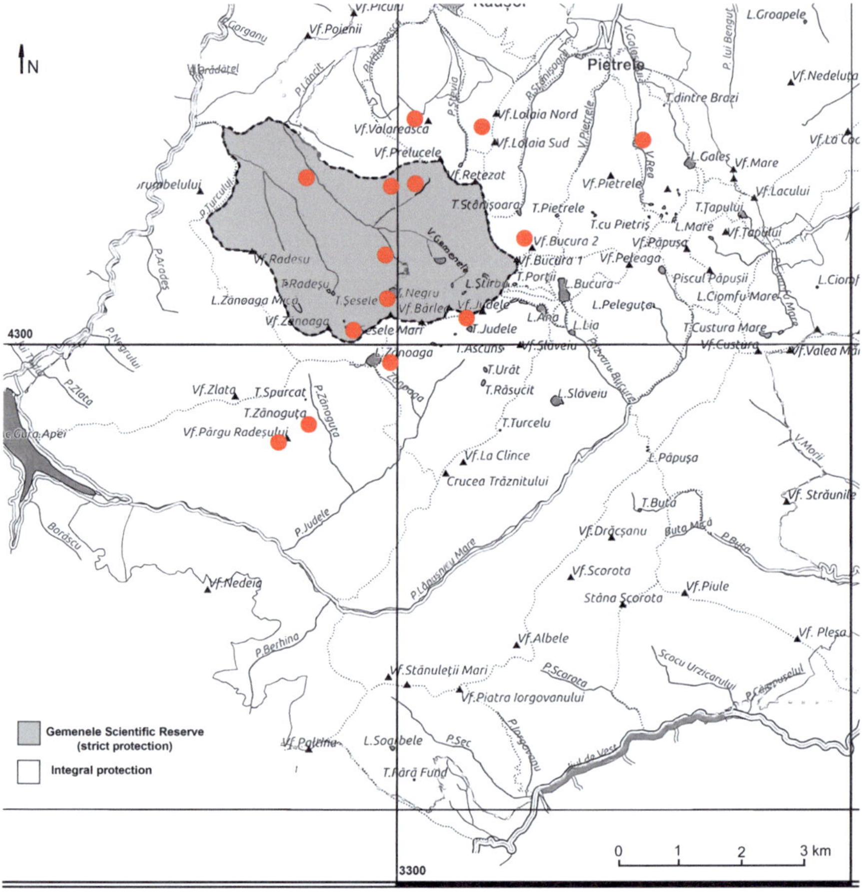

Fig. 4.40 Distribution of *Laserpitium krapfii* (Heuff.) Zahn in the Retezat National Park

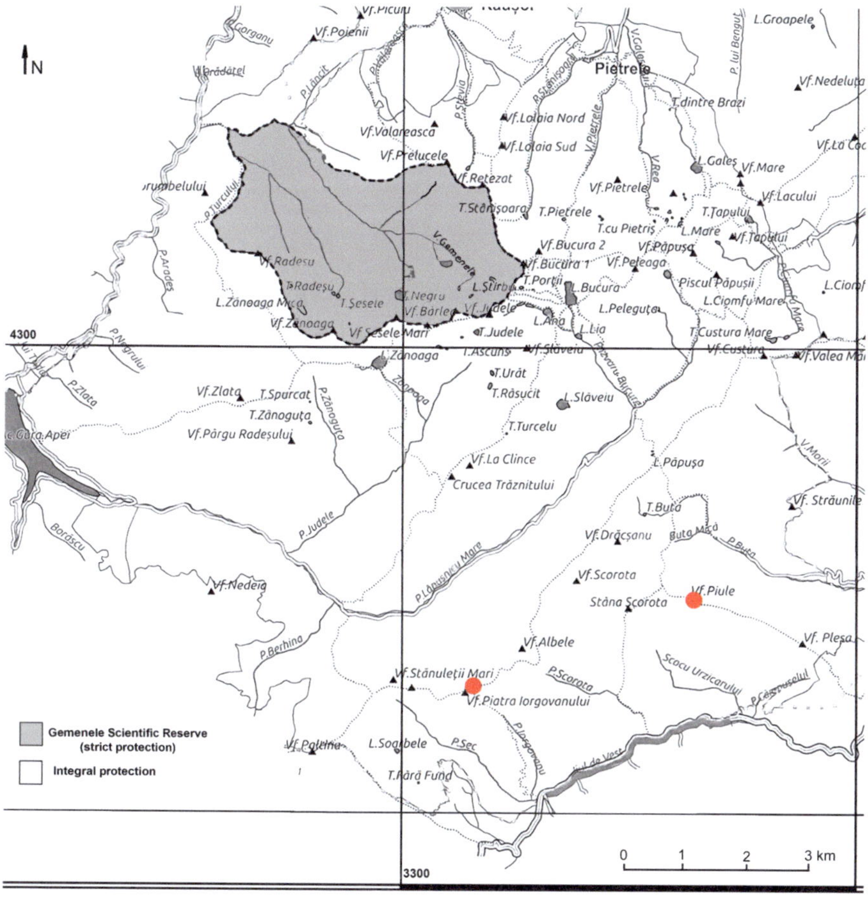

Fig. 4.41 Distribution of *Leontodon croceus* subsp. *rilaensis* in the Retezat National Park

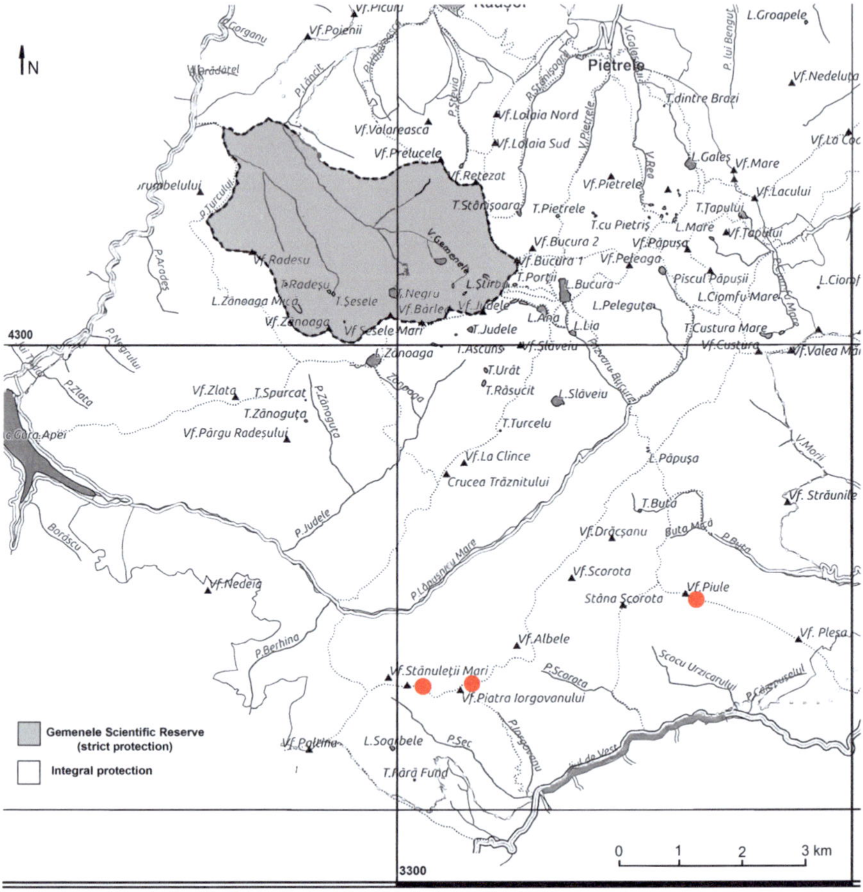

Fig. 4.42 Distribution of *Leontodon montanum* subsp. *pseudotaraxaci* in the Retezat National Park

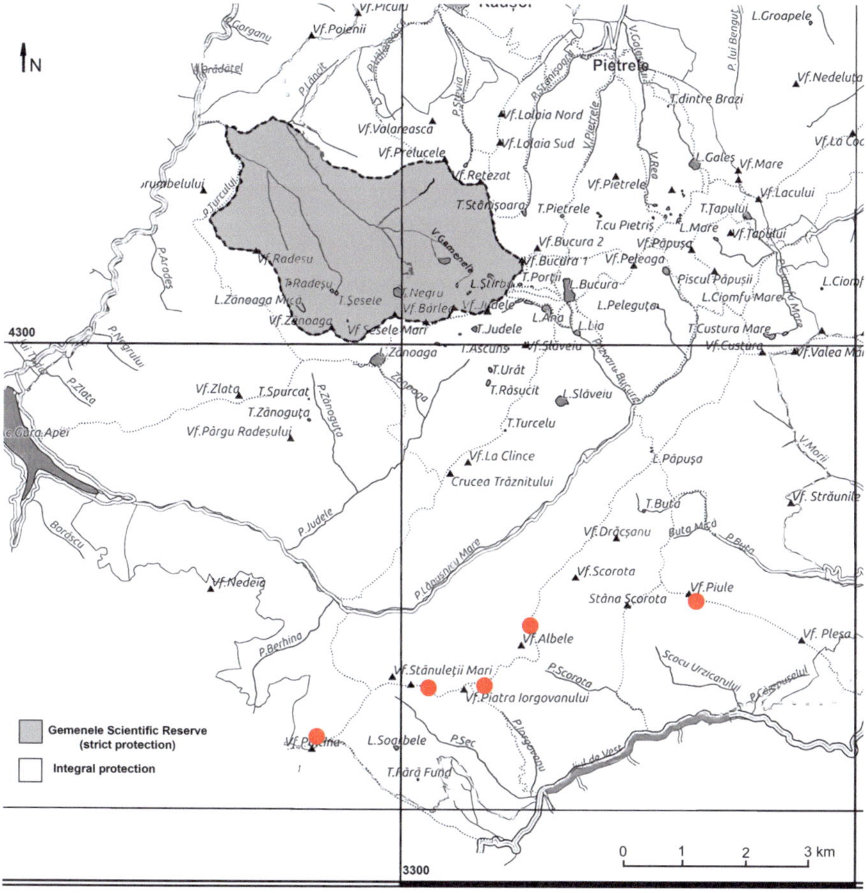

Fig. 4.43 Distribution of *Leontopodium alpinum* in the Retezat National Park

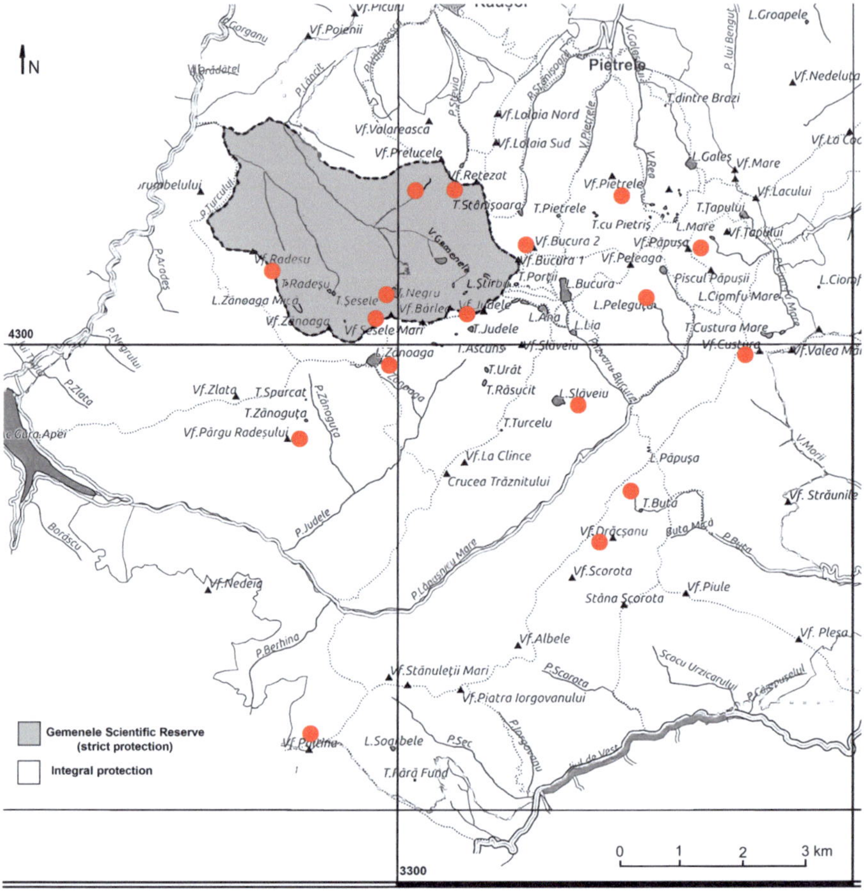

Fig. 4.44 Distribution of *Leucanthemopsis alpina* in the Retezat National Park

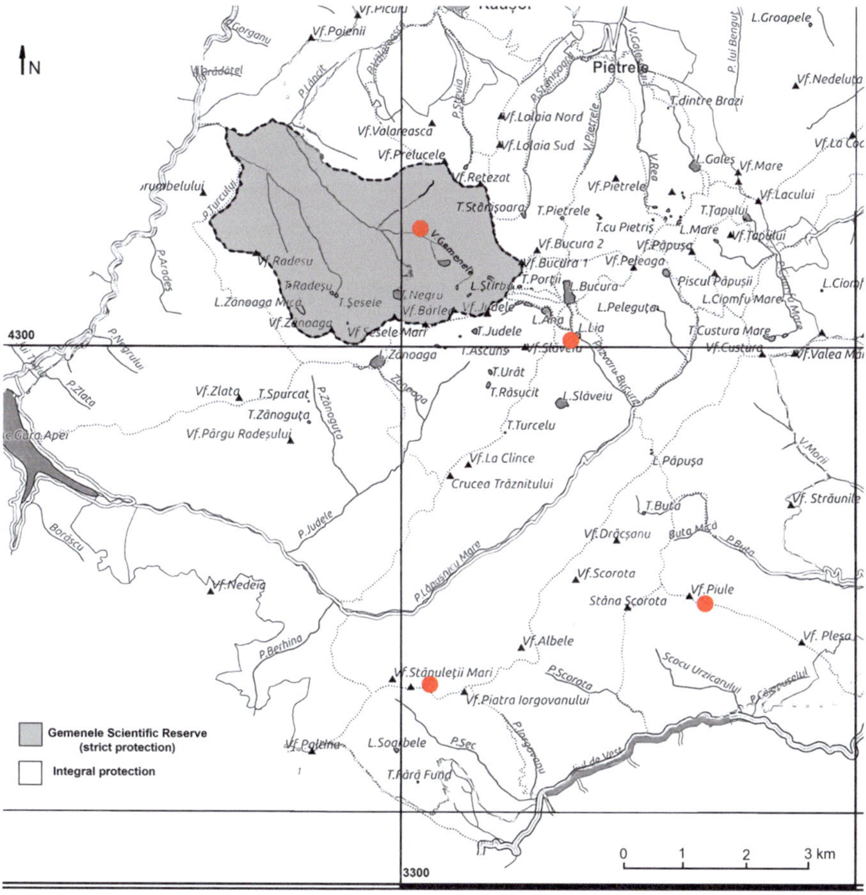

Fig. 4.45 Distribution of *Lilium jankae* in the Retezat National Park

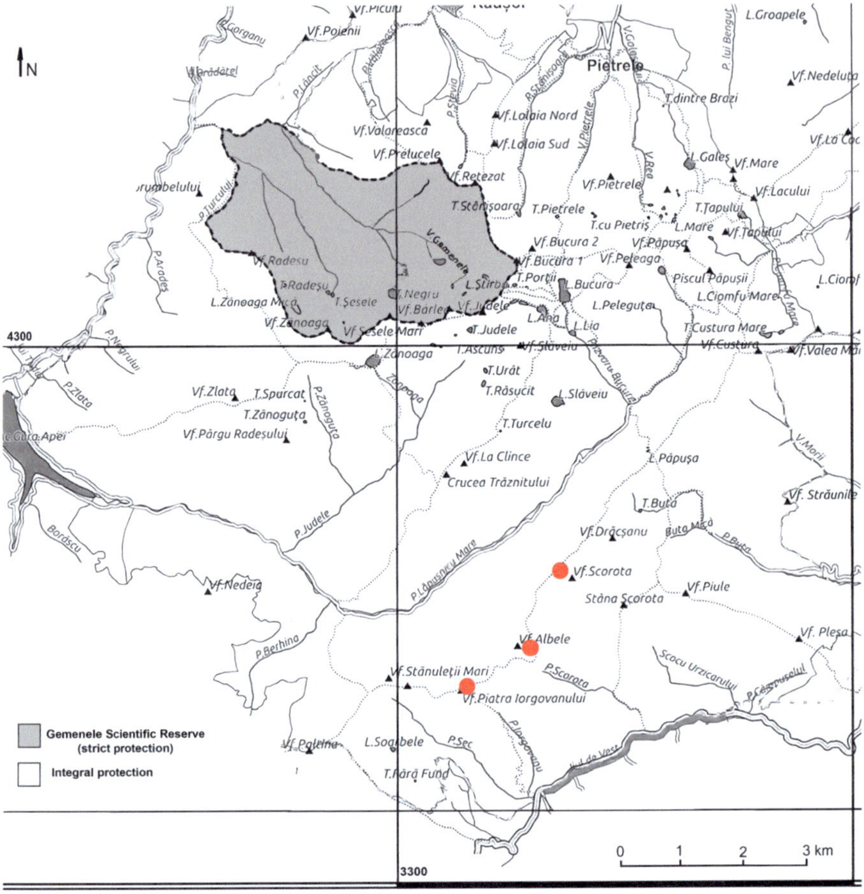

Fig. 4.46 Distribution of *Nigritella nigra* in the Retezat National Park

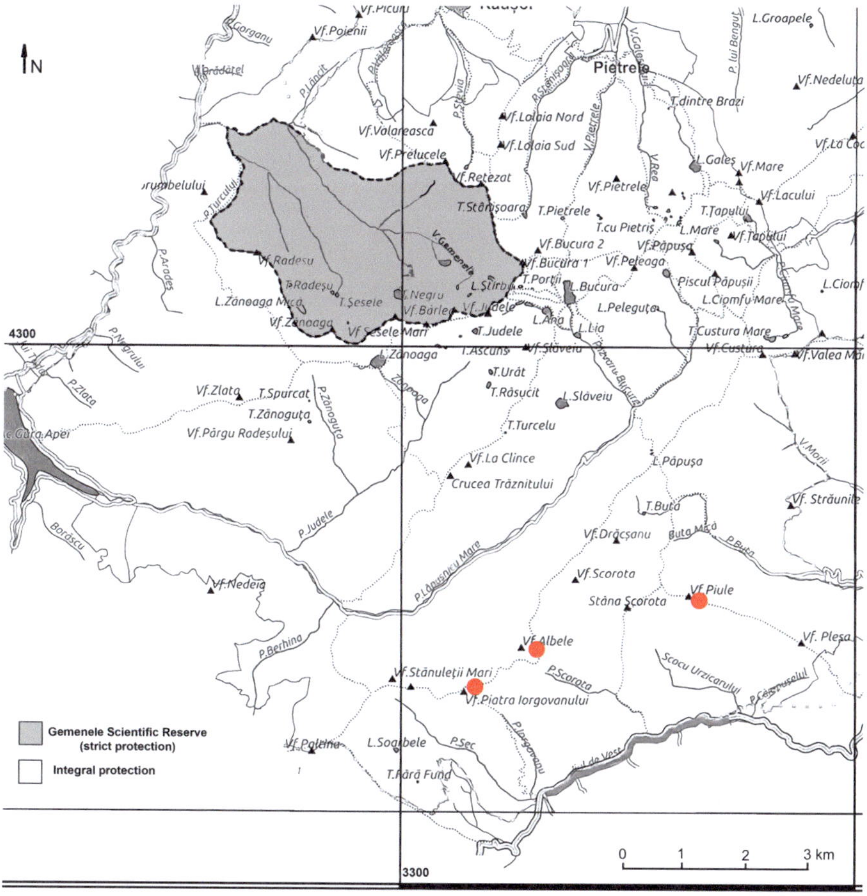

Fig. 4.47 Distribution of *Oxytropis pyrenaica* in the Retezat National Park

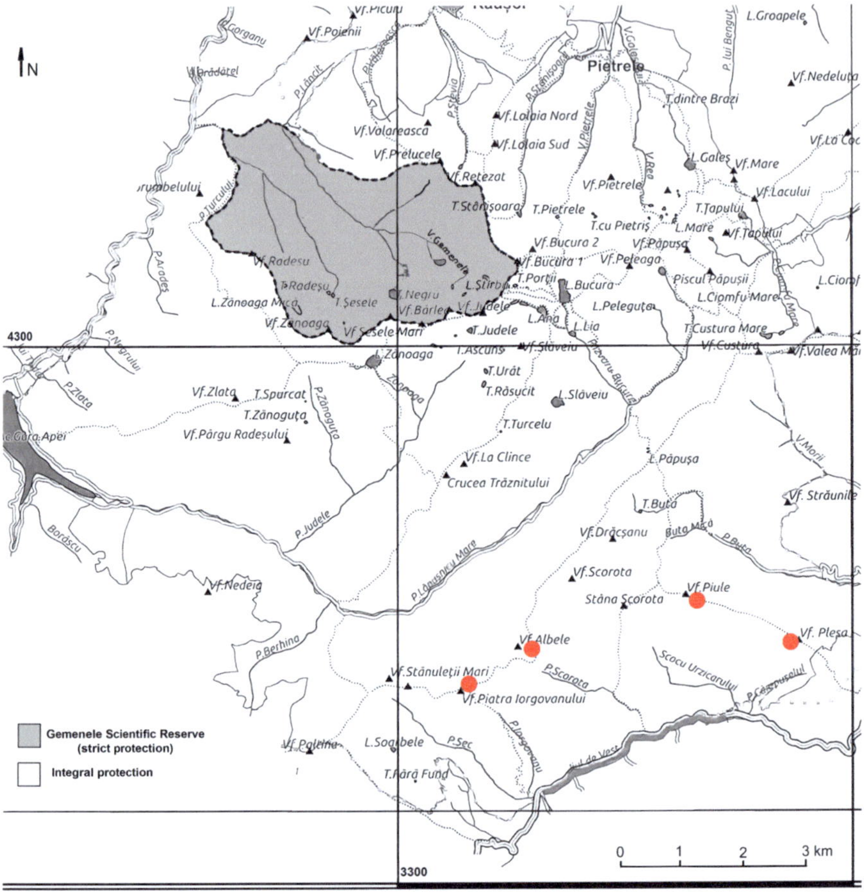

Fig. 4.48 Distribution of *Papaver alpina* subsp. *corona-sancti-stephani* in the Retezat National Park

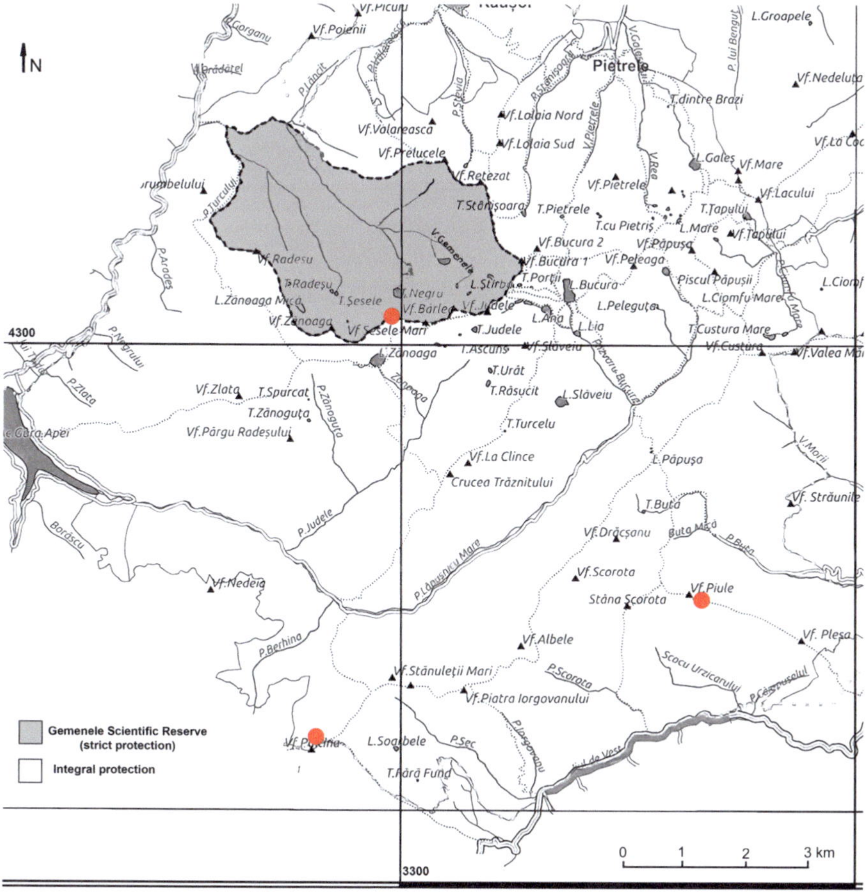

Fig. 4.49 Distribution of *Pedicularis baumgartenii* in the Retezat National Park

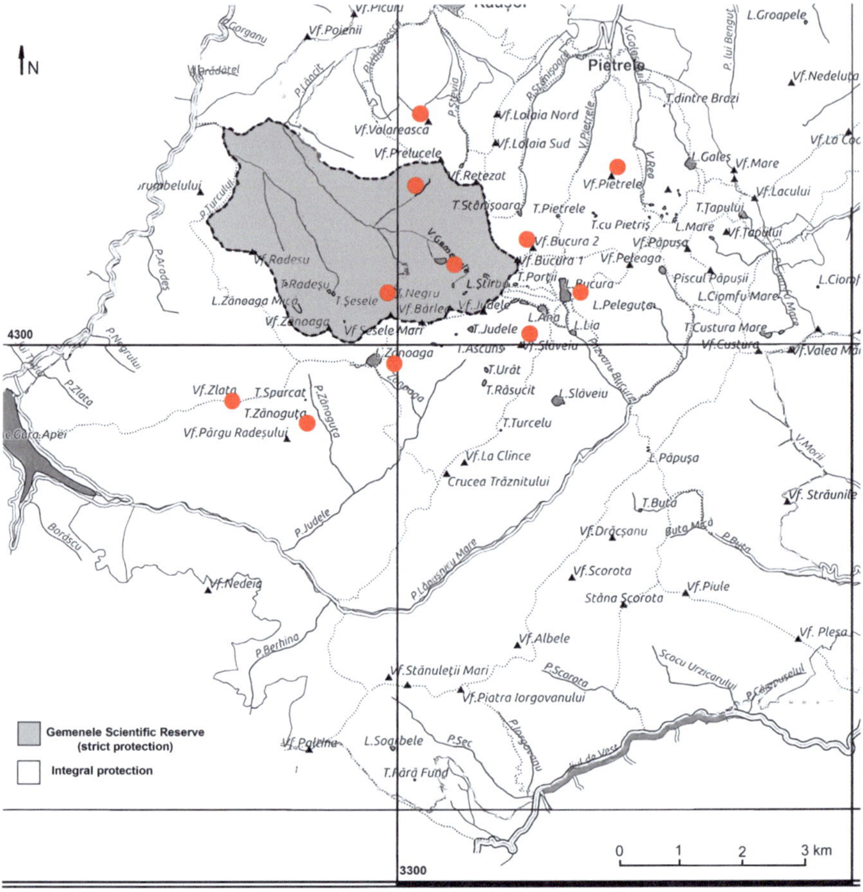

Fig. 4.50 Distribution of *Phyteuma vagneri* in the Retezat National Park

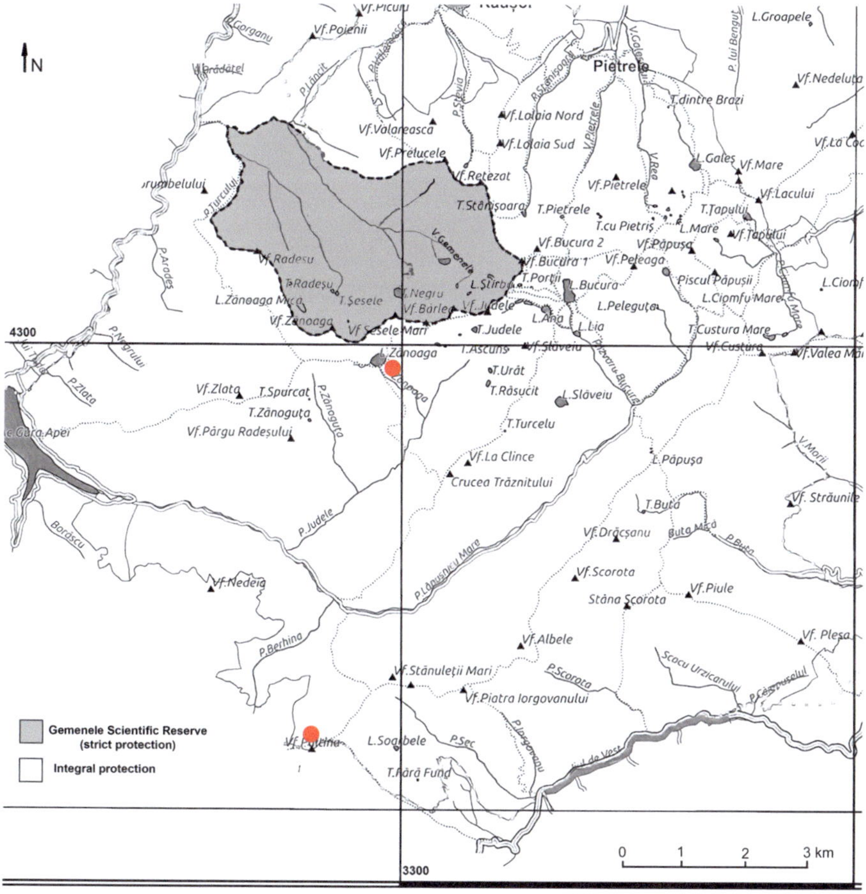

Fig. 4.51 Distribution of *Plantago atrata* subsp. *carpatica* in the Retezat National Park

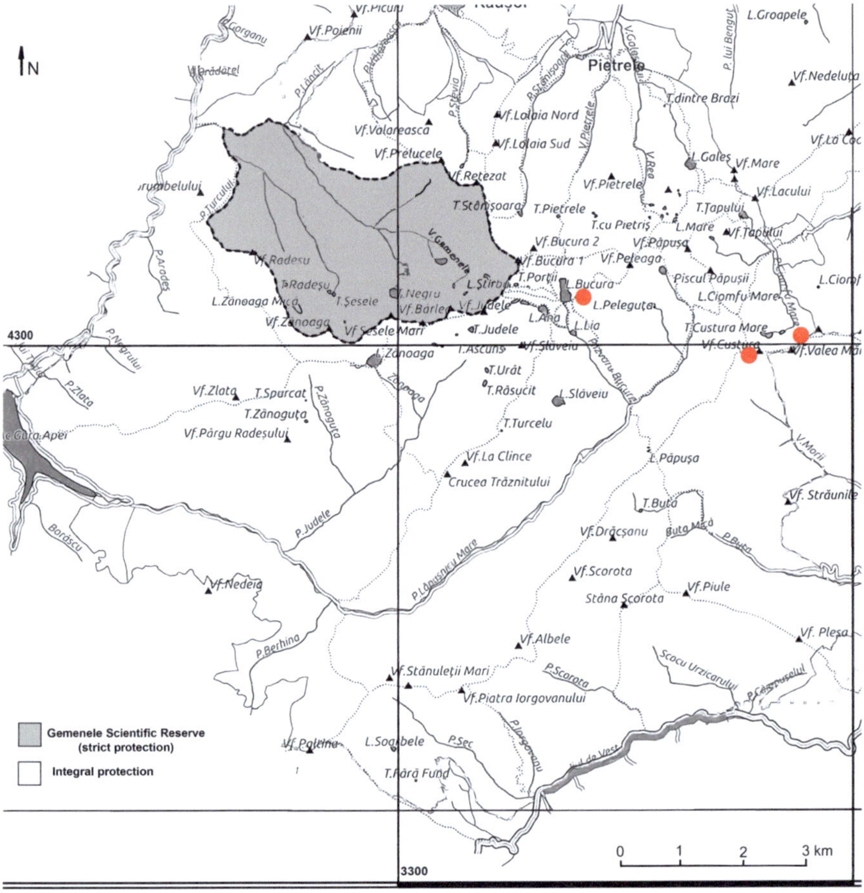

Fig. 4.52 Distribution of *Poa cenisia* subsp. *contracta* in the Retezat National Park

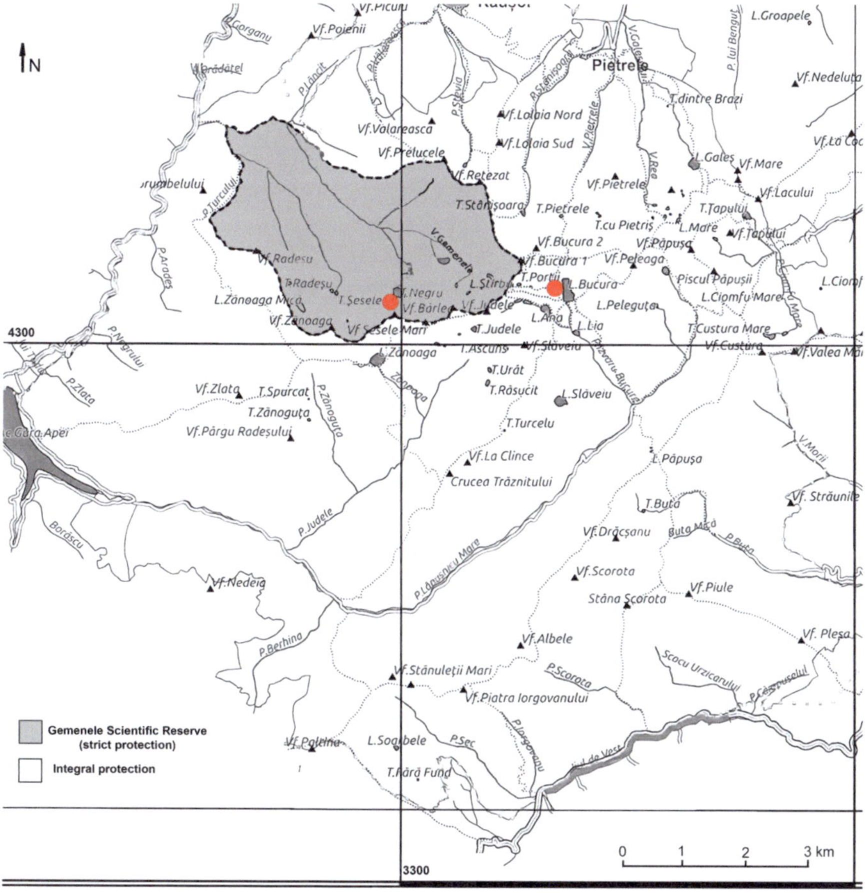

Fig. 4.53 Distribution of *Poa granitica* subsp. *disparilis* in the Retezat National Park

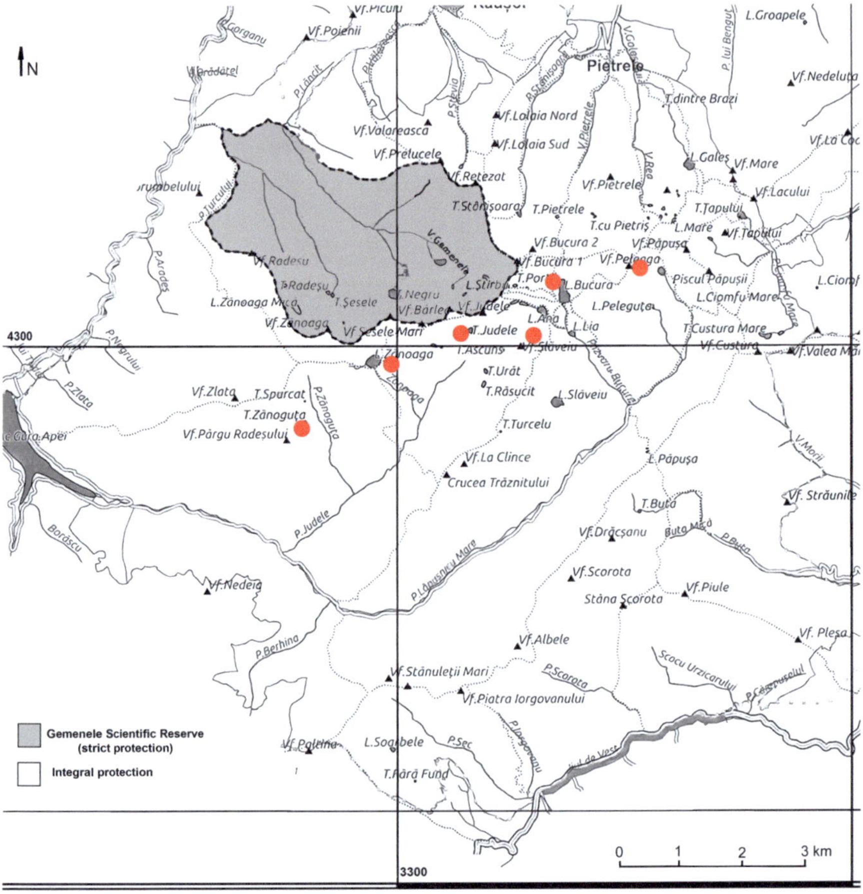

Fig. 4.54 Distribution of *Pseudorchis frivaldii* in the Retezat National Park

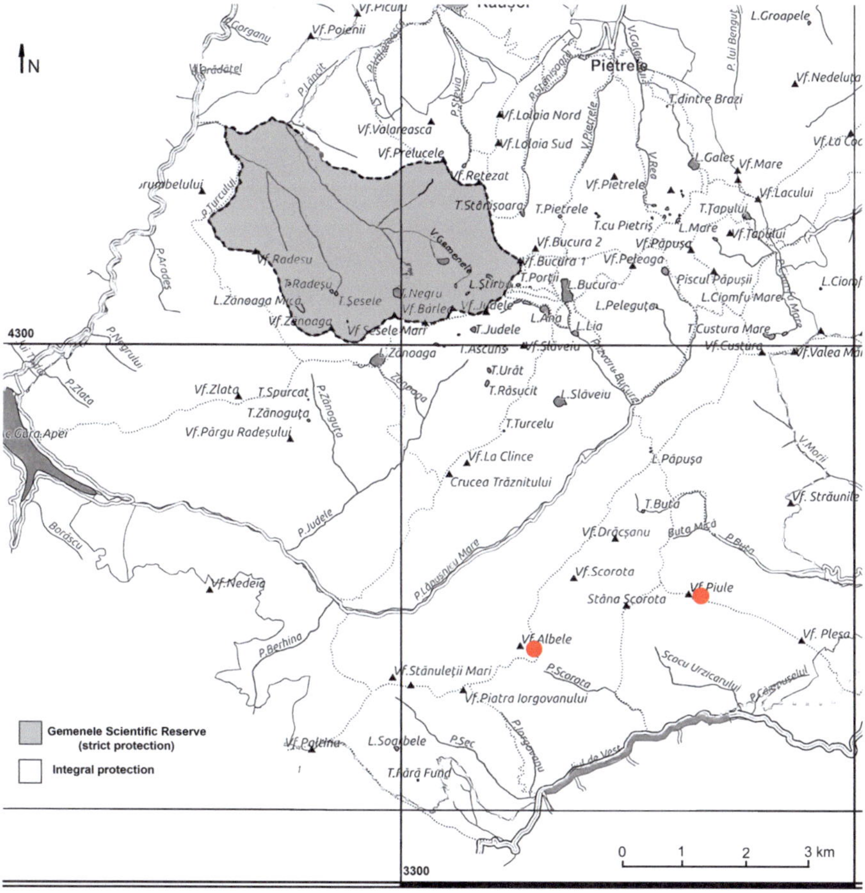

Fig. 4.55 Distribution of *Pulsatilla vulgaris* subsp. *grandis* in the Retezat National Park

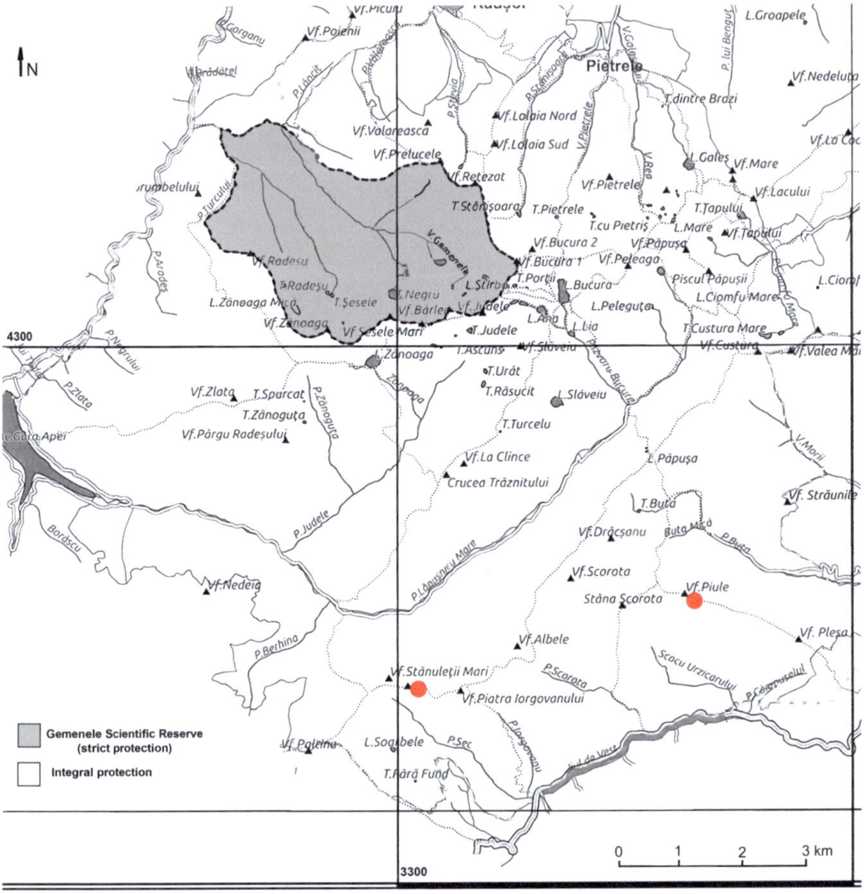

Fig. 4.56 Distribution of *Saxifraga corymbosa* in the Retezat National Park

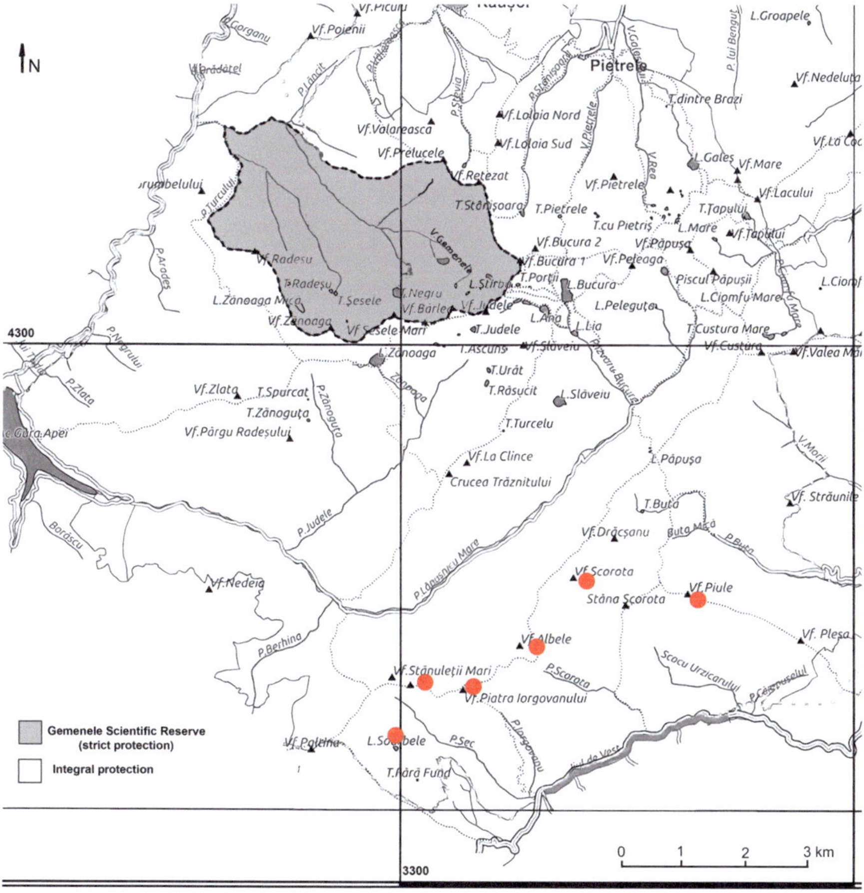

Fig. 4.57 Distribution of *Saxifraga marginata* in the Retezat National Park

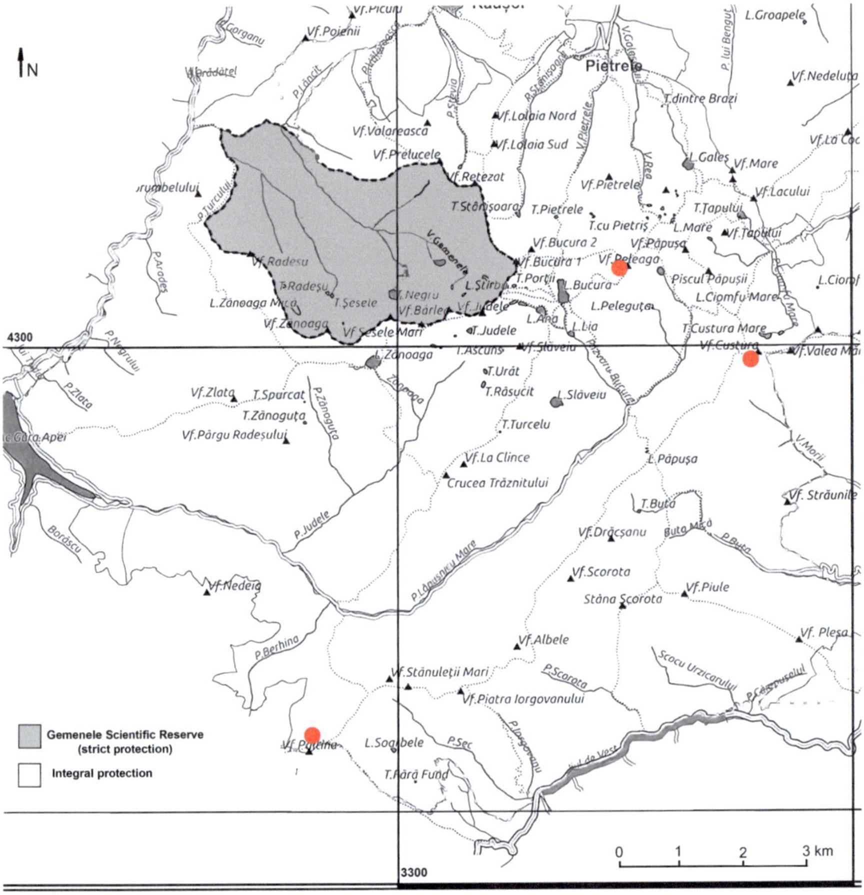

Fig. 4.58 Distribution of *Senecio abrotanifolius* subsp. *carpaticus* in the Retezat National Park

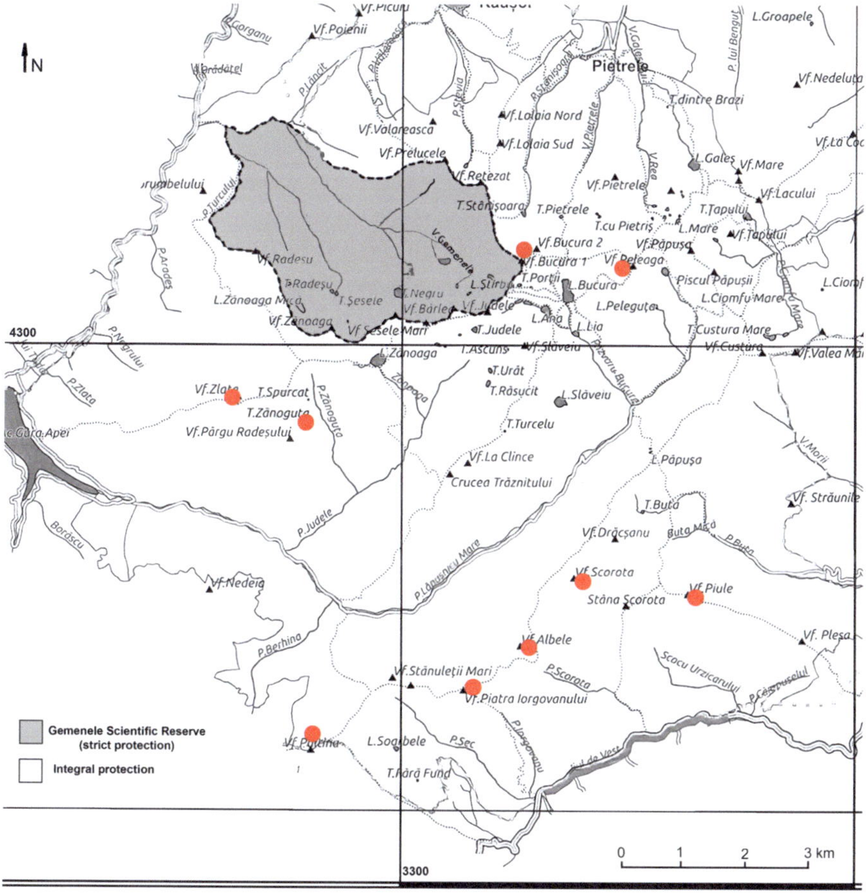

Fig. 4.59 Distribution of *Thlaspi dacicum* in the Retezat National Park

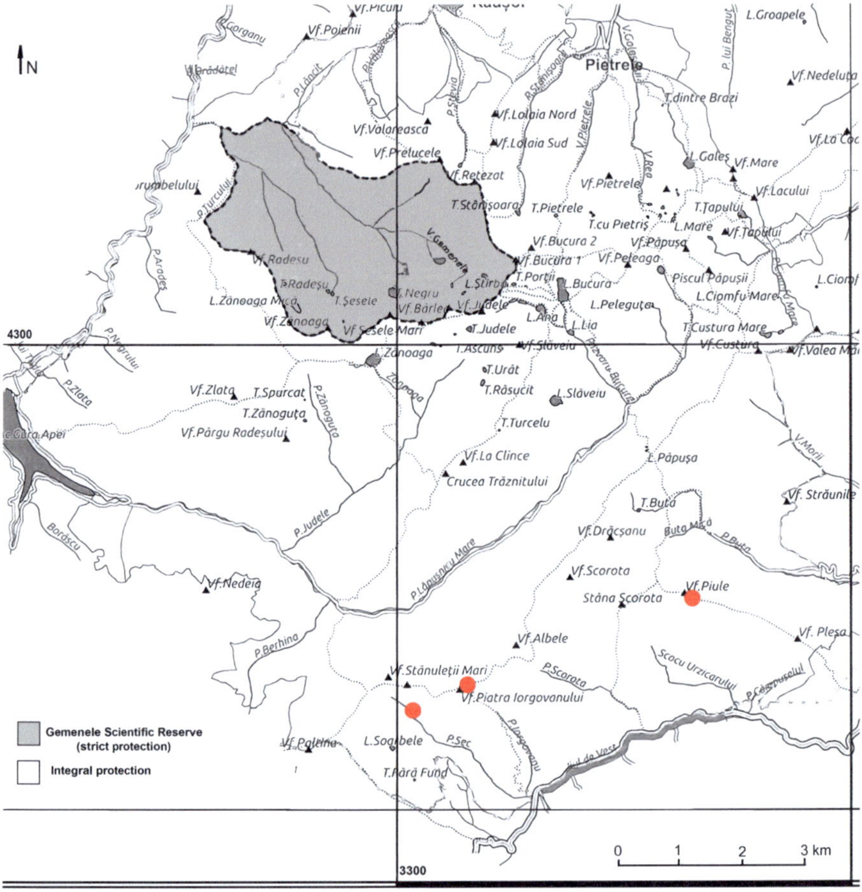

Fig. 4.60 Distribution of *Tozzia alpina* subsp. *carpathica* (Wol.) Dostal in the Retezat National Park

Chapter 5
Vegetation of the Retezat Mountains

5.1 History of Late- and Post-glacial Vegetation

Spore-pollen analysis of the lake sediment from Tăul Zănoguța (1840 m), situated in the southeastern part of the Retezat Mountains, was carried out by Pop et al. (1970) and continued later by Fărcaș et al. (1999). ^{14}C-dating of pollen sequences revealed the late- and post-glacial forest phases in the area and their placement in time.

Zone 1, located at the lowest level in the Tăul Zănoguța (565–555 cm) profile revealed the post-glacial vegetation succession indicated by the high frequency of *Pinus, Artemisia* (50%), and *Chenopodiaceae* pollen. These results suggest that the lowlands surrounding the Retezat Mountains were covered by steppic vegetation dominated by *Artemisia* and some species of *Chenopodiaceae*. The slopes below Zănoguța glacial cirque were covered by more or less compact mountain pine (*Pinus mugo*) shrublands (Pop et al. 1970).

Zone 2, located immediately above the first within the profile (555–465 cm) showed increasing *Pinus* pollen content and a slight increase in *Betula* pollen. This indicated a late glacial interstage which, according to ^{14}C-dating, corresponded to 11,140 years B.P. The steppic vegetation and the *Poaceae* species remained abundant while the juniper shrublands covered the slopes up to an altitude of 1400 m.

Zone 3, in the upper profile (465–445 cm), revealed a slight recovery of *Artemisia* pollen, a high level of *Pinus* (30%), and regression in *Picea* pollen, indicating a deterioration of the climate in the young Dryas period (about 10,000 years B.P.).

Zone 4 of the profile (445–365 cm) revealed increased *Alnus viridis* and *Pinus mugo* pollen, marking the beginning of Holocene, when the expansion of *Picea* pollen initiated.

Zone 5 of the profile (363–305 cm) was characterized by the continuous increase in *Picea* pollen and decrease in *Pinus mugo* pollen, estimated to 9000 years B.P., based on its ^{14}C analysis.

G. Coldea, V. Cristea, *The Vascular Plant Communities of the Retezat National Park (Southern Carpathians)*, Geobotany Studies,
https://doi.org/10.1007/978-3-031-05618-5_5

Zone 6 of the profile (305–205 cm) was assessed to 8500 years B.P. according to ^{14}C-dating. It is characterized by the development of mesophytic vegetation at low altitudes. This coincided with a significant increase in *Corylus* pollen and the important presence of *Quercus* and *Ulmus* pollen in the palynological diagram. This forest phase corresponds to the climatic optimum of the Boreal.

Zone 7 of the profile (205–145 cm) is marked by an important increase in *Carpinus* pollen in the Retezat Mountains. This occurred 5275 years B.P. based on ^{14}C analyzes and corresponded to the Subboreal.

Zone 8, the last one in the profile (145–20 cm), showed a visible increase in *Fagus* pollen towards the end of the Subboreal, that intensified gradually throughout the whole Subatlantic period. The start in the *Fagus* pollen spreading was estimated at 3000 years B.P. (Pott 2000).

These palynological studies by Fărcaş et al. (1999) confirmed the existence of glacial refugia of *Picea abies* populations in the Southern Romanian Carpathians, which expanded northwards in the Eastern Romanian Carpathians in the Holocene.

5.2 Methodology for the Study and Presentation of Current Vegetation

Phytocoenological research in the Retezat National Park was carried out based on the phytosociological approach by Braun-Blanquet (1964), adapted by Borza (1934) to the particularities of the Romanian vegetation. Phytosociological relevés were carried out in the field on representative surfaces as close as possible to the minimum area of each vegetation type. The relevés with similar floristic composition were grouped in distinct plant associations, based on characteristic (diagnostic) species and constant species (Jarolimek and Šibik 2008), also considering the dominant ones, which indicate the local pedo-climatic conditions. In analytical tables, the first digit of the species scores represents the combined estimation of abundance-dominance, expressed by the scale proposed by Braun-Blanquet (r, +, 1, 2, 3, 4, 5), whereas the second digit represents the local frequency of the species in the relevé's area, according to the method proposed by Borza (Borza 1934, 1959). By using this notation system, the local frequency of species individuals within the relevé's area is expressed, resulting in a scale with five classes (1, 2, 3, 4, 5). The species frequencies were calculated in percentage but reported in tables by means of five classes defined as follows: I (<20%); II (21–40%); III (41–60%); IV (61–80%); V (81–100%). The identified plant associations were assigned to higher-ranking coenotaxonomic units (alliances, orders, and vegetation classes) based on the presence of characteristic species designated by several foreign phytocoenologists (Oberdorfer 2001; Chytrý 2007, 2011, 2013; Valachovič 2001; Jarolimek and Šibik 2008), but also by Romanians (Borza 1959; Boşcaiu 1971; Dihoru 1975; Coldea 1990, 1991; Cristea et al. 2015). The regional, Carpathian, and Carpathian-Balkan species present in the composition of the phytocoenoses were the main floristic arguments for the distinction and description of new syntaxa in the region.

5.3 Vegetation Belts of the Retezat Mountains

Anthropo-zoogenic factors have produced, over time, profound changes in the composition of the natural vegetation of the Retezat Mountains, affecting particularly the extension of the forested area (Geanana 1975), although since 1935 these mountains have been included in the list of protected areas as National Park by A. Borza (JCM 593/1935). In many parts of the massif, such as the upper basins of Pietrele, Bărbat, and Buta valleys, in the northeastern Retezat, the human activities, inducing deforestation of woody vegetation and intensive grazing, resulted in the artificial lowering of the upper forest limit below the natural climatic-determined treeline. Mesophilic, secondary grasslands, settled at altitudes over 1600 m on the deforested and continuously grazed lands and dominated by *Nardus stricta* and *Festuca nigrescens*, outline a misleading image of "alpine area", without including the typical alpine phytocoenosis. Subsequently, we present the distribution of forest, shrub and natural (potential) herbaceous formations in the area of the Retezat Mountains in relation to the orographic factors (altitude, aspect, and slope).

The mountain belt, well-individualized, is sub-divided into the lower mountain belt (500–800 m), middle (800–1250 m), and upper mountain belt (1250–1800 m). The lower mountain belt is characterized by mixed sessile oak (*Quercus petraea*) and hornbeam (*Carpinus betulus*), linden (*Tilia cordata*), and Scots pine (*Pinus sylvestris*) forests. The middle mountain belt consists of pure beech (*Fagus sylvatica*) forests as well as mixed forests of beech, fir (*Abies alba*) and spruce (*Picea abies*). The upper mountain belt is mostly covered with spruce (*Picea abies*) forests and mixed spruce-Swiss stone pine (*Pinus cembra*) woodlands.

The subalpine belt (1800–2200 m) includes mountain pine (*Pinus mugo*), alpine juniper (*Juniperus communis* subsp. *alpina*), and mountain alder (*Alnus viridis*) scrubs, which include boreal species typical of spruce forests (*Homogyne alpina*, *Soldanella major*, *Vaccinium vitis-idaea,* etc.) as well as microthermal species found in alpine grasslands (*Agrostis rupestris*, *Festuca supina*, *Campanula alpina*, *Calamagrostis villosa*).

The alpine belt (2200–2509 m) is differentiated from the subalpine one by the lower thermal regime and longer duration of the snow cover above 2200 m altitude. The dwarf scrub and herbaceous plant communities typical of the lower alpine belt are those grouped in the *Loiseleurio-Vaccinion*, *Salicion herbaceae* and *Veronicion baumgartenii* alliances. The microthermal herbaceous communities typical to the alpine belt itself are those grouped in the *Caricion curvulae* alliance, i.e. *Primulo-Caricetum curvulae*, *Phyteumo-Juncetum trifidi,* and *Potentillo ternatae-Festucetum supinae* associations.

The distinction between these vegetation belts in the field is easily noticeable as one climbs from the Gura Zlata lodge to the Gemenea chalet and then to the Faţa Retezatului and Retezat peak (2485 m). The main sigma-associations observed on this route were described in a previous paper by Coldea (1982).

Chapter 6
Description of the Plant Associations Distinguished in the Retezat National Park

Class *Asplenietea trichomanis* Br.-Bl. in Meier and Br.-Bl. 1934 corr. Oberd. 1977

This class groups saxicolous, pioneer phytocoenoses occurring on rocky walls and rock cracks, mainly composed of crystalline schists and occasionally Jurassic and Cretaceous limestone reefs of the montaine and alpine belts. The saxicolous phytocoenoses are poor in species, and their floristic structure differs according to the geological substrate (basic or acidic). They are grouped into two different orders: *Potentilletalia caulescentis*, comprising basophilic coenoses, and *Androsacetalia vandellii*, containing acidophilic ones.

Order *Potentilletalia caulescentis* Br.-Bl. in Br.-Bl. et Jenny 1926

This order groups calcophilic communities of sunny cliffs and shaded rocky cracks from the montane and subalpine belts. The order includes two alliances, well individualized floristically and phytogeographically: *Gypsophilion petraeae* and *Cystopteridion* (Boşcaiu et al. 1978; Coldea 1991).

Alliance *Gypsophilion petraea* Borhidi et Pócs 1957
Natura 2000: habitat type 8210

The sunny limestone cliffs from the Calcareous Retezat (Piatra Iorgovanului) are frequently populated by the xero-mesophilous communities of some Carpathian endemic species such as *Gypsophila petraea, Androsace villosa* subsp. *arachnoidea, Asperula capitata, Eritrichium nanum* subsp. *jankae,* and *Campanula kladniana* that define this regional alliance for the South-Eastern Carpathians (Boşcaiu et al. 1978).

Ass. *Saxifrago rochelianae-Gypsophiletum petraeae* Boşcaiu et al. 1977 (Table 6.1)

The calcophilic, saxicolous, xero-mesophilous coenoses of the regional species *Gypsophila petraea, Saxifraga marginata* var. *rocheliana* and *Edraianthus graminifolius*, frequent in the Piatra Iorgovanului peak area (1650–1720 m), were

G. Coldea, V. Cristea, *The Vascular Plant Communities of the Retezat National Park (Southern Carpathians)*, Geobotany Studies,
https://doi.org/10.1007/978-3-031-05618-5_6

Table 6.1 Ass. *Saxifrago rochelianae-Gypsophiletum petraeae* Boşcaiu et al. 1978

Relevé No.	1	2	3	4	5	
Altitude (m a.s.l.)	1640	1650	1650	1710	1720	
Aspect	SW	W	SE	S	S	
Slope (degrees)	70	80	80	80	80	
Herb cover (%)	10	25	10	10	25	
Sample area (sq. m)	4	4	4	4	4	K
Char. ass.						
Saxifraga marginata var. *rocheliana*	1.4	2.5	1.3	1.5	2.5	V
Gypsophila petraea	1.5	1.5	1.4	1.5	1.5	V
Edraianthus graminifolius	+	+	.	+.3	+	IV
Gypsophilion petraeae						
Androsace villosa subsp. *arachnoidea*	+	+	+	+	+	V
Campanula kladniana	+.4	+.4	+	+	+	V
Eritrichium nanum subsp. *jankae*	+	+	+	.	+.3	IV
Asperula capitata	+	+	+	.	.	III
Androsace lactea	.	+	.	+	+	III
Potentilletalia caulescentis						
Asplenium viride	+	+	+	.	.	III
Kernera saxatilis	+	+	+	.	.	III
Asplenium ruta-muraria	.	+	+	.	.	II
Cystopteris fragilis	.	+	+	.	.	II
Saxifraga paniculata	.	+	.	.	+	II
Seslerietalia s.l.						
Sesleria rigida subsp. *haynaldiana*	+	+	+	+.3	+	V
Leontopodium alpinum	+.4	+.3	+.4	+.2	+.3	V
Carex sempervirens	+	+	+	.	+	IV
Helianthemum alpestre	+	+	.	+	+	IV
Dianthus petraeus	+	.	.	.	.	I
Aster alpinus	.	.	+	.	.	I
Festuca versicolor	.	.	.	.	+	I
Companion						
Seseli libanotis	+	.	.	.	.	I

Relevé: 1–5. Piatra Iorgovanului (26.06.1971)

allocated to this regional South-Carpathian association. They populate the sunny, southern, and western slopes with superficial soils, such as the skeletal rendzinas. The floristic structure of the associations includes a lower number of species (14 in average) that achieve 10–25% coverage of the limestone substrate. From a floristic point of view, these phytocoenoses partially resemble the ones assigned to the association *Saxifragetum rochelianae* Blečić 1958, described in Montenegro and included in the alliance *Micromerion croaticae* Horv. 1931 (Horvat et al. 1974).

Alliance *Cystopteridion* (Norhag. 1936) Rich. 1972

This alliance includes hygro-mesophilous phytocoenoses of calcophilic fern species, that inhabit the basal part of the semi-shaded cliffs from the montane and subalpine belts from southwestern and southeastern Europe (Richard 1972). The characteristic species for the alliance are *Asplenium viride, Moehringia muscosa,* and *Cystopteris fragilis* (Oberdorfer 2001).

Ass. *Asplenio-Cystopteridetum fragilis* Oberd. 1949 (Table 6.2)

Representative phytocoenoses for this association were identified in the southern part of the Retezat Mountains, in the area called "Retezatul Calcaros" (the Calcareous Retezat), which includes the peaks Stănuleți, Piatra Iorgovanului, Albele, and Piule. The dominant species for the association are *Asplenium viride* and *Cystopteris fragilis,* along with several characteristic species for the alliance and order. The plant association's composition also includes a high number of species characteristic of the order *Seslerietalia,* and *Saxifraga moschata, Ranunculus oreophilus, Poa molinerii, Veronica aphylla, Acinos alpinus,* and *Carex sempervirens* are frequent. The hygro-mesophilous character is strengthened by some hygrophilous companion species such as *Selaginella selaginoides, Saxifraga aizoides,* and *Silene pusilla.*

Order *Androsacetalia vandellii* Br.-Bl. in Meier and Br.-Bl. 1934
Natura 2000: habitat type 8220

This order includes petrophilous and crack-dwelling phytocoenoses of siliceous cliffs of montane and subalpine belts. Apart from saxicolous species with a wider European distribution, these communities also include in their structure Carpathian-Balkan acidophilic species such as *Galium kitaibelianum, Dianthus tenuifolius, Silene nutans* subsp. *dubia, Veronica bachofenii,* and *Silene lerchenfeldiana,* offering a regional specificity (Boşcaiu 1971). The order comprises the following two alliances: *Asplenion septentrionalis* and *Silenion lerchenfeldianae.*

Alliance *Asplenion septentrionalis* Oberd. 1938

Within this alliance we include the chasmophytic phytocoenoses that populate the siliceous, submontane, and montane cliffs from the basin of the Râul Mare valley in the Retezat Mountains. The diagnostic species for the alliance are *Asplenium septentrionale, Sedum telephium,* and *Epilobium colinum.*

Ass. *Asplenio trichomanes-Poetum nemoralis* (Borza 1959) Boşcaiu 1971 (Table 6.3)
– *veronicetosum bachofenii* Borza ex Boşcaiu 1971
Nomenclatural type: Borza 1959, Table 1, rel. 4, lectotype *hoc loco*

The saxicolous phytocoenoses dominated by *Asplenium trichomanes* and *Poa nemoralis,* with scattered distribution in the basin of Râul Mare valley in the Retezat Mountains, are included in this association. The regional species *Veronica bachofenii, Galium kitaibelianum,* and *Dianthus tenuifolius* are constantly present in the association, providing the basis for Boşcaiu (Boşcaiu 1971) to describe the sub-association *veronicetosum bachofenii,* typical for the Southern Romanian Carpathians.

Table 6.2 Ass. *Asplenio-Cystopteridetum fragilis* Oberd. 1949

Relevé No.	1	2	3	4	5	6	
Altitude (m a.s.l.)	1880	1870	1850	1860	1800	1840	
Aspect	NE	E	NE	N	N	N	
Slope (degrees)	90	80	90	90	90	90	
Herb cover (%)	30	30	20	40	40	30	
Sample area (sq. m)	4	4	4	4	4	4	K
Char. ass.							
Asplenium viride	2.4	+.3	1.3	2.5	2.5	1.3	V
Cystopteris fragilis	1.3	2.5	1.4	2.5	2.5	1.4	V
Cystopteridion **et** ***Potentilletalia caulescentis***							
Draba lasiocarpa	+	.	+	+	+	.	IV
Saxifraga paniculata	.	2.5	+	.	.	2.5	III
Doronicum columnae	.	+	+	.	.	.	II
Asplenium ruta-muraria	.	.	.	.	+.3	.	I
Androsace lactea	.	.	.	.	.	+.4	I
Moehringia muscosa	.	.	.	.	.	+	I
Seslerietalia s.l.							
Saxifraga moschata	1.3	+	.	1.5	1.3	.	IV
Ranunculus oreophilus	+	+	.	+.3	.	+	IV
Poa molinerii	+	.	+	+	.	.	III
Veronica aphylla	+	.	.	+	.	+.5	III
Acinos alpinus subsp. *alpinus*	.	+	+	+	.	.	III
Selaginella selaginoides	.	.	+	+	+	.	III
Carex sempervirens	.	.	+	+	.	+	III
Polygonum viviparum	.	.	+	+	.	+	III
Ditrichum flexicaule	1.3	.	+	.	1.3	.	III
Timmia austriaca	+	.	1.1	.	1.1	.	III
Pohlia cruda	+	.	+	.	+	.	III
Leiocolea müllleri	1.2	.	+	.	1.1	.	III
Thymus pulcherrimus	1.3	.	1.3	.	.	.	II
Sesleria rigida subsp. *haynaldiana*	.	.	.	.	+	+	II
Scapania aequiloba	+	.	.	.	+	.	II
Saxifraga adscendens	+	.	.	.	.	.	I
Galium anisophyllon	+	.	.	.	.	.	I
Sedum atratum	.	.	+	.	.	.	I
Myosotis alpestris	.	.	.	+	.	.	I

Companion species with one occurrence: *Senecio rupestris* 2: +; *Potentilla thuringiaca* 2: +; *Scrophularia vernalis* 2: +; *Saxifraga androsacea* 3: +; *Viola alpina* 5: +; *Campanula kladniana* 5: +; *Saxifraga aizoides* 6: +; *Silene pusilla* 6: +

Relevé: 1. Mt. Stănuleți—Piatra Iorgovanului (23.06.1971); 2. Mt. Stănuleți (23.06.1971); 3-5. Mt. Albele (24.06.1971); 6. Mt. Piule (6.07.1972)

Table 6.3 Ass. *Asplenio trichomanis-Poetum nemoralis* (Borza 1959) Boşcaiu 1971

Relevé No.	1	2	3	4	5	
Altitude (m a.s.l.)	640	690	810	815	830	
Aspect	SE	V	SW	SW	SW	
Slope (degrees)	85	80	80	75	80	
Herb cover (%)	75	40	60	80	50	
Sample area (sq. m)	25	25	25	25	25	K
Char. ass.						
Asplenium trichomanes subsp. *trichomanes*	+	+	+	+	+	V
Poa nemoralis	4.5	2.5	3.5	3.5	2.5	V
Diff. subass.						
Veronica bachofenii	+.4	+	+	+.3	+	V
Dianthus tenuifolius	.	+	+	+	+	IV
Galium kitaibelianum	.	+	.	+	+	III
Androsacion **et** ***Androsacetalia vandellii***						
Sedum telephium subsp. *maximum*	1.5	+.3	+	+	+	V
Asplenium septentrionale	.	.	+	.	2.5	II
Symphyandra wanneri	.	+	.	.	.	I
Epilobium collinum	.	+	.	.	.	I
Asplenietea trichomanis						
Jovibarba heuffelii	+	+	+	+	+	V
Moehringia pendula	+	+	.	+	+	IV
Polypodium vulgare	.	+	+	+	+	IV
Silene nutans subsp. dubia	+	+	.	.	+	III
Cystopteris fragilis	.	+	.	.	.	I
Companion						
Potentilla thuringiaca	+	+	+	+	+	V
Mycelis muralis	.	+	+	+	+	IV
Seseli libanotis	+	+	+	.	.	III
Sedum album	+	+	.	.	.	II
Digitalis grandiflora	.	+	.	+	.	II
Genista tinctoria	.	.	+	.	+	II
Hypericum perforatum	.	.	.	+	+	II
Thymus pulegioides s.l.	.	.	.	+	+	II
Hypnum cupressiforme	1.5	2.5	2.5	3.5	2.5	V
Thuidium tamariscinum	.	.	+	.	.	I
Leucodon sciuroides	.	.	.	2.5	.	I

Companion species with one occurrence: *Melica ciliata* 1: +.3; *Centaurea stoebe* subsp. *australis* 1: +; *Euphorbia cyparissias* 1: +; *Campanula rapunculoides* 1: +; *Allium carinatum* subsp. *pulchellum* 1: +; *Trifolium medium* s.l. 2: +; *Achillea distans* 2: +; *Lathyrus hallersteinii* 2: +; *Luzula luzuloides* 2: + *Cardaminopsis arenosa* 2: +; *Saxifraga rotundifolia* 2: +; *Geranium robertianum* 2: +; *Cnidium silaifolium* 2: +; *Selaginella helvetica* 3: +; *Coronilla varia* 4: +; *Veronica urticifolia* 4: +; *Viscaria vulgaris* 5: +

Relevé: 1–2. Râul Mare (14.09.1970); 3–5. Râul Mare—Gura Zlata (11.09.1970)

Table 6.4 Ass. *Senecio glaberrimi-Silenetum lerchenfeldianae* Boşcaiu et al. 1978

Relevé No.	1	2	3	4	5	6	
Altitude (m a.s.l.)	1890	2020	1920	1930	2040	2100	
Aspect	W	S	W	SW	S	W	
Slope (degrees)	90	90	90	90	90	80	
Herb cover (%)	5	10	10	10	5	10	
Sample area (sq. m)	4	4	4	4	4	10	K
Char. ass.							
Silene lerchenfeldiana	+.3	1.3	1.3	1.4	+.3	1.3	V
Senecio doronicum subsp. *transylvanicus*	+	+	+	+	+.3	+	V
Asplenietea s.l.							
Asplenium viride	.	+	+	.	.	+	III
Cystopteris fragilis	.	+	.	.	+	+	III
Campanula kladniana	.	+.2	.	+	.	+	III
Silene nutans subsp. *dubia*	.	.	.	.	.	+	I
Draba donneri	.	.	.	.	.	1.3	I
Companion							
Juncus trifidus	1.4	1.3	1.3	1.4	+.2	.	V
Solidago virgaurea subsp. *minuta*	+	+	+	+	.	.	IV
Gentiana punctata	.	+	.	.	.	.	I

Companion species with one occurrence: *Cardaminopsis hallerii* subsp. *ovirensis* 6: +; *Euphorbia stricta* 6: +; *Achillea oxyloba* subsp. *schurii* 6: +; *Laserpitium krapfii* 6: +; *Moehringia pendula* 6: +
Relevé: 1. Casa Gemenele (21.07.1972); 2. Stâncărie Lacul Ana (10.08.1970); 3–4. Valea Judelui (13.07.1970); 5. Muchia Ascuţită (9.07.1971); Piciorul Colţului (21.07.1972)

Alliance *Silenion lerchenfeldianae* Simon 1958

This Balkan alliance described from the Pirin Mountains by Simon (1958) is well individualized by several regional species such as *Potentilla haynaldiana, Silene lerchenfeldiana, Saxifraga pedemontana* subsp. *cymosa,* and *Symphyandra wanneri.* We include in this alliance the saxicolous phytocoenoses that populate the granite rock walls of the subalpine belt in the Retezat Mountains.

Ass. *Senecio glaberrimi-Silenetum lerchenfeldianae* Boşcaiu et al. 1978 (Table 6.4)

This plant association includes the saxicolous, species-poor, pioneer communities that colonize the sunny granite cliffs of the subalpine and alpine belts in the scientific reserve of the Retezat National Park (Boşcaiu et al. 1978). The dominant characteristic species for the association is *Silene lerchenfeldiana*, along with the frequent Carpathian-Balkan species *Senecio doronicum* subsp. *transsilvanicum (S. glaberrimus).* The petrophilous species *Juncus trifidus* and *Solidago virgaurea* subsp. *minuta* are also frequent in this association, emphasizing its acidophilic character (Ellenberg 1979) (Photo 6.1).

Class *Thlaspietea rotundifolii* Br.-Bl. 1948

Photo 6.1 *Daraba dorneri* (photo B. Hurdu)

This class includes communities that populate both mobile and semi-fixed screes of the subalpine and alpine belts of the Retezat Mountains. Scree formation was mostly favored by the cryogenic processes during the Pleistocene glaciation, leaving its mark on the alpine landscape of Retezat. Over time, particularly the plant species adapted to extreme environmental conditions were preserved in the structure of these phytocoenoses, conferring them an important conservative value. The floristic diversity of these phytocoenoses is determined by the petrographic nature of geological substrate and orographic factors. Taking into account the ecopedological factors, such communities are grouped into two different orders: *Thlaspietalia rotundifolii* and *Androsacetalia alpinae*. The characteristic species for the class are *Rumex scutatus, Oxyria digyna, Arabis alpina* and *Saxifraga carpathica* (Jarolimek and Šibik 2008).

Order *Thlaspietalia rotundifolii* Br.-Bl. et Jenny 1926
Natura 2000: habitat type 8120

In this order we include the basophilic associations that populate the limestone screes in the subalpine belt of the Calcareous Retezat. They are mostly located on sunny, steep slopes with superficial rendzinas. Among the characteristic species of the order we mention *Sedum atratum, Pritzelago alpina* subsp. *brevicaulis* and *Cystopteris alpina*. The order comprises a single alliance: *Papavero-Thymion pulcherrimi.*

Alliance *Papavero-Thymion pulcherrimi* Pop 1968

The subalpine phytocoenoses on the limestone screes from the South-Eastern Carpathians, with different floristic composition from those in the Alps assigned to *Thlaspion rotundifolii,* were grouped into this regional alliance (Pop 1968). Among the regional characteristic species, we mention *Papaver alpinum* subsp. *corona-sancti-stephani, Thymus pulcherrimus, Alyssum repens,* and *Cerastium lerchenfeldianum.*

Ass. *Calamintho baumgartenii-Galietum anisophylii* Beldie 1967 (Table 6.5)

The phytocoenoses grouped into this association populate the small calcareous screes situated at the base of cliffs from the Calcareous Retezat (Piatra Iorgovanului, Albele, Stănuleți, Piule), on moderately steep, south-facing slopes. The soil type in these areas is a superficial, discontinuous rendzina. The characteristic species for the association are *Galium anisophyllum* and *Acinos alpinus* agg., achieving on average 20–25% coverage. Among the Carpathian regional species with high presence in the association, we mention *Thymus pulcherrimus, Alyssum repens, Draba lasiocarpa,* and *Festuca stricta* subsp. *rumelica*. The high number of basophilic species typical of the order *Seslerietalia* reveals the syndynamic tendency towards grasslands, as the soil evolves and increases its edaphic volume enough to support the development of *Poaceae* species (Beldie 1967; Boşcaiu 1971).

Ass. *Cerastio lerchenfeldiani-Papavaretum coronae-sancti-stephani* Boşcaiu et al. 1978 (Table 6.6)

The communities dominated by *Papaver alpinum* subsp. *corona-sancti-stephani* and *Cerastium lerchenfeldianum* mainly populate the coarse, semi-mobile calcareous screes of the Piule peak from the Calcareous Retezat. Due to the poor development of rendzinas and their, these phytocoenoses cover a smaller area than those belonging to the association *Calamintho-Galietum anisophylii* (Beldie 1967). Also, the number of characteristic species for the order *Seslerietalia* present in the association is smaller. These structural peculiarities of the *Papaver* communities reveal, on one hand, the unfavorable eco-pedological conditions for grassland development in these locations and the petrophilous character of the phytocoenoses, on the other hand. This regional, south-eastern Carpathian association is a geographical variant of the *Cerastio latifolii-Papaveretum tatrici* Pawł. et Stecki ex Valachovič 1995, described in the Western Carpathians (Valachovič 1995).

Ass. *Doronico columnae-Rumicetum scutati* Boşcaiu et al. 1978 (Table 6.7)

Representative coenoses with *Rumex scutatus* were identified in the Retezat Mountains on the coarse, semi-mobile screes from Piatra Iorgovanului and Piule Mountain. They are located in the subalpine belt, on moderate-steep sunny slopes with southern and western aspects. *Rumex scutatus* is the dominant species in the association, achieving on average 30% coverage. It is frequently accompanied by the calcophilic, meso-hygrophilous preferential species *Doronicum columnae*, underlining the eco-pedological specificity of the association (Ellenberg 1996; Oberdorfer 2001). The inclusion of the association in the alliance *Papavero-Thymion* and order *Thlaspietalia rotundifolii* is supported by the high presence of species characteristic of these syntaxa, similarly to those described from Southeast Europe (Horvat et al. 1974; Mucina 2016). Floristically, such phytocoenoses differ

Table 6.5 Ass. *Calamintho baumgartenii-Galietum anisophylli* Beldie 1967

Relevé No.	1	2	3	4	5	6	7	8	
Altitude (m a.s.l.)	1840	1890	1880	1850	1800	1850	1860	2020	
Aspect	E	SE	SE	S	SE	S	S	S	
Slope (degrees)	15	45	35	15	10	15	40	45	
Herb cover (%)	70	75	40	60	40	50	60	60	
Sample area (sq. m)	25	25	25	25	4	25	25	25	K
Char. ass.									
Galium anisophyllon	2.5	1.5	1.5	+.4	+.3	+.4	2.5	1.5	V
Acinos alpinus subsp. *alpinus*	1.4	2.5	2.5	2.5	1.4	1.4	2.5	2.5	V
Papavero-Thymion pulcherrimae									
Thymus pulcherrimus	+.4	1.5	1.5	1.5	2.5	+	+	+	V
Alyssum repens	+	+	1.3	+	.	+	1.3	+.5	V
Saxifraga moschata	3.5	2.5	1.3	2.5	.	+	.	.	IV
Thlaspietalia rotundifolii									
Sedum atratum	+	+	+	+	.	+	+	+	V
Cystopteris alpina	+	+	.	.	.	.	.	+	II
Arabis alpina	.	.	.	.	.	.	+	+	II
Galium album subsp. *album*	.	.	.	.	.	.	+	+	II
Senecio rupestris	.	.	.	.	.	.	+	+	II
Hutchinsia brevicaulis	.	.	.	.	.	.	+	.	I
Selerietalia s.l.									
Ranunculus oreophilus	1.4	1.5	+.5	1.5	+	+	+	+	V
Poa molinerii	1.5	+	1.5	2.5	1.5	1.5	+	2.5	V
Festuca stricta subsp. *rumelica*	.	1.5	+	+	+.4	2.5	+	1.5	V
Saxifraga adscendens	+	+	.	.	+	+	+	+	IV
Carex sempervirens	+.3	1.4	.	.	+	.	1.5	1.5	IV
Helianthemum alpestre	.	2.5	.	.	2.5	2.5	1.4	+	IV
Alchemilla flabellata	.	+	+	+.3	.	.	.	1.4	III
Veronica aphylla	+.3	+	+	.	.	.	+	.	III
Myosotis alpestris	.	+		+.5	.	.	+	+	III
Minuartia verna	.	.	+	.	+	+	.	+	III
Botrychium lunaria	+	+	.	.	+	.	.	.	II
Gentiana verna	+	.	.	.	.	.	+	+	II
Polygonum viviparum	.	+	.	+	.	.	.	+	II
Trifolium pallescens	.	+	.	.	+	+	.	.	II
Scabiosa lucida	.	.	+	.	.	.	.	+	II
Androsace villosa subsp. *arachnoidea*	.	.	.	+.3	.	.	+	.	II
Seleria rigida subsp. *haynaldiana*	.	.	+	.	.	.	.	.	I
Euphrasia salisburgensis	+	.	.	.	.	.	.	.	I

(continued)

Table 6.5 (continued)

Potentilletalia caulescentis									
Draba lasiocarpa	+.3	+	+	+	+	+	+	+	V
Saxifraga paniculata	.	.	.	.	.	.	1.3	+.3	II
Asplenium viride	+	.	.	.	.	.	.	.	I
Saxifraga marginata	.	+	.	.	.	.	.	+	I
Companion									
Luzula multiflora	+	+	.	.	.	.	.	+	II
Phyteuma confusum	.	.	.	.	+	+	+	.	II

Companion species with one or two occurrences: *Polygala alpestris* 1, 5: +; *Dianthus petraeus* 5: +; *Dryas octopetala* 6: +; *Edraianthus graminifolius* 2, 6: +; *Androsace lactea* 3: +; *Saxifraga androsacea* 3: +; *Taraxacum panalpinum* 6: +; *Viola alpina* 7: +; *Saxifraga aizoides* 8: +; *Thlaspi dacicum* 8: +; *Biscutella laevigata* 8: +

Relevé: 1–2. Curmătura Soarbele (23.06.1971); 3. Mt. Stănuleți (23.06.1971); 4. Mt. Piatra Iorgovanului (23.06.1971); 5–6. Mt. Albele (26.06.1971); 7. Mt. Piule (25.06.1971); 8. Mt. Piule (6.07.1972)

markedly from those described in Germany (Pott 1995) and Slovakia (Valachovič 1995) within the alliance *Stipion calamagrostis*.

Ass. *Saxifragetum moschatae-aizoidis* Boşcaiu 1971 (Table 6.8)

This association was initially described in the Țarcu Mountains (Boşcaiu 1971) and later identified also in the Făgăraş Mountains (Voik and Schneider-Binder 1978). Such phytocoenoses were identified on moist, low-mobility screes of the subalpine belt from the Piatra Iorgovanului, Albele, and Stănuleți limestone massifs in the Calcareous Retezat. The hygro-mesophilous specificity of the association is denoted by the presence and high coverage of *Saxifraga aizoides* (Ellenberg 1979). Unlike the phytocoenoses described in Țarcu and Făgăraş Mountains, those from Retezat lack the chionophilous-hygrophilous species in their floristic structure, but instead include several mesophylous species characteristic of the order *Seslerietalia*. This structural peculiarity indicates the syndynamic trajectory of the *Saxifraga moschata* coenoses from Retezat toward the alpine calcareous grasslands of the alliance *Seslerio bielzii*.

Order *Androsacetalia alpinae* Br.-Bl. et Jenny 1926
Natura 2000: habitat type 8110

The order includes phytocoenoses developed on mobile and semi-mobile screes resulted from disaggregation of siliceous cliffs in the subalpine and alpine belts of European mountain ranges. The characteristic species for the order are *Cardamine resedifolia, Saxifraga bryoides,* and *S. hieracifolia* (Matuszkiewicz 2008; Jarolimek and Šibik 2008).

Alliance *Veronicion baumgartenii* Coldea 1991

We include in this alliance phytocoenoses on siliceous screes from the South-Eastern Carpathians, lacking the main characteristic species for the alliance *Androsacion*

Table 6.6 Ass. *Cerastio lerchenfeldiani-Papaveretum coronae-sancti-stephani* Boşcaiu et al. 1978

Relevé No.	1	2	3	4	5	6	7	8	9	10	
Altitude (m a.s.l.)	1820	1830	1840	1850	1860	1860	1840	2100	1840	2050	
Aspect	SW	SW	SW	S	SW	SW	SW	S	N	N	
Slope (degrees)	45	45	45	55	45	45	45	45	45	45	
Herb cover (%)	40	30	30	30	30	20	20	50	40	40	
Sample area (sq. m)	4	10	25	4	25	4	25	25	25	16	K
Char. ass.											
Papaver alpinum subsp. *corona-sancti-stephani*	1.5	+.4	1.4	1.5	+.3	1.4	1.5	2.5	2.5	2.4	V
Cerastium lerchenfeldianum	2.5	2.5	2.5	2.5	1.4	1.5	1.4	+	.	.	IV
Papavero-Thymion pulcherrimi											
Thymus pulcherrimus	.	+.4	1.3	1.4	.	+	+.4	1.5	.	+	IV
Alyssum repens	1.4	+.4	+	.	+	+	+	+	.	.	IV
Acinos alpinus subsp. *alpinus*	2.5	1.5	+	+	.	+	.	+	.	.	III
Galium anisophyllon	.	+	.	.	.	.	.	+	+	+	II
Thlaspietalia rotundifolii											
Arabis alpina	.	+	+	+.3	2.5	1.3	.	1.5	2.5	2.4	IV
Senecio rupestris	.	+.3	+	+.3	+	+	+	+	+	.	IV
Sedum atratum	.	+	+	+	.	.	+	+	.	.	III
Hutchinsia brevicaulis	.	.	.	.	.	1.4	1.5	+	+	.	II
Thlaspi dacicum	.	.	.	.	+	.	+	.	+	.	II
Doronicum columnae	.	.	.	.	.	.	.	.	+.3	1.3	I
Cystopteris alpina	.	.	.	.	.	.	.	.	+	+	I
Seslerietalia s.l.											
Festuca stricta subsp. *rumelica*	1.5	+	+	1.5	.	+	+	1.5	+.5	.	IV
Myosotis alpestris	+.4	+	+	+	.	.	.	+	+	+	IV
Ranunculus oreophilus	+	+	.	+.3	.	.	.	+	+.5	+	III

(continued)

Table 6.6 (continued)

Poa molinerii	+	+	.	.	1.4	+	+	+	.	.	III
Biscutella laevigata	+	+	.	.	.	.	.	.	.	+	II
Saxifraga adscendens	+	.	+	.	.	.	.	.	.	.	I
Sesleria rigida subsp. *haynaldiana*	.	.	.	.	.	.	+	1.5	.	.	I
Asperula capitata	.	.	.	.	.	+	.	.	.	.	I
Festuca versicolor	.	.	.	.	.	.	+	.	.	.	I
Anemone narcissiflora	.	.	.	.	.	.	.	+	.	.	I
Minuartia verna	.	.	.	.	.	.	.	+	.	.	I
Veronica aphylla	.	.	.	.	.	.	.	.	+	.	I
Potentilletalia caulescentis											
Draba lasiocarpa	.	.	.	+	+	+	+	+	.	.	III
Saxifraga paniculata	.	.	.	.	.	.	.	+	.	.	I
Saxifraga marginata	.	.	.	.	.	.	.	+	.	.	I
Companion											
Taraxacum panalpinum	+	+	.	.	+	.	.	.	.	.	II
Silene pusilla	.	.	+	.	1.3	.	.	+	+	.	II
Saxifraga aizoides	.	.	+	.	+	.	.	.	+	.	II

Companion species with one occurrence**:** *Cerastium alpinum* subsp. *lanatum* 8: +; *Aconitum napellus* subsp. *tauricum* 9: +
Relevé: 1–8. Mt. Piule (25.06.1971); 9–10. Mt. Piule (6.07.1972)

Table 6.7 Ass. *Doronico columnae-Rumicetum scutati* Boşcaiu et al. 1978

Relevé No.	1	2	3	4	5	6	7	8	9	10	
Altitude (m a.s.l.)	1650	1670	1715	1720	1730	1740	1750	1750	1730	1700	
Aspect	W	SW	S	S	S	S	S	W	W	W	
Slope (degrees)	45	50	40	40	30	30	30	40	35	40	
Herb cover (%)	40	70	60	50	40	30	50	40	45	45	
Sample area (sq. m)	25	25	25	4	4	25	25	10	10	15	K
Char. ass.											
Rumex scutatus	3.5	3.5	3.5	2.5	2.5	2.5	3.5	2.5	3.5	3.5	V
Doronicum columnae	1.3	2.5	2.5	2.5	1.5	1.3	+.3	.	+	.	IV
Papavero-Thymion pulcherrimi											
Thymus pulcherrimus	+.4	.	+	+	+.3	+.3	+	+	+	1.2	V
Alyssum repens	+	+	+	.	.	+	+	+	+	+	IV
Cerastium lerchenfeldianum	.	2.5	.	+.4	+	+	.	.	+	+	III
Acinos alpinus subsp. *alpinus*	+	.	.	+	+	.	.	1.3	+.3	1.2	III
Papaver alpinum subsp. *corona-sancti-stephani*	.	.	+	.	.	1.3	+.3	.	.	.	II
Saxifraga moschata	.	.	.	+	.	+	1.3	.	.	.	II
Galium anisophyllon	.	+	.	.	.	.	.	+.2	1.5	+.4	II
Thlaspietalia rotundifolii											
Galium album subsp. *album*	+	+	+	+.4	+	+	+	+	+	.	V
Senecio rupestris	1.5	+.4	+.4	1.4	2.5	1.5	1.5	+.3	+.2	+.3	V
Cystopteris alpina	+	+	+	1.3	+	+	.	+	+	.	IV
Arabis alpina	+	+.2	+	.	.	+	.	2.2	1.2	1.3	IV
Seslerietalia s.l.											
Myosotis alpestris	+	+	+	.	+	+	+	.	.	+	IV
Poa molinerii	.	+	+	+	+	.	+	+	+	+	IV
Ranunculus oreophilus	.	+	+	+	.	+	+	+	+	+	IV

(continued)

Table 6.7 (continued)

Sesleria rigida subsp. *haynaldiana*	.	+.5	.	.	.	+	+	.	.	.	II
Alchemilla flabellata	.	.	+	+	.	.	.	.	.	.	I
Carduus kerneri	+	.	.	.	.	.	.	.	.	.	I
Festuca versicolor	.	.	.	.	.	.	+	.	.	.	I
Companion											
Taraxacum panalpinum	.	+	+	+	+	+	+	.	.	+	IV
Silene pusilla	.	.	.	.	.	.	.	+	+	+	II
Saxifraga aizoides	.	.	.	.	.	+	.	+	.	.	I
Parnassia palustris	.	.	.	.	.	.	.	.	+	+	I

Companion species with one occurrence: *Corydalis solida* 1: +.3; *Scrophularia vernalis* 1: +; *Polystichum lonchitis* 3: +
Relevé: 1–7. Mt. Piatra Iorgovanului (26.06.1971); 8–10. Vf. Piule (1.08.1986)

Table 6.8 Ass. *Saxifragetum moschatae-aizoidis* Boşcaiu 1971

Relevé No.	1	2	3	4	5	
Altitude (m a.s.l.)	1940	1860	1880	1820	1830	
Aspect	S	N	N	W	W	
Slope (degrees)	45	30	25	45	30	
Herb cover (%)	50	50	60	25	40	
Sample area (sq. m)	25	4	4	4	4	K
Char. ass.						
Saxifraga moschata	1.4	1.4	2.5	1.4	2.4	V
Saxifraga aizoides	3.5	3.5	3.5	2.4	2.4	V
Papavero-Thymion pulcherrimi						
Cerastium arvense subsp. *molle*	.	+	+	+	+	IV
Thymus pulcherrimus	+	.	.	+	+	III
Cerastium lerchenfeldianum	+	.	.	.	.	I
Acinos alpinus subsp. *alpinus*	+	.	.	.	.	I
Galium anisophyllon	.	.	+	.	.	I
Thlaspietalia rotundifolii						
Cystopteris alpina	+	.	+	.	+	III
Sedum atratum	+	.	.	.	+	II
Cardaminopsis halleri subsp. *ovirensis*	.	.	+	.	+	II
Arabis alpina	.	.	.	+	+	II
Senecio rupestris	+	.	.	.	.	I
Seslerietalia s.l.						
Poa molinerii	+	.	1.3	+	+	IV
Polygonum viviparum	.	+	+	+	+	IV
Ranunculus oreophilus	+	.	+	+	.	III
Sesleria rigida subsp. *haynaldiana*	+	+	.	.	.	II
Myosotis alpestris	+	.	+	.	.	II
Festuca stricta subsp. *rumelica*	.	+	.	.	+	II
Androsace lactea	.	.	.	+	+	II
Helianthemum alpestre	.	+	.	.	.	I
Gentiana verna	.	+	.	.	.	I
Minuartia verna	.	+	.	.	.	I
Potentilletalia caulescentis						
Draba lasiocarpa	+	.	.	+	+	III
Saxifraga paniculata	.	+	.	.	.	I
Asplenium viride	.	.	+	.	.	I
Companion						
Taraxacum panalpinum	.	.	+	+	+	III
Alchemilla glabra	+	.	+	.	.	II
Saxifraga androsacea	.	.	+	+	.	II
Selaginella selaginoides	.	+	.	.	.	I

Relevé: 1. Mt. Stănuleţi (23.06.1971); 2-3. Mt. Piatra Iorgovanului (24.06.1971); 4-5. Mt. Albele (24.06.1971)

alpinae (Oberdorfer 1977; Grabherr 1993). Their floristic structure includes Carpathian-Balkan regional species, such as *Veronica baumgartenii, Saxifraga pedemontana* subsp. *cymosa, Poa cenisia* subsp. *contracta,* and *Poa granitica* subsp. *disparilis,* which markedly distinguished them from phytocoenoses described in the Alps and Pyrenees (Coldea 1991).

Ass. *Veronico baumgartenii-Saxifragetum bryoides* Boşcaiu et al. 1978 (Table 6.9)

The pioneer phytocoenoses consisting of *Saxifraga bryoides* and *Veronica baumgartenii* on slightly turfed, mobile screes with superficial, skeletal regosol from Retezat were grouped into these syntaxa (Boşcaiu et al. 1978). The characteristic species for the alliance *Veronicion baumgartenii* and order *Androsacetalia* are well represented in the structure of the association, supporting its assignment to these syntaxa. The alpine feature of the association is also shown by the presence of microthermal species, typical for the alpine grasslands, attributed to the alliance *Caricion curvulae* and by some chionophilic species characteristic of the alliance *Salicion herbaceae.*

Ass. *Saxifrago bryoidis-Silenetum acaulis* Boşcaiu et al. 1978 (Table 6.10)

The petrophilic phytocoenoses developed on consolidated, alpine screes with wet, strongly acidic soils from Retezat were assigned to this association (Boşcaiu et al. 1978). The ecological peculiarities of the areas occupied by these phytocoenoses are indicated by the species *Primula minima, Soldanella pusilla, Salix herbacea,* and *Polygonum viviparum,* with high presence in the association (Ellenberg 1979). The reduced presence of the *Poaceae* and *Juncaceae* species in the association denotes the initial serial stage of these phytocoenoses developing toward alpine grasslands.

Ass. *Poo contractae-Oxyrietum digynae* Horvat et al. 1937 (Table 6.11)

This association, initially described from the Rila Planina Mountains (Bulgaria), was found and described in the Ţarcu-Godeanu Mountains by Boşcaiu (1971), taking into account seven relevées. We identified such phytocoenoses in the eastern and northeastern part of the Retezat Mountains (Custura Păpuşii and Custura Mountains). The floristic composition of these phytocoenoses in the Ţarcu-Godeanu and Retezat Mountains is similar to that described from Rila Planina, except for some Balkan species, absent from the Southern Romanian Carpathians (Boşcaiu 1971). Compared to the first two associations described in this alliance, the phytocoenoses of this association are in a more advanced successional stage toward the alpine grasslands, suggested by the presence and high coverage of the species *Poa cenisia* subsp. *contracta* and *Luzula alpinopilosa* subsp. *obscura,* similarly to those described in the Tatra Mountains (Valachovič 1995).

Class *Salicetea herbaceae* Br.-Bl. 1947

We included in this vegetation class the chionophilic coenoses in the alpine belt of the Retezat Mountains, developed on shaded areas with northern aspects, where snow persists until summer. The first specific phytocoenoses for this vegetation type in Retezat were published by Borza (1934). Subsequently, Boşcaiu (1971) published

Table 6.9 Ass. *Veronico baumgartenii-Saxifragetum bryoidis* Boşcaiu et al. 1978

Relevé No.	1	2	3	4	5	6	7	8	9	10	
Altitude (m a.s.l.)	2360	2350	2340	2240	2100	2260	2370	2500	2340	2250	
Aspect	NW	E	N	N	N	N	W	S	SW	S	
Slope (degrees)	30	50	30	80	80	45	30	30	45	20	
Herb cover (%)	25	50	20	20	20	30	20	20	50	20	
Sample area (sq. m)	4	4	10	10	4	4	4	4	4	4	K
Char. ass.											
Saxifraga bryoides	2.4	3.4	1.3	1.3	.	2.4	1.3	1.4	1.3	1.3	V
Veronica baumgartenii	.	1.4	1.3	1.4	1.3	+	+.3	1.5	+	+	V
Veronicion baumgartenii* et *Androsacetalia											
Saxifraga pedemontana subsp. *cymosa*	1.3	1.3	.	1.3	1.3	1.3	.	.	2.5	2.5	IV
Poa laxa	+	.	1.3	+	+	+	.	.	.	.	III
Senecio incanus subsp. *carniolicus*	+	.	+	.	.	.	.	.	+	.	II
Luzula spicata	.	.	.	.	.	+	.	.	.	.	I
Gentiana frigida	.	.	.	.	.	.	.	.	+	.	I
Thlaspietea rotundifolii											
Doronicum carpaticum	+	+	.	+	+	.	.	.	.	+	III
Saxifraga androsacea	.	.	.	.	.	.	.	.	+	+	I
Salicion herbaceae											
Leucanthemopsis alpina	+	+	+	.	+	+	.	+	+	1.3	IV
Luzula alpinopilosa subsp. *obscura*	.	1.3	+	.	1.3	.	1.3	1.3	.	+	III
Soldanella pusilla	.	+	.	.	+	.	.	+	.	.	II
Festuca picturata	.	.	+	.	+	.	.	+	.	.	II
Taraxacum panalpinum	.	.	.	.	+	.	.	.	.	+	I
Caricion curvulae											
Cerastium alpinum	+	+	+	+	.	.	+	.	+	+	IV

(continued)

Table 6.9 (continued)

Oreochloa disticha	1.3	+.3	+	+	+	.	.	.	2.5	.	III
Primula minima	.	.	+	+	.	1.3	+	.	+	1.3	III
Hieracium alpinum	+	.	.	+	.	+	.	+	+	.	III
Sesleria bielzii	+	.	.	.	.	.	+	+	.	+	II
Festuca supina	+	.	.	.	.	+	.	.	1.3	.	II
Cetraria nivalis	.	.	.	+	.	.	.	.	+	+	II
Thamnolia vermicularis	.	.	.	.	.	.	.	+	+	+	II
Juncus trifidus	.	.	.	+	.	+	.	.	.	.	I
Pedicularis verticillata	.	.	.	.	+	.	.	+	.	.	I
Phyteuma confusum	.	.	.	.	.	+	+	.	.	.	I
Carex atrata	.	.	.	.	.	+	.	.	.	+	I

Companion species with one or two occurrences: *Pulsatilla alba* 1: +; *Avenula versicolor* 2: +; *Huperzia selago* 4: +; *Ligusticum mutellina* 8: +; *Senecio abrotanifolius* susbp. *carpathicus* 8: +; *Polytrichum sexangulare* 1: +; *Alectoria ochroleuca* 2, 9: +; *Solorina crocea* 2: +
Relevé: 1. Vf. Judele (12.07.1970); 2. Vf. Muchia Ascuțită (12.07.1970); 3. Vf. Bucura (8.07.1970); 4. Custura Pietrele (8.07.1970); 5. V. Pietrele-Curmătura Bucurii (6.08.1970); 6–8. Vf. Peleaga (7.08.1970); 9–10. Vf. Păpușa (11.07.1971)

Table 6.10 Ass. *Saxifrago bryoidis-Silenetum acaulis* Boşcaiu et al. 1978

Relevé No.	1	2	3	4	5	6	7	8	9	10	
Altitude (m a.s.l.)	2350	2310	2320	2330	2320	2325	2330	2315	2240	2250	
Aspect	W	S	SW	SW	N	N	N	N	W	N	
Slope (degrees)	20	35	40	40	40	45	45	45	40	70	
Herb cover (%)	50	70	70	75	75	75	75	80	60	60	
Sample area (sq. m)	4	4	4	4	4	4	4	4	4	4	K
Char. ass.											
Silene acaulis subsp. *bryoides*	3.5	3.5	3.5	3.5	4.5	4.5	3.5	4.5	3.5	3.5	V
Veronicion baumgartenii **et** ***Androsacetalia***											
Saxifraga bryoides	.	1.3	1.4	2.5	1.3	1.4	3.5	2.5	2.5	1.3	V
Saxifraga pedemontana subsp. *cymosa*	1.3	.	.	+	+	.	+	+	.	.	III
Poa cenisia subsp. *contracta*	1.5	.	.	.	1.5	+.4	1.5	1.5	.	.	III
Geum reptans	.	.	.	.	.	1.3	.	+	.	1.3	II
Gentiana frigida	.	.	.	+	.	.	.	.	.	.	I
Luzula spicata	.	.	.	.	.	.	.	.	+	.	I
Veronica baumgartenii	+	.	+	.	.	.	.	.	.	.	I
Thlaspietea rotundifolii											
Saxifraga androsacea	.	.	.	+	+	+.4	+.3	+	+.4	+.4	IV
Oxyria digyna	.	.	.	.	+	+	.	.	.	2.5	II
Saxifraga moschata	.	.	+	.	+	.	.	.	.	+	II
Saxifraga carpathica	.	.	.	+	+	.	.	.	.	.	I
Doronicum carpaticum	.	.	.	.	.	.	.	+	.	.	I
Salicion herbaceae											
Soldanella pusilla	.	.	.	.	1.5	+.5	+.5	+.3	+	.	III
Salix herbacea	1.3	+	+	+.3	.	.	.	.	.	.	II
Sedum alpestre	+	+	.	.	.	.	.	.	+	.	II

(continued)

Table 6.10 (continued)

Leucanthemopsis alpina	+.3	.	.	.	.	.	.	.	.	.	I
Luzula alpinopilosa subsp. *obscura*	.	.	.	.	.	.	+	.	.	.	I
Caricion curvulae											
Primula minima	.	2.5	2.5	1.5	.	+.4	+	+	1.5	.	IV
Polygonum viviparum	+.4	+.4	+.5	+.5	.	.	.	.	.	.	III
Cerastium alpinum	+	.	.	+	.	.	.	.	+	.	II
Carex curvula	.	1.3	.	.	.	.	.	.	.	+.4	I
Festuca supina	.	.	.	1.4	.	.	.	.	1.4	.	I

Companion species with one or two occurrences: *Oreochloa disticha* 1, 2: +; *Sesleria bielzii* 3: +; *Campanula alpina* 4: +; *Phyteuma confusum* 9: +; *Cetraria nivalis* 3: +, 4: 1.3

Relevé: 1. Vf. Custura (11.07.1971); 2–10. Vf. Gruniu (12.07.1971)

Table 6.11 Ass. *Poo contractae-Oxyrietum digynae* Horv., Pawł., Wal. 1937 nom. invers.

Relevé No.	1	2	3	4	5	6	7	8	9	
Altitude (m a.s.l.)	2350	2165	2170	2175	2180	2190	2200	2280	2290	
Aspect	N	N	N	N	N	N	NW	NW	N	
Slope (degrees)	45	45	45	45	45	45	30	15	95	
Herb cover (%)	40	75	60	50	30	60	70	60	30	
Sample area (sq. m)	25	25	25	25	25	25	25	25	25	K
Char. ass.										
Poa cenisia subsp. *contracta*	1.5	2.5	2.5	1.4	2.5	1.5	1.4	+.3	+	V
Oxyria digyna	1.4	2.5	1.5	2.5	1.3	2.4	1.3	.	1.4	V
Veronicion baumgartenii										
Saxifraga pedemontana subsp. *cymosa*	+.3	2.5	1.3	2.4	1.3	1.3	3.5	2.4	1.3	V
Saxifraga bryoides	2.5	1.3	+	+	.	1.3	+	3.5	+	V
Veronica baumgartenii	+	+.4	+	+	+	+	+	.	.	V
Cardamine resedifolia	+	.	.	+	+	.	.	.	.	II
Geum reptans	.	.	.	.	.	.	1.3	.	1.3	II
Thlaspietea rotundifolii										
Doronicum carpaticum	2.4	1.3	+	+	+	+	1.5	.	2.5	V
Saxifraga androsacea	1.3	+	.	.	.	.	.	.	.	II
Saxifraga carpathica	.	.	.	.	.	.	.	.	+	I
Salicion herbaceae s.l.										
Luzula alpinopilosa subsp. *obscura*	.	1.4	2.5	2.5	1.3	1.4	+	1.3	.	IV
Salix herbacea	.	+	+	+	+	1.3	+	+	.	IV
Leucanthemopsis alpina	+	.	1.5	+.4	+	+.3	.	.	+	IV
Soldanella pusilla	.	.	1.5	+.5	+.4	1.5	1.3	+	.	IV
Taraxacum panalpinum	.	.	+	+.4	.	+	+	+	.	III
Gnaphalium supinum	.	+	+	+	.	.	.	.	.	II
Cerastium cerastioides	.	.	.	+	+	.	.	.	.	II
Sedum alpestre	.	.	.	+	.	.	.	.	+	II
Companion										
Primula minima	+	1.5	+	+.4	+	.	.	+	.	IV
Polygonum viviparum	.	+.5	+.4	+.5	.	+.4	1.5	+.4	.	IV
Oreochloa disticha	+	.	+	.	.	+	+	.	.	III

(continued)

Table 6.11 (continued)

Geum montanum	.	1.3	.	+	.	1.3	.	.	.	II
Cerastium alpinum	.	.	.	.	.	.	+	+	+	II

Companion species with one or two occurrences: *Saxifraga moschata* 1: +; *Campanula alpina* 1: +; *Ligusticum mutellina* 2: +.4, 3: +; *Homogyne alpina* 2, 4: +; *Saxifraga rotundifolia* subsp. *heucherifolia* 5: +; *Festuca supina* 7: +, 8: 1.3; *Potentilla ternata* 8: +; *Sedum roseum* 9: +
Relevé: 1. Custura Păpușii (11.07.1971); 2–9. Mt. Custura (11.07.1971)

several chionophile associations representative for the region, based on his own research in the neighboring Țarcu-Godeanu Mountains. The occurrence of these microthermal phytocoenoses is largely conditioned by the presence of glacial landscape. The characteristic species of the class are *Sedum alpestre, Poa granitica* subsp. *disparilis, Cerastium cerastioides, Veronica alpina, Taraxacum alpinum, T. nigricans,* and *Sagina saginoides.* This vegetation type is grouped into two different orders.

Order *Salicetalia herbaceae* Br.-Bl. in Br.-Bl. et Jenny 1926
Natura 2000: habitat type 6150

This order includes the microthermal vegetation that develops in the area of the glacial cirques, on small terraces or microdepressions with northern aspects, where snow persists up to 7–9 months (Braun-Blanquet and Jenny 1926; Braun-Blanquet 1930). The geological substrate consists of siliceous cliffs, covered by superficial, acid soils, usually lithosols. Although the landscape configuration and soil properties influence the structure of these microthermal coenoses, the differences between them are small and we therefore assigned them to a single alliance, similarly to those described in the Țarcu-Godeanu Mountains (Boşcaiu 1971). The species characteristic of the order are identical with those of the alliance.

Alliance *Salicion herbaceae* Br.-Bl. in Br.Bl. et Jenny 1926

This alliance, which groups the European chionophile coenoses of siliceous cliffs, has the following characteristic species: *Salix herbacea, Gnaphalium supinum, Luzula alpinopilosa* subsp. *obscura, Soldanella pussilla, Leucanthemopsis alpina, Carex pyrenaica, Polytrichum sexangulare*. The Carpathian-Balkan regional species *Ranunculus crenatus* and *Plantago gentianoides*, requiring identical ecological conditions join the previous species (Borza 1934; Boşcaiu 1971), along with some other alpine species with wider ecological preferences, such as *Ligusticum mutellina* and *Geum montanum* (Jarolimek and Šibik 2008; Oberdorfer 2001).

Ass. *Salicetum herbaceae* Br.-Bl. 1913 (Table 6.12)

Representative coenoses for this chionophile association have been identified in the alpine belt of the Retezat Mountains, especially on the highest peaks, Păpușa, Peleaga, Bucura, Custura, and Gruniu. They populate the small peneplains and microdepressions with superficial soil, rich in gravel and with acidic reaction (pH = 4.5–5) of the northern slopes, where snow persists for a long time. The dominant species for the association is *Salix herbacea*, achieving an average

Table 6.12 Ass. *Salicetum herbaceae* Br.-Bl. 1913

Relevé No.	1	2	3	4	5	6	7	8	9	10	
Altitude (m a.s.l.)	2460	2370	2370	2480	2270	2260	2260	2250	2240	2200	
Aspect	N	N	W	N	N	N	S	W	NW	N	
Slope (degrees)	35	10	3	40	20	30	25	30	45	45	
Herb cover (%)	50	60	60	80	70	50	50	80	60	60	
Sample area (sq. m)	25	25	4	25	25	25	4	4	25	25	K
Char. ass.											
Salix herbacea	3.5	3.5	3.5	4.5	3.5	3.5	3.5	3.5	3.5	3.5	V
Salicion herbaceae et ***Salicetalia herbaceae***											
Leucanthemopsis alpina	+	+.3	.	+	+	+	.	+	+	+.4	IV
Luzula alpinopilosa subsp. *obscura*	+	+.4	.	.	+	+	+	.	1.5	.	III
Soldanella pusilla	.	.	.	.	+	+	+	.	+	+	III
Geum montanum	.	.	.	.	+	.	+	.	.	+	II
Gnaphalium supinum	.	.	.	.	.	+	+	.	+.3	.	II
Polytrichum sexangulare	.	1.5	.	.	1.3	.	.	.	+	.	II
Ligusticum mutellina	.	.	.	.	.	.	.	.	+	.	I
Salicetea herbaceae											
Sedum alpestre	.	+	.	.	+	+	.	.	+	.	II
Taraxacum panalpinum	.	.	.	.	+	.	.	.	.	+	I
Caricon curvulae											
Primula minima	+	+	+	+	2.5	+	+	+	+	+	V
Phyteuma confusum	+	.	+	.	+	.	+	.	+	+	III
Festuca supina	.	+	1.5	.	.	.	+	+.3	.	+.5	III
Cetraria islandica	.	1.5	+.4	+	.	.	.	2.5	.	1.5	III
Sesleria bielzii	–	+	+	.	.	.	.	+	.	.	II
Campanula alpina	–	.	.	+	+	.	.	+	.	.	II

(continued)

Table 6.12 (continued)

Oreochloa disticha	.	.	+	+	.	.	+	.	+	.	II
Thamnolia vermicularis	+	.	+	+	.	.	.	.	.	.	II
Hieracium alpinum	.	+	.	+	.	1.5	.	.	.	.	II
Saxifraga bryoides	.	+	.	.	.	+	.	.	+	.	II
Luzula spicata	.	.	.	+	1.3	.	.	.	.	+	II
Cetraria nivalis	+	.	+	+	.	.	.	.	.	.	II
Cerastium alpinum	+	.	.	+	.	.	.	.	.	.	I

Companion species with one or two occurrences: *Rhododendron myrtifolium* 8: +; *Festuca picturata* 8: 2.5; *Carex curvula* 7: +; *Carex curta* 8: +.3; *Cerastium alpinum* subsp. *lanatum* 1, 4: +; *Homogyne alpina* 3: +.; *Poa molinerii* 1: +; *Polygonum viviparum* 2, 7: +; *Senecio abrotanifolius* subsp. *carpathicus* 4: +; *Senecio incanus* subsp. *carniolicus* 8: +; *Veronica baumgartenii* 1, 4: +; *Potentilla ternata* 7: +; *Vaccinium gaulterioides* 9: +; *Silene acaulis* 7: +; *Loiseleuria procumbens* 8: +

Relevé: 1. Vf. Peleaga (8.08.1970); 2. Vf. Custura-Gruniu (12.07.1971); 3. Mt. Peleaga (8.08.1971); 4. Vf. Peleaga (8.08.1970); 5. Vf. Gruniu (12.07.1971); 6. Mt. Păpuşa (8.07.1971); 7. Mt. Păpuşa (8.07.1971); 8. Custura Păpuşei (7.07.1971); 9. Peleaga-Bucura (8.07.1970); 10. Muchea Custurei (8.07.1970)

coverage of 35%. The chionophilic species characteristic of the alliance *Salicion* and order *Salicetalia herbaceae* are also present, similarly to those described in Germany (Oberdorfer 1977). The *Salix herbacea* coenoses come into contact with the *Festuca supina* alpine grasslands, revealing that under the current climatic conditions, in some places, the grassland coenoses tend to replace the willow communities as soil development advances.

Ass. *Luzuletum obscurae* Szfer et al. 1927 corr. Dúbracová 2007 (Table 6.13)

The *Luzula alpinopilosa* subsp. *obscura* coenoses, specific to the Carpathians, form compact communities on areas of 50–100 m^2, on the gutters of the slopes of some peaks in Retezat (Şesele, Custura, Păpuşa) and on the humus-rich colluvia at the scree base of some frontal moraines near the Bucura, Galeşu, Tău Negru, and Valea Rea glacial lakes. The dominant species, *Luzula alpinopilosa* subsp. *obscura,* achieves between 35% and 65% of coverage. The chionophilic species characteristic of the alliance and order are well represented in the composition of this association. The presence of some species characteristic of the alliance *Adenostylion*, such as *Adenostyles alliaria, Saxifraga rotundifolia,* and *Veratrum album,* denotes the more pronounced hygrophilous character of the association as compared to *Salicetum herbaceae.*

Ass. *Soldanello pusillae-Plantaginetum gentianoides* Boşcaiu 1971 (Table 6.14)
Nomenclatural type: Boşcaiu 1971, Table 23, rel. 4, lectotype *hoc loco*

This Dacian-Balkan regional association was found and described for the first time in the Ţarcu-Godeanu Mountains (Boşcaiu 1971) in the glacial cirques and nival niches from flat areas, where water from the melting snow stagnates for a long time. We identified such phytocoenoses near some glacial lakes (Bucura, Galeşu, Judele, Zănoaga) in the alpine belt of the Retezat Mountains, where they populate wet, slightly lumpy areas with fluvisols rich in humic colluvium and having acidic reaction (pH = 5). These soils often evolve into histosols over time and favor the development of mesotrophic swamp phytocoenoses, grouped into the association *Carici dacicae-Plantaginetum* Boşcaiu et al. 1972. The hygrophilous specificity of the association is revealed by the dominant species *Soldanella pusilla* and *Plantago gentianoides*, that achieve on average 50% coverage, and also by *Carex pyrenaica.*

Ass. *Soldanello pussillae-Ranunculetum crenati* (Borza 1963) Boşcaiu 1971 (Table 6.15)

This chionophilic-petrophilous association reported by Borza (1963) from the Southern Romanian Carpathians was also identified on the long-lasting snow screes of the Radeşu, Zănoaga, Judele, Huta high peaks, located at the southern limit of the Retezat scientific reserve. The coenoses of this association have a more mesophilous character because the water resulting from the melting snow is easily drained by the underlying semi-fixed screes. This ecological specificity of the association is revealed by the reduced presence of *Plantago gentianoides* and the absence of hygrophilous *Carex pyrenaica* (Sârbu et al. 2013).

Table 6.13 Ass. *Luzuletum obscurae* Szafer et al. 1927 corr. Dúbravcová 2007

Relevé No.	1	2	3	4	5	6	7	8	9	10	
Altitude (m a.s.l.)	2300	2260	2240	2120	2110	2070	2100	2010	2090	1980	
Aspect	NW	E	NW	NW	W	NE	N	NW	W	E	
Slope (degrees)	45	45	45	30	30	20	25	30	5	45	
Herb cover (%)	40	80	45	65	65	70	40	50	70	70	
Sample area (sq. m)	10	25	25	25	25	16	16	4	25	25	K
Char. ass.											
Luzula alpinopilosa subsp. *obscura*	3.4	4.5	3.5	3.5	3.5	4.5	2.5	3.5	4.5	4.5	V
Salicion herbaceae **et** ***Salicetalia herbaceae***											
Soldanella pusilla	+.4	+	+	2.5	2.5	1.5	+	1.3	+	+	V
Ligusticum mutellina	+	+	+	1.5	.	.	1.5	1.5	+	+	IV
Geum montanum	+	1.3	+	+.3	+.3	+	.	.	+	.	IV
Gnaphalium supinum	+	+.5	+	+	.	.	+	.	.	+	III
Leucanthemopsis alpina	.	.	+	.	.	.	.	.	+	+	II
Plantago gentianoides	.	.	.	1.5	1.3	+	.	.	.	.	II
Polytrichum sexangulare	.	.	.	.	.	.	.	.	.	+	I
Salicetea herbaceae											
Sedum alpestre	+	+	.	.	.	.	.	+	.	.	II
Taraxacum panalpinum	.	+	.	+	.	.	.	.	.	.	I
Poa granitica subsp. *disparilis*	.	.	.	.	.	+	.	+	.	.	I
Companion											
Homogyne alpina	+	+	1.5	.	+	+	+	.	+	+	IV
Senecio incanus subsp. *carniolicus*	+	.	+	.	.	.	+	.	.	.	II
Doronicum carpaticum	.	+	.	.	.	.	.	+	.	+	II
Primula minima	.	.	+.5	.	.	.	+	.	.	+	II
Agrostis rupestris	.	.	.	.	.	.	1.5	+	.	1.4	II

Veronica baumgartenii	.	.	+	.	.	.	+	.	.	+	II
Deschampsia flexuosa	.	.	+	.	.	.	.	.	+	.	I
Campanula alpina	.	.	+	.	.	.	.	.	.	+	I

Companion species with one or two occurrences: *Adenostyles alliariae* 9: 1.3; *Anthoxanthum alpinum* 3: +; *Epilobium alpinum* 4: +; *Festuca picturata* 3, 10: +; *Hieracium alpinum* 3, 10: +; *Gentiana punctata* 3, 10: +; *Phyteuma confusum* 7, 10: +; *Pulsatilla alba* 3, 7: +; *Pedicularis verticillata* 3, 10: +; *Potentilla ternata* 3: +; *Saxifraga rotundifolia* subsp. *heucherifolia* 6: +; *Oreochloa disticha* 6: +; *Solidago virgaurea* subsp. *minuta* 6: +; *Senecio doronicum* subsp. *transylvanicus* 10: +; *Veratrum album* 8: +; *Rododendron myrtifolium* 10: +; *Vaccinium myrtillus* 10: +; *Cetraria islandica* 7, 9: +; *Dicranum scoparium* 7, 8: +; *Polytrichum piliferum* 7: +; *Racomitrium sudeticum* 7: +

Relevé: 1. Muchia Ascuţită (10.07.1971); 2. Mt. Custura (10.07.1972); 3. Peleaga-Bucura (8.08.1970); 4. Deasupra Galeşului (17.07.1971); 5. V. Rea sub Păpuşa (8.08.1971); 6. Sub poarta Bucurei (5.07.1969); 7. Mt. Şesele (5.07.1969); 8. Tăul Negru (5.07.1969); 9. V. Judelui (13.07.1970); 10. Căldarea Şeselor (12.07.1967)

Table 6.14 Ass. *Soldanello pusillae-Plantaginetum gentianoidis* Boşcaiu 1971

Relevé No.	1	2	3	4	5	6	7	8	9	10	
Altitude (m a.s.l.)	2180	2150	2115	2070	2130	2130	2150	2150	2010	1840	
Aspect	S	N	W	N	–	N	–	–	E	–	
Slope (degrees)	10	25	20	20	0	10	0	0	10	0	
Herb cover (%)	80	40	35	80	45	35	50	50	45	100	
Sample area (sq. m)	25	25	25	25	25	25	25	25	25	25	K
Char. ass.											
Soldanella pusilla	+.4	1.5	2.5	1.5	+.4	1.5	+.4	+.4	1.3	2.4	V
Plantago gentianoides	3.5	3.5	2.5	4.5	3.5	2.5	3.5	3.5	3.5	4.5	V
Salicion herbaceae **et** ***Salicetalia herbaceae***											
Ligusticum mutellina	+	1.5	+	1.5	1.5	1.4	+.5	+	.	1.5	V
Luzula alpinopilosa subsp. *obscura*	+	+	+	1.5	+	+	+	+	.	.	IV
Geum montanum	.	+.4	.	+.3	.	+	+	+	+.3	+	IV
Carex pyrenaica	3.5	.	1.3	.	1.3	.	.	.	.	.	II
Leucanthemopsis alpina	+	.	.	.	.	.	.	.	.	.	I
Ranunculus crenatus	.	.	.	+	.	.	.	.	.	.	I
Salicetea herbaceae											
Taraxacum panalpinum	.	+	.	.	+	.	.	.	.	+	II
Sagina saginoides	.	.	.	.	.	.	+	.	+	.	I
Companion											
Nardus stricta	2.5	.	+.3	1.5	+	2.5	2.5	2.5	2.5	+	V
Homogyne alpina	+	.	+	1.5	+	+	+	+	+	+	V
Potentilla ternata	1.5	.	.	+.5	.	+.4	+.5	+.5	+	.	III
Cetraria islandica	.	.	.	+	.	+	.	.	+	.	II
Crocus vernus	.	.	.	.	.	.	+	+	.	.	I
Poa alpina	.	+	.	.	.	.	.	.	.	+	I

Poa media	+	.	.	.	.	.	.	.	.	2.5	I
Pulsatilla alba	.	.	.	.	+	.	+	.	.	.	I
Dicranum spadiceum	.	.	.	+	.	+	.	.	.	.	I
Polytrichum alpinum	.	.	.	.	.	+	.	.	+	.	I

Companion species with one occurrence: *Agrostis rupestris* 1: +; *Campanula alpina* 6: +; *Festuca nigrescens* 10: +; *Hieracium alpinum* 1: +; *Poa supina* 10: +; *Primula minima* 3: +; *Vaccinium myrtillus* 3 +; *Veratrum album* 10: +

Relevé: 1. Vf. Bucura (8.07.1970); 2. V. Galeșului (17.07.1971); 3. V. Rea-Păpușa (13.07.1971); 4. Mt. Huta (9.07.1970); 5. Lacul Bucura (8.07.1970); 6. Tăul Judelui (13.07.1970); 7. V. (12.07.1970); 8. Tăul Ascuns (8.07.1970); 9. Lacul Zănoaga (10.07.1969); 10. V. Pietrele (6.08.1970)

Table 6.15 Ass. *Soldanello pusillae-Ranunculetum crenati* (Borza 1934) Boşcaiu 1971

Relevé No.	1	2	3	4	5	6	7	8	
Altitude (m a.s.l.)	2030	2000	2010	2010	2010	1845	2010	1845	
Aspect	N	NW	NW	NW	N	N	NW	NW	
Slope (degrees)	60	60	60	15	15	10	60	5	
Herb cover (%)	90	70	70	100	100	60	100	50	
Sample area (sq. m)	10	25	25	25	25	4	25	4	K
Char. ass.									
Ranunculus crenatus	4.5	3.5	3.5	+.3	3.5	3.4	4.5	2.5	V
Soldanella pusilla	1.5	1.5	1.5	4.5	2.5	1.3	1.5	1.4	V
Salicion herbaceae **et** ***Salicietalia herbaceae***									
Geum montanum	+.3	+	1.5	1.5	1.5	1.3	1.3	+	V
Luzula alpinopilosa subsp. *obscura*	+	+	1.4	2.5	1.4	.	+	.	IV
Leucanthemopsis alpina	.	+	+	.	+	+	+	.	IV
Ligusticum mutellina	.	+.5	+.5	.	+.5	+	.	1.3	IV
Plantago gentianoides	+	.	+	.	+.5	.	.	+	III
Gnaphalium supinum	+	.	+	.	.	.	+	.	II
Salicetea herbaceae									
Polytrichum sexangulare	+	.	.	+	+	.	.	.	II
Taraxacum nigricans	.	.	.	+	.	+	.	.	II
Sedum alpestre	.	.	.	.	.	.	.	+	I
Companion									
Nardus stricta	1.5	2.5	2.5	1.3	2.5	2.3	2.5	1.3	V
Homogyne alpina	+	.	1.5	+	1.5	+	.	+	IV
Cetraria islandica	+.5	+	1.5	+	+.5	.	.	+	IV
Potentilla ternata	.	+	+	+	+	.	+	.	III
Poa alpina	.	+	+	.	+	.	+	1.3	III
Primula minima	.	+	+	.	+	.	+	.	III
Dicranum scoparium	.	+	+	.	+	.	.	.	II
Campanula alpina	.	+	.	.	.	+	.	.	II

Companion species with one or two occurrences: *Agrostis rupestris* 1: +; *Anthoxanthum alpinum* 1: +; *Avenula versicolor* 1: +; *Juncus trifidus* 7: +.5; *Pulsatilla alba* 6: +; *Drepanocladus uncinatus* 2: +; *Polytrichum juniperinum* 2: +

Relevé: 1. Mt. Huta (9.07.1970); 2. Obârşia Radeşului (8.07.1970); 3. Vf. Huta-Judele (10.07.1970); 4. Sub Curmătura Zănoagei (7.07.1970); 5. Obârşia Radeşului (8.07.1970); 6. Mt. Huta (9.07.1970); 7. Vf. Judele-Huta (9.07.1970); 8. Mt. Huta (9.07.1970)

Ass. *Poo supinae-Cerastium cerastioides* (Söry 1954) Oberd. 1957 (Table 6.16)

The meso-hygrophilous coenoses built up by *Cerastium cerastioides* and *Poa supina* from Retezat mostly vegetate on rock debris in the nivation niches and semi-mobile scree cones from the basal part of subalpine-alpine torrents. The chionophilic-hygrophilous feature of these pioneer phytocoenoses is revealed by the presence of microthermal species typical for the alliance, order, class and by the very small

Table 6.16 Ass. *Poo supinae-Cerastietum cerastioidis* (Söry. 1954) Oberd. 1957

Relevé No.	1	2	3	4	5	
Altitude (m a.s.l.)	1780	1780	1980	1980	1750	
Aspect	E	E	–	–	–	
Slope (degrees)	5	3	0	0	0	
Herb cover (%)	90	90	70	90	80	
Sample area (sq. m)	10	10	25	10	10	K
Char. ass.						
Cerastium cerastioides	3.5	3.5	3.5	3.5	3.5	V
Poa supina	4.5	4.5	3.5	4.5	3.5	V
Salicion herbaceae **et** ***Salicetum herbaceae***						
Gnaphalium supinum	+	+	+	1.3	+	V
Ligusticum mutellina	+	+	.	.	1.5	III
Sedum alpestre	.	.	+	+	+	III
Geum montanum	1.1	+	.	.	+	III
Salicetea herbaceae						
Taraxacum panalpinum	.	+	1.5	1.5	+	IV
Veronica alpina	+	+	.	.	+	III
Companion						
Phleum alpinum	+	+.3	+	+	+	V
Poa alpina	+	+	+	+	.	IV
Agrostis rupestris	+	+	.	.	1.5	III
Potentilla ternata	+	+	.	.	+	III
Alchemilla vulgaris	+	+	.	.	.	II
Phyteuma confusum	+	+	.	.	.	II
Taraxacum fontanum	1.5	+	.	.	.	II
Veronica serpyllifolia s.l.	+	+	.	.	.	II
Primula minima	.	.	.	+	+.2	II
Poa media	.	+	.	.	.	I

Relevé: 1–2. Mt. Piule (6.07.1972); 3–4. Mt. Drăgşanu (6.07.1972); 5. Mt. Albele (7.07.1972)

number of co-occurring species, similarly to those described in the Ţarcu Mountains (Boşcaiu 1971).

Ass. *Nardo-Gnaphalietum supinae* Bartsch 1940 (Table 6.17)

The association groups pioneer chionophilic-terricolous phytocoenoses developed on the solifluction terraces with sandy-clay fluvisol (Florea and Munteanu 2003) of the subalpine-alpine belt from Retezat (Zănoaga, Şesle, Judele, Bucura). In these places the snow persists for 6–8 months a year, favoring the development of chionophilic species typical for *Salicion* and *Salicetalia herbaceae*, as well as acidophilic and oligotrophic species characteristic of the alliance *Potentillo-Nardion*, towards which these phytocoenoses evolve in time (Boşcaiu 1971). The ecological, alpine peculiarity of these areas is revealed by the species *Festuca*

Table 6.17 Ass. *Nardo-Gnaphalietum supinae* Bartsch 1940

Relevé No.	1	2	3	4	5	6	7	
Altitude (m a.s.l.)	2100	2100	2050	2060	1920	1940	1890	
Aspect	E	–	W	NE	SE	–	S	
Slope (degrees)	15	0	15	50	5	0	60	
Herb cover (%)	60	60	50	40	60	60	50	
Sample area (sq. m)	4	4	4	4	4	4	4	K
Char. ass.								
Gnaphalium supinum	2.5	1.5	1.4	1.3	3.5	3.5	+	V
Salicion herbaceae **et** ***Salicetalia herbaceae***								
Luzula alpinopilosa subsp. *obscura*	+.3	+.3	.	+	.	.	+	III
Polytrichum sexangulare	+	2.5	.	.	1.5	+	.	III
Plantago gentianoides	1.3	1.3	+	.	.	.	.	III
Soldanella pusilla	+	+	+	.	.	.	.	III
Ligusticum mutellina	1.3	1.3	.	+	.	.	.	III
Salicetea herbaceae								
Taraxacum nigricans	+	+	.	.	.	.	.	II
Plantago atrata subsp. *carpatica*	.	.	.	+	.	.	.	I
Sedum alpestre	.	.	.	.	.	.	+	I
Potentillo-Nardion								
Nardus stricta	3.5	3.5	3.4	3.4	2.3	2.5	2.5	V
Potentilla ternata	+	1.5	+	.	+	+	+	V
Poa alpina	+	+	+.3	.	+	+	.	IV
Campanula serrata	.	.	.	.	+	+	+	III
Luzula sudetica	.	.	.	.	+	+	.	II
Companion								
Deschampsia flexuosa	.	.	.	+	.	.	+	II
Phyteuma confusum	.	.	.	+	.	.	+	II
Festuca supina	.	.	.	.	+	+	.	II
Hieracium alpinum	.	.	.	.	+	+	.	II

Companion species with one occurrence: *Anthoxanthum alpinum* 1: +; *Campanula alpina* 3: +; *Avenula versicolor* 4: +; *Primula minima* 2: +; *Agrostis rupestris* 7: +; *Juncus trifidus* 7: +
Relevé: 1. Circul glaciar Zănoaga (8.07.1970); 2. Şeselor Ridge (8.07.1970); 3. V. Judelui (10.07.1970); 4–5. Lacul Zănoaga (8.07.1970); 6. Mt. Zănoaga (8.07.1970); 7. Gura Bucurei (7.08.1970)

supina, Hieracium alpinum, Campanula alpina, Primula minima, and *Agrostis rupestris*, sporadically present in this association.

Order *Arabidetalia caeruleae* Rübel 1933

In this order are grouped the alpine, chionophilic phytocoenoses from calcareous cliffs where snow persists 7–9 months a year. The soil in these places is superficial, a discontinuous rendzinic leptosol. The phytocoenoses are located on the weather-excavated shelves in the limestone wall and on the screes at the base of the cliffs where snow persists longer. The characteristic species of the order are identical to those of the alliance (Oberdorfer 1977).

Alliance *Arabidion caeruleae* Br.-Bl. in Br.-Bl. et Jenny 1926

The alliance includes the microthermal, calcophilic phytocoenoses of the alpine areas in Europe. Horvat described the vicariant alliance *Salicion retusae* in the Balkans and included in it the coenoses with *Salix retusa* and *Salix reticulata* (Horvat et al. 1974). The absence of *Arabis caerulea* in the Carpathians determined the Romanian phytocoenologists to include such phytocoenoses in the alliance *Salicion retusae* (Boşcaiu 1971; Coldea 1990; Mihăilescu 2001). Taking into account that such microthermal coenoses described in the Tatra Mountains are also grouped into the alliance *Arabidion*, we consider it justified to also include the coenoses from Retezat within the same syntaxon. The characteristic species of the alliance are *Saxifraga androsacea, Polygonum viviparum, Sedum roseum, Veronica aphylla,* and *Selaginella selaginoides* (Jarolimek and Šibik 2008).

Ass. *Soldanello pusillae-Salicetum retusae* (Boşcaiu 1971) Coldea 1993 (Table 6.18)

The coenoses built up by *Salix retusa* (including *S. kitaibeliana*) from the Retezat Mountains develop on Jurassic limestone shelves (Albele, Piule, Piatra Iorgovanului, Custura) and on their basal screes, where snow persists for 6–7 months a year. The soils in these areas are skeletal, humus-rich leptosols. Through its preference for base rich soil (Florea and Munteanu 2003), the species *Soldanella pusilla* gives a distinct ecological specificity to this association in comparison to *Soldanello alpinae-Saliceteum retusae* Horvat 1933, reported in the Dinaric Alps (Horvat et al. 1974). The presence of a large number of basophilic species in the association, characteristic of the order *Seslerietalia,* reveals the direct contact between the chionophilic coenoses and the alpine basophilic vegetation. The coenoses of this association evolve, in time, towards communities of the alliance *Seslerion bielzii* (Boşcaiu 1971), as humus accumulates and progressive decalcification occurs in the upper soil horizon.

Class *Montio-Cardaminetea* Br.-Bl. et Tx. ex Klika et Hadač 1944

This class includes the spring and mountain stream phytocoenoses, well individualized floristically by several hygrophilic and microthermal species, adapted to the continuous flow of cold and well oxygenated water. The water temperature in these springs and streams remains around 5–6 °C throughout the growing season. The local microclimatic conditions favor the development of both hygrophilous cormophyte and bryophyte synusiae. Taking into account their floristic composition, we grouped the montane and subalpine fontinal phytocoenoses of Retezat in a single order, *Montio-Cardaminetalia*. The characteristic species of the class and order are *Cardamine amara, Epilobium alsinifolium, Saxifraga stellaris, S. aizoides, Stellaria nemorum, Silene pulsilla, Philonotis seriata, Brachythecium rivulare,* and *Scapania undulata.*

Table 6.18 Ass. *Soldanello pusillae-Salicetum retusae* (Boşcaiu 1971) Coldea 1993

Relevé No.	1	2	3	4	5	6	7	8	9	10	
Altitude (m a.s.l.)	2280	2240	2200	2100	2150	2000	2000	1880	1850	2240	
Aspect	E	E	W	N	W	NNE	WWE	NE	N	E	
Slope (degrees)	70	40	45	30	50	40	30	30	15	40	
Herb cover (%)	50	50	50	50	95	100	100	100	100	100	
Sample area (sq. m)	25	10	25	25	25	10	25	25	25	25	K
Char. ass.											
Salix retusa	4.5	3.5	3.5	3.5	3.5	3.5	5.5	2.5	4.5	4.5	V
Soldanella pusilla	+	+	+	+	.	+	.	.	.	+	III
Arabidion caeruleae et ***Arabidetalia***											
Polygonum viviparum	+	+	+	+	.	+	+	+	1.5	+	V
Sedum roseum	+	+	+	+	.	.	.	.	.	+	III
Myosotis alpestris	.	+	+	.	+	.	.	.	.	+	II
Veronica aphylla	.	.	+	.	+	+	.	.	.	.	II
Selaginella selaginoides	.	.	.	.	.	+	.	.	+	.	I
Saxifraga androsacea	.	.	.	.	.	.	+	.	.	.	I
Salicetea herbaceae											
Luzula alpinopilosa subsp. *obscura*	.	.	+	+	+	+	.	.	.	.	II
Veronica alpina	.	.	.	.	.	+	.	.	+	.	I
Festuca picturata	.	+	.	.	.	.	.	.	.	+	I
Ligusticum mutellina	.	+	.	.	.	+	.	.	.	.	I
Salix herbacea	+	.	.	.	.	.	.	.	.	.	I
Viola alpina	.	+	.	.	.	.	.	.	.	.	I
Seslerion et ***Seslerietalia***											
Bartsia alpina	+	+	+.3	+	+	+	+	+	+.2	.	V
Ranunculus oreophilus	.	+	+.3	.	+	.	+	+	.	+	III

Silene acaulis	.	+	.	+	.	+	.	.	.	.	II
Dryas octopetala	.	.	.	.	+	.	.	3.5	2.5	.	II
Hedysarum hedysaroides	.	.	.	.	+	.	.	.	+.4	+	I
Saxifraga moschata	.	+	+	.	.	.	.	.	.	.	I
Cardamine resedifolia	.	+	+	.	.	.	.	.	.	.	I
Artemisia petrosa	.	+	.	.	+	.	.	.	.	.	I
Carex sempervirens	.	.	.	.	+	.	.	2.5	.	.	I
Lloydia serotina	.	+	.	.	.	.	.	.	.	.	I
Galium anisophyllon	.	.	.	.	.	.	.	+	.	.	I
Sesleria rigida subsp. *haynaldiana*	.	.	.	.	.	.	.	.	+	.	I
Companion											
Rhododendron myrtifolium	1.3	2.4	1.5	+	+	+	+	.	.	1.3	IV
Pulsatilla alba	+	+	+	+	.	+	.	.	.	+	III
Campanula alpina	+	.	.	+	+	+	.	.	.	+	III
Oreochloa disticha	.	+	+	+	+	+	.	.	.	.	III
Homogyne alpina	.	+	.	+	+	+	.	.	.	+	III
Cerastium alpinum	+	.	.	+	+	.	.	.	.	+	II
Vaccinium gaultherioides	+	.	.	+	.	+	.	.	.	+	II
Carex atrata	.	+	+	.	+	+	.	.	.	.	II
Gentiana punctata	.	+	.	+	.	+	.	.	.	+	II
Doronicum columnae	.	.	+	.	+	+	+	.	.	.	II
Huperzia selago	.	.	.	+	+	+	.	.	+	.	II
Coeloglossum viride	.	.	+	+	.	+	.	.	.	.	II
Pedicularis verticillata			+		+					+	II

(continued)

Table 6.18 (continued)

Saxifraga paniculata	.	.	+	.	.	.	+	.	+	.	II
Primula minima	.	.	.	+	+	+	.	.	.	.	II
Vaccinium myrtillus	.	.	.	+	+	+	.	.	.	.	II

Companion species with one or two occurrences: *Agrostis alpina* 2: +; *Aconitum licoctonum* subsp. *moldavicum* 2: +; *Anthoxanthum alpinum* 3, 4: +; *Centaurea nervosa* 2: +; *Doronicum carpaticum* 2: +; *Festuca supina* 1: +; *Hieracium alpinum* 1 +; *Loiseleuria procumbens* 4, 6: +; *Geum reptans* 2: +; *Galium album* subsp. *album* 3: +; *Phyteuma confusum* 3: +; *Sesleria coerulans* 1: 1.3, 10: +; *Senecio incanus* subsp. *carniolicus* 1: +; *Veronica baumgartenii* 1: +; *Saxifraga pedemontana* subsp. *cymosa* 2: +; *Thymus pulcherrimus* 3: +; *Gentiana verna* 9: +; *Vaccinium vitis-idaea* 5, 8: +; *Sesleria rigida* 9: +; *Anthemis carpatica* 6: +; *Carex curvula* 6: +

Relevé: 1–2. Mt. Custura (10.07.1971); 3. Stâna de Râu (17.071971); 4. Sub Custura (14.07.1971); 5. Vf. Cionfu (17.07.1971); 6. Stîna de Râu (17.07.1971); 7. Mt. Piule (25.06.1971); 8. Mt. Albele (29.06.1972); 9. Mt. Piatra Iorgovanului (24.06.1971); 10. Mt. Custura (10.07.1971)

Order *Montio-Cardaminetalia* Pawłowski 1928

The fontinal coenoses from Central Europe were usually included in this single order (Boşcaiu 1971; Coldea 1990; Matuszkiewicz 2008; Oberdorfer 1977; Oprea and Sîrbu 2009; Zechmeister 1993). After Hinterlang (1992) published the new order *Cardamino-Chrysosplenietalia,* typical for the vegetation springs of low-altitude mountains in Central Europe, several phytocoenologists accepted it (Ellenberg 1996; Jarolimek and Šibik 2008; Mucina 2016; Rodwell et al. 2002; Valachovič 2001) and assigned such phytocoenoses to this syntaxon. Since montane fontinal coenoses are very narrowly distributed in the Retezat Mountains and because they lack some characteristic species of the order *Cardamino-Chrysosplenietalia,* we included all spring associations into the order *Montio-Cardaminetalia.* This order comprises two alliances with very well individualized phytocoenoses from floristic and eco-pedological point of view.

Alliance *Cardamino-Montion* Br.-Bl. 1926.
Natura 2000: habitat type 3220

We include in this alliance all montane and subalpine spring and stream phytocoenoses on siliceous substrates, with cold and well-oxygenated waters and having the following diagnostic species: *Cardamine amara* subsp. *amara, C. amara* subsp. *opicii, Chrysosplenium alternifolium, C. alpinum, Caltha palustris, Epilobium nutans, Stellaria uliginosa, Myosotis scorpioides, Pellia epiphylla, Mnium punctatum, Plagiomnium undulatum* (Ellenberg 1996; Oberdorfer 1977).

Ass. *Chrysosplenio-Cardaminetum amarae* (Tx. 1937) Mass 1959 (Table 6.19)

We include in this association the spring and stream coenoses of the mixed (beech-spruce and beech-fir) forests from the Retezat Mountains. The dominant species in this association is *Cardamine amara* subsp. *amara*, achieving on average 60% coverage. Apart from it, the hygrophilous and montane microthermal species characteristic of the alliance and order are also frequent, such as: *Stellaria nemorum, S. uliginosa, Epilobium nutans, Myosotis scorpioides, Mnium punctatum,* and *Scapania undulata*. The composition of these phytocoenoses above 1000 m altitude also includes some transgressive, hygrophilic species of *Adenostylion alliariae* (*Cicerbita alpina, Leucanthemum rotundifolium, Doronicum austriacum, Milium effusum, Senecio nemorensis, Adenostyles alliariae*), occasionally found in mixed beech-coniferous forests.

Ass. *Cardaminetum opicii* Szafer et al. 1923 (Table 6.20)

We assigned to this association, originally described from the Tatra Mountains (Szafer et al. 1923), the fontinal coenoses of the subalpine belt from the Retezat Mountains, dominated by *Cardamine amara* subsp. *opicii*. The floristic composition of this association comprises a large number of characteristic species of the alliance *Cardamino-Montion* and order *Montio-Cardaminetalia*, but they have low coverage. These phytocoenoses come into contact with the hygrophilous coenoses of tall forbs of the order *Adenostyletalia*, similarly to those described by Valachovič (1995)

Table 6.19 Ass. *Chrysosplenio-Cardaminetum* (Tüxen 1937) Mass 1959

Relevé No.	1	2	3	4	5	6	
Altitude (m a.s.l.)	1050	1200	840	1050	1150	960	
Aspect	SW	N	NE	W	N	N	
Slope (degrees)	0	25	10	10	20	5	
Herb cover (%)	90	90	75	75	90	90	
Sample area (sq. m)	25	25	25	25	25	25	K
Char. ass.							
Cardamine amara	4.5	4.5	3.5	3.5	4.5	4.5	V
Chrysosplenium alternifolium	.	+.3	.	+.5	+.5	.	III
Cardamino-Montion et ***Montio-Cardaminetalia***							
Stellaria nemorum	+.3	1.3	1.3	+.2	1.5	.	V
Myosotis scorpioides	1.5	.	.	+	.	+	IV
Mnium punctatum	.	+.5	+	+	+	.	IV
Epilobium nutans	.	+.5	+	.	+	+.2	IV
Impatiens noli-tangere	.	.	+	+	+	1.3	IV
Scapania undulata	.	2.5	.	3.5	2.5	.	III
Plagiomnium undulatum	.	.	+	1.3	+	.	III
Caltha palustris	.	.	+.3	.	.	+	II
Chaerophyllum hirsutum	+	.	.	+	.	.	II
Brachytecium rivulare	.	1.3	.	.	+.5	.	II
Stellaria uliginosa	.	.	+	.	.	.	I
Pellia epiphylla	.	.	.	.	.	+	I
Petasition							
Petasites hybridus	.	+.3	.	1.3	1.3	+	IV
Rumex obtusifolius subsp. *subalpinus*	+	.	.	.	+	+	III
Stachys palustris	.	+	.	.	+	.	II
Lamium maculatum	.	+	.	.	+	.	II
Angelica sylvestris	+	.	.	.	+	.	II
Valeriana officinalis	+	.	.	.	.	.	I
Companion							
Urtica dioica	.	+	+	+	+	+	V
Oxalis acetosella	.	+	+	+	+	.	IV
Plagiochila asplenioides	.	2.5	1.3	+	2.5	.	IV
Plagiothecium ruthei	.	1.3	+	+	+	.	IV
Brachythecium rutabulum	.	.	+	+	1.3	1.5	IV
Senecio nemorensis	1.3	+	.	.	+	.	III
Ranunculus repens	1.5	.	+.3	.	.	1.3	III
Milium effusum	.	+	+	+	.	.	III
Doronicum austriacum	.	+	.	+	+	.	III
Geranium robertianum	.	.	+	.	+	+	III
Veronica beccabunga	+	.	.	.	.	+	II
Calamagrostis arundinacea	.	.	.	+	+	.	II

Companion species with one occurrence: *Cirsium oleraceum* 1: +; *Geum rivale* 1: +; *Adenostyles alliariae* 1: +; *Leucanthemum rotundifolium* 1: +; *Cicerbita alpina* 2: +; *Gentiana asclepiadea* 2: +; *Poa trivialis* 3: +; *Athyrium distentifolium* 4: +; *Chiloscyphus polyanthus* 5: +; *Bryum pseudotriquetrum* 6: +

Relevé: 1. V. Lăpuşnicului (15.07.1970); 2. V. Şesele pe pârîul Pârgului (9.07.1968); 3. V. Zlătuia (18.06.1968); 4. V. Şesele (21.06.1968); 6. V. Turcului (9.09.1969)

Table 6.20 Ass. *Cardaminetum opicii* Szafer et al. 1923

Relevé No.	1	2	3	4	5	
Altitude (m a.s.l.)	1800	209	1650	1750	1760	
Aspect	S	S	–	SW	SW	
Slope (degrees)	5	5	0	30	30	
Herb cover (%)	80	100	75	50	60	
Sample area (sq. m)	25	25	25	25	25	K
Char. ass.						
Cardamine amara subsp. *opicii*	4.5	5.5	4.5	3.5	2.5	V
Philonotis seriata	3.5	2.4	+.5	+	+	V
Cardamino-Montion **et** ***Montio-Cardaminetalia***						
Saxifraga stellaris	+	+	.	1.3	+	IV
Epilobium alsinifolium	+	.	+	+	+	IV
Caltha palustris	+	+.3	+	.	.	III
Stellaria nemorum	.	.	+	+	+	III
Chrysosplenium alternifolium	.	.	+	+	+	III
Mnium punctatum	.	.	.	+	+	II
Saxifraga rotundifolia subsp. *heucherifolia*	.	.	.	+	1.4	II
Silene pusilla	.	.	.	+	+	II
Brachythecium rivulare	.	.	+	+	.	II
Myosotis scorpioides	.	.	+	.	.	I
Adenostyletalia						
Deschampsia caespitosa	+	+	+	.	+	IV
Chaerophyllum hirsutum	.	.	+	+	2.5	III
Epilobium alpestre	.	+	+	.	.	II
Viola biflora	.	.	.	+	+	II
Aconitum napellus subsp. *firmum*	.	.	.	+	+	II
Leucanthemum rotundifolium	.	.	.	.	+	I
Companion						
Poa alpina	+	.	+	.	+	III
Plantago gentianoides	+	+	.	.	.	II
Rumex alpinus	.	.	.	+	1.5	II

Companion species with one occurrence: *Poa laxa* 1: +; *Carex nigra* subsp. *dacica* 2: +; *Senecio subalpinus* 3: +; *Taraxacum panalpinum* 3: +; *Geum rivale* 4: +; *Athyrium distentifolium* 4: +; *Rumex alpestris* 5: +; *Crepis paludosa* 5: +

Relevé: 1. Căldarea Zănoaga (10.07.1970); 2. V. Pelegii (7.08.1968); 3–4. Ridge Drăgşanu (17.07.1967); 5. Drăgşanul-Albele (18.07.1967)

in the Tatra Mountains. The coenoses from Retezat differ floristically from those described in the Western Carpathians (Hájková and Hájek 2011; Valachovič 1995) by several regional species such as *Plantago gentianoides, Carex nigra* subsp. *dacica,* and *Aconitum napellus* subsp. *firmum.*

Ass. *Philonotido seriatae-Calthetum laetae* (Krajina 1933) Coldea 1991 (Table 6.21)
Syn.: *Philonotidetum seriatae* Borza 1934 (Art. 37)

The subalpine spring and stream phytocoenoses from Retezat, dominated by *Caltha laeta,* with alluvial soils and rich in coarse gravel are included in this association. The subalpine species *Philonotis seriata* reveals the microthermal character of these coenoses, compared to the ones with *Caltha laeta* from the montane and submontane belt in the Carpathians, mostly with mesothermal species (Dihoru 1975; Oprea and Sîrbu 2009; Oroianu 1998; Sămărghițan 2005). The microthermal specificity of the coenoses assigned to this association is denoted by the presence of some transgressive species from *Salicion herbaceae*, such as *Luzula alpinopilosa* subsp. *obscura, Plantago gentianoides,* and *Cerastium cerastioides.* Some hygrophilous species from *Adenostylion alliariae* are present in the association in some areas of Retezat.

Ass. *Chrysosplenio alpini-Saxifragetum stellaris* Pawl. et Walas 1949 (Table 6.22)

The springs and streams on siliceous substrates from the alpine and subalpine belts of the Retezat Mountains are frequently populated by the phytocoenoses assigned to this association, similarly to those from the Țarcu-Godeanu Mountains (Boșcaiu 1971). It develops luxuriantly at the edge of streams and springs with cold, fast-flowing water on alluvial soils, rich in fine gravel. The dominant species of the association are *Saxifraga stellaris* subsp. *alpigena* and *Philonotis seriata*, achieving on average 50% coverage. The endemic species *Chrysosplenium alpinum* is present but has a scattered distribution, as well as the characteristic species of the *Cardamino-Montion* and *Montio-Cardaminetalia* syntaxa. We separated the new sub-association *caricetosum dacicae* (rel.: 1–5) on the basis of differential species *Carex nigra* subsp. *dacica* and *Juncus filiformis* var. *transsilvanicus,* occurring on gentle slopes, rich in fine alluvia and prone to swamping. This sub-association reveals the syndynamic trend of *Saxifraga stellaris*-dominated coenoses toward the association *Carici dacicae-Plantaginetum gentianoides,* as soil peat formation advances in time.

Alliance *Cratoneurion commutatae* W. Koch 1926

In this alliance are included hygrophilous phytocoenoses developed around mountain springs and streams on calcareous substrate. Such fontinal phytocoenoses are rare in Retezat because the calcareous substrate they populate facilitates water infiltration into the substrate, preventing the settlement of hygrophilous coenoses. The characteristic species of this alliance are *Saxifraga aizoides, Doronicum carpaticum, Silene pusilla, Philonotis calcarea, Bryum pseudotriquetrum,* and *Cratoneuron filicinum.*

Ass. *Doronico carpatici-Saxifragetum aizoidis* Coldea 1990 (Table 6.23)

In this association, we group the fontinal, basophilic coenoses identified in the southern and eastern part of Retezat, where they populate the edges of subalpine springs and streams, with limestone-rich geological substrate. The dominant species

Table 6.21 Ass. *Philonotido seriatae-Calthetum laetae* (Krajina 1933) Coldea 1991

Relevé No.	1	2	3	4	5	6	7	8	9	10	
Altitude (m a.s.l.)	1950	1910	1850	2040	1870	2000	1750	1650	2180	2070	
Aspect	SW	NE	N	W	–	–	NE	–	S	S	
Slope (degrees)	10	25	30	5	0	0	15	0	10	10	
Herb cover (%)	80	80	95	95	90	100	100	75	60	80	
Sample area (sq. m)	25	10	25	25	25	10	25	25	25	25	K
Char. ass.											
Caltha palustris	4.5	4.5	4.5	3.5	4.5	4.5	4.5	3.5	3.5	4.5	V
Philonotis seriata	+	+	1.3	1.3	1.4	1.3	+	.	1.3	1.3	V
Cardamino-Montion et ***Montio-Cardaminetalia***											
Saxifraga stellaris	1.5	1.4	+	+	1.3	1.3	.	1.3	.	.	IV
Saxifraga rotundifolia subsp. *heucherifolia*	+	.	2.5	.	.	.	+	.	+	1.3	III
Stellaria nemorum	.	+	+	+	.	.	+	.	+	+	III
Senecio subalpinus	.	.	+	.	.	.	+	+	.	.	II
Chrysosplenium alpinum	.	.	.	.	+	.	.	.	.	.	I
Chrysosplenium alternifolium	.	.	.	.	.	.	+	.	.	.	I
Epilobium alsinifolium	.	.	.	.	.	.	.	+.4	.	.	I
Silene pusilla	.	.	.	.	.	.	.	+	.	.	I
Mnium punctatum	.	+	.	.	.	+	.	.	.	.	I
Pellia fabbroniana	.	.	.	.	.	.	1.3	.	+	.	I
Scapania undulata	.	.	.	.	.	.	+	.	.	.	I
Adenostyletalia											
Deschampsia caespitosa	+	.	+	3.5	.	2.5	+	1.5	1.3	1.4	IV
Aconitum napellus subsp. *tauricum*	.	+	+	+	+	.	.	.	+	+	III
Angelica archangelica	.	.	.	.	.	+	2.5	.	+	.	II
Viola biflora	.	.	.	.	+	.	+	.	.	.	I

(continued)

Table 6.21 (continued)

Epilobium alpestre	.	.	.	+	.	.	.	.	.	.	I
Chaerophyllum hirsutum	.	.	.	.	.	.	+	.	.	.	I
Adenostyles alliariae	.	.	.	.	.	.	.	.	.	+	I
Caricetalia fuscae											
Carex curta	+	.	.	+	.	.	.	+	.	.	II
Carex nigra subsp. *dacica*	.	.	.	.	+	.	.	+	.	+	II
Carex stellulata	.	.	.	.	.	.	.	1.3	.	.	I
Agrostis canina	.	.	.	.	.	.	.	+	.	.	I
Companion											
Plantago gentianoides	+	.	+	.	+	+	+	.	+	+	IV
Luzula alpinopilosa subsp. *obscura*	.	+	.	+	.	.	1.5	.	1.3	1.3	III
Rumex alpinus	+	+	1.3	.	.	.	.	.	1.3	.	II
Taraxacum panalpinum	.	.	.	+	+	.	.	.	.	+	II
Sphagnum giergensohnii	.	.	.	.	1.3	1.3	+	.	.	.	II
Rumex alpestris	.	+	.	.	.	.	.	.	+	.	I
Poa annua	.	.	.	+	.	.	.	.	+	.	I
Poa alpina	.	.	.	+	.	.	.	.	.	1.3	I
Veronica serpyllifolia s.l.	.	.	.	+	.	.	.	.	.	+	I
Veratrum album	.	.	.	.	.	+	.	.	.	+	I
Cerastium cerastioides	.	.	.	2.5	.	.	.	.	.	.	I
Ligusticum mutellina	.	.	.	.	.	.	1.5	.	.	.	I
Cratoneurum filicinum	.	.	.	.	.	.	.	2.5	.	.	I

Companion species with one occurrence: *Geum montanum* 1: +; *Heracleum palmatum* 2: +; *Dactylorhiza cordigera* 6: +; *Scirpus sylvaticus* 7: +; *Galium uliginosum* 7: +; *Juncus alpinus* 8: +; *Athyrium distentifolium* 9: +; *Aulaconium palustre* 8: +

Relevé: 1. Mt. Zlata (10.07.1970); 2. Lacul Gemenea (16.07.1967); 3. Mt. Zlata (9.07.1970); 4. Obârșia Radeșului (10.07.1970); 5. Căldarea Zănoaga-V. Judele-Slăvei (14.07.1970); 6. Lacul Lia (6.07.1969); 7. Mt. Șesele (8.07.1969); 8. Ridge Pelegii (7.08.1970); 9. Lacul Viorica (8.08.1970); 10. Lacul Bucura (9.08.1970)

Table 6.22 Ass. *Chrysosplenio alpini-Saxifragetum stellaris* Pawł. et Walas 1949

Relevé No.	1	2	3	4	5	6	7	8	9	10	
Altitude (m a.s.l.)	1750	1880	1870	2050	2010	2000	2010	1800	2050	1900	
Aspect	NE	SE	E	NE	S	SW	E	W	W	–	
Slope (degrees)	3	5	5	3	10	10	5	5	5	0	
Herb cover (%)	95	85	70	90	70	95	95	80	90	60	
Sample area (sq. m)	10	4	25	25	25	4	10	10	4	10	K
Char. ass.											
Saxifraga stellaris	3.5	4.5	2.4	4.5	3.5	2.3	3.5	4.5	3.5	2.5	V
Philonotis seriata	4.5	1.4	1.3	1.5	2.4	1.4	1.3	1.3	4.5	2.5	V
Chrysosplenium alpinum	.	+	.	+	.	.	.	.	.	+	II
Diff. subass.											
Carex nigra subsp. *dacica*	1.3	1.3	1.3	+.3	+	.	.	.	.	.	III
Juncus filiformis var. *transsilvanicus*	+	+	.	+	+	.	.	.	.	.	II
Cardamino-Montion **et** ***Montio-Cardaminetalia***											
Caltha palustris	.	.	+.3	+	2.5	.	+.2	.	.	+.3	III
Senecio subalpinus	+	.	+	.	+	.	.	1.3	.	.	II
Saxifraga rotundifolia subsp. *heucherifolia*	.	.	.	+	+	.	.	.	.	+	II
Scapania undulata	.	.	3.4	.	.	+	4.5	.	.	.	II
Stellaria nemorum	.	1.4	.	+	.	.	.	.	.	.	I
Epilobium anagallidifolium	.	+	.	+	.	.	.	.	.	.	I
Scapania dentata	.	.	.	.	.	4.5	.	.	.	.	I
Pellia fabbroniana	.	+	.	.	.	.	.	.	.	.	I
Betulo-Adenostyletea											
Deschampsia caespitosa	.	.	.	.	1.5	2.4	1.3	1.3	1.3	1.5	III
Aconitum napellus subsp. *tauricum*	+	.	+	.	+	.	.	.	+	.	II

(continued)

Table 6.22 (continued)

Chaerophyllum hirsutum	.	.	+	.	+	.	.	.	.	.	I
Viola biflora	.	.	+	.	.	.	.	.	.	.	I
Companion											
Plantago gentianoides	+	+.3	1.3	+	+	+.3	+	.	+	1.5	V
Luzula alpinopilosa subsp. *obscura*	+	.	.	.	+	.	+	.	+	.	II
Eriophorum vaginatum	+	.	.	.	.	.	.	+	.	1.5	II
Sphagnum girgensohnii	.	1.4	1.3	.	.	.	.	2.5	.	1.5	II
Poa alpina	.	.	.	.	+	.	.	.	+	.	I
Soldanella major	.	.	.	.	.	+	.	.	.	+	I
Carex curta	+	.	.	.	.	.	.	.	.	1.5	I
Calliergon stramineum	.	.	.	.	.	.	.	1.3	.	.	I
Drepanocladus exannulatus	.	.	.	3.5	.	.	.	.	.	.	I

Companion species with one occurrence: *Cerastium cerastioides* 4: +; *Allium schoenoprasum* subsp. *sibiricum* 6: +; *Veratrum album* 7: +; *Rumex alpestris* 5: +; *Taraxacum panalpinum* 5: +

Relevé*:* 1. V. Pietrele (6.08.1970); 2. Lacul Zănoaga (7.07.1971); 3. Slăvei-Muchia ascuțită (14.07.1970); 4. V. Zănoagei (8.07.1971); 5. Lacul Lia (11.08.1970); 6. Slăvei-Muchia Ascuțită (9.07.1971); 7. Lacul Lia (6.07.1969); 8. Casa Gemenea (9.07.1969); 9. Lacul Galeșu (13.07.1971); 10. Mt. Zlata Căldarea Zănoguței (9.07.1970)

Table 6.23 Ass. *Doronico carpatici-Saxifragetum aizoidis* Coldea 1990

Relevé No.	1	2	3	4	5	
Altitude (m a.s.l.)	1850	1900	1990	1950	1850	
Aspect	W	W	W	SW	SE	
Slope (degrees)	40	65	60	38	30	
Herb cover (%)	40	75	80	30	30	
Sample area (sq. m)	4	4	4	4	4	K
Char. ass.						
Saxifraga aizoides	2.4	3.5	3.5	2.3	2.2	V
Doronicum carpaticum	.	.	+	+	+	III
Pseudorchis frivaldii	+	+	.	.	.	II
Cratoneurion commutati						
Silene pusilla	1.4	+.3	2.5	+	+	V
Philonotis calcarea	.	.	1.5	+	+	III
Bryum pseudotriquetum	.	+	+	.	.	II
Cratoneurum filicinum	.	.	2.4	.	.	I
Montio-Cardaminetalia						
Saxifraga rotundifolia subsp. *heucherifolia*	.	+	+	1.2	1.1	IV
Viola biflora	1.5	2.5	1.3	.	.	III
Saxifraga stellaris	.	+	1.3	+	.	III
Deschampsia caespitosa	+	.	.	+	.	II
Epilobium alsinifolium	.	+	.	.	+	II
Caltha palustris	.	+	.	.	+	II
Cardamine amara	.	.	.	+	+	II
Stellaria nemorum	.	.	.	.	+	I
Companion						
Poa alpina	.	+	+	+	+	IV
Aconitum napellus subsp. *tauricum*	.	+	+	+	+	IV
Plantago gentianoides	+	+.3	.	.	.	II
Luzula alpinopilosa subsp. *obscura*	1.3	.	+.3	.	.	II
Ranunculus oreophilus	.	+	+	.	.	II
Soldanella pusilla	.	.	+.3	.	.	I

Relevé: 1. V. Pelegii (7.08.1968); 2. Vf. Pelegii-Tăul Ghimpelui (12.07.1971); 3. Mt. Custura (12.07.1971); 4. Mt. Piatra Iorgovanului (10.07.1971); 5. Mt. Stănuleții (15.07.1970)

of the association is *Saxifraga aizoides*, populating both the river beds and stream edges and achieving on average 20% coverage. The regional species *Doronicum carpaticum* and *Pseudorchis frivaldii* also occur sporadically in the marginal areas of these sites, conferring to this association a Carpathian character. The association comprises several hygrophilous species characteristic of the alliance and order, among which *Silene pusilla, Saxifraga rotundifolia, Viola biflora, Saxifraga stellaris,* and *Philonotis calcarea* have higher presence. The current distribution of *Saxifraga aizoides* in Europe and the different ecological conditions where this species grows are evidences of the preglacial origin of these coenoses (Boşcaiu 1971).

Class *Scheuchzerio-Caricetea fuscae* (Nordh. 1936) Tx. 1937

In this class we group the hygrophilous phytocoenoses that develop on the swampy areas surrounding the subalpine springs (Quellmoor) and montane and subalpine lakes in Retezat, partially or completely clogged with acidic, peaty soil (histosols). Swamp genesis in the massif has an obvious topogenic determinism. Some swamps of the alpine belt formed in areas with small concavities, where water persists longer. Although such coenoses are widespread in the massif, there are only a few characteristic species of the class. Among these (Chytrý 2011; Pott 1995; Valachovič 2001) the following are present in the phytocoenoses identified in Retezat: *Carex nigra, Carex echinata, Carex rostrata, Parnassia palustris, Sphagnum recurvum, Sphagnum teres, Sphagnum subsecundum,* and *Aulacomnium palustre.*

Order *Caricetalia fuscae* Kock 1926 in Nordh. 1936

The swampy phytocoenoses on limestone-poor terrains (Neidermoor) of the Euro-Siberian region are grouped into this order. In the Retezat Mountains such phytocoenoses were found more frequently in the subalpine belt, where they form small mesotrophic swamps (Boşcaiu et al. 1972). The characteristic species of the order are *Carex curta, Juncus filiformis, Dactylorhiza cordigera, Pseudorchis frivaldii,* and *Sphagnum compactum.*

Alliance *Caricion fuscae* Koch 1926

We include in this alliance all eu-mesotrophic swamp coenoses formed near the glacial lakes and alpine and subalpine springs in Retezat. Their floristic composition includes a low number of species, among which some Carpathian-Balkan ones such as *Dactylorhiza cordigera, Pseudorchis frivaldii,* and *Plantago gentianoides*. These species confer a regional specificity to these associations described in the Southern Romanian Carpathians (Boşcaiu et al. 1972).

Ass. *Carici daciae-Plantaginetum gentianoidis* Boşcaiu et al. 1972 (Table 6.24)

The coenoses built up by *Carex nigra* subsp. *dacica* and *Plantago gentianoides* are found more frequently in the subalpine belt of the Retezat Mountains, on slightly uneven lands where water from snow melting collects in spring. The moss synusiae are largely absent in their initial development stages, whereas some chionophilic species characteristic of the order *Salicetalia herbaceae* persist. The moss synusia, represented by *Sphagunm teres, Sphagnum compactum, Polytrichum strictum,* and *Aulacomium palustre,* develops as the swamping progresses.

Ass. *Carici dacicae-Drepanocladetum exannulati* Boşcaiu et al. 1972 (Table 6.25)

The phytocoenoses of this association develop more frequently by the clogging of small lacustrine formations of the subalpine and alpine belt from the Retezat Mountains. The initial stages of the association are represented only by submerged moss synusiae of *Drepanocladus exannulatus* and *Drepanocladus fluitans*. Their centripetal development finally results in a compact layer across the lakes' surfaces. As the submerged moss layer turns into peat, conditions become favorable for the

Table 6.24 Ass. *Carici dacicae-Plantaginetum gentianoidis* Boşcaiu et al. 1972

Relevé No.	1	2	3	4	5	6	7	8	9	
Altitude (m a.s.l.)	2000	2100	1750	2130	1880	2202	2120	1990	2000	
Aspect	–	–	S	–	–	–	–	–	–	
Slope (degrees)	0	0	10	0	0	0	0	0	0	
Herb cover (%)	100	100	100	100	100	100	100	100	100	
Sample area (sq. m)	25	25	25	25	25	25	25	25	25	K
Char. ass.										
Carex nigra ssp. *dacica*	3.5	3.5	2.5	4.5	3.5	3.5	3.5	3.5	2.5	V
Plantago gentianoides	3.5	2.5	1.4	2.5	2.5	3.5	1.5	2.5	3.5	V
Caricion* et *Caricetalia fuscae										
Juncus filiformis	+	.	+	.	+	.	.	1.3	.	III
Carex curta	.	.	1.5	.	+	.	.	+	.	II
Dactylorhiza cordigera	.	.	.	.	1.5	.	.	.	+.3	II
Sphagnum teres	.	.	2.3	.	.	.	.	+	.	II
Carex echinata	.	.	.	.	+	.	.	.	.	I
Eriophorum scheuchzeri	.	.	.	.	.	+	.	.	.	I
Pseudorchis frivaldii	.	.	.	.	+.3	.	.	.	.	I
Parnassia palustris	.	.	.	.	+	.	.	.	.	I
Aulacomnium palustre	+	.	.	.	.	.	.	.	.	I
Oxycocco-Sphagnetea										
Eriophorum vaginatum	.	+	+	+.3	.	.	+	+.3	.	III
Salicetalia herbaceae										
Carex pyrenaica	+.3	+	3.5	+	.	.	.	.	.	III
Taraxacum panalpinum	+	.	.	+	+	+	.	.	1.5	III
Soldanella pusilla	.	+.3	.	.	+.5	+.5	.	.	.	II
Luzula alpinopilosa subsp. *obscura*	.	.	.	+.3	+	+	.	.	.	II
Geum montanum	.	+	.	.	+	+	.	.	.	II
Companion										
Ligusticum mutellina	+	+.4	+	.	+.5	+.5	+.4	.	+	IV
Anthoxanthum alpinum	+	+	.	.	.	.	+	+.4	+.5	III
Nardus stricta	3.5	.	.	.	2.5	2.5	3.5	.	+.3	III
Potentilla ternata	+	+	.	.	+.5	+.5	.	.	.	III
Sphagnum girgensohnii	.	3.4	.	.	.	.	1.3	4.5	4.5	III
Deschampsia caespitosa	+	1.3	+	.	.	.	.	.	1.3	III

(continued)

Table 6.24 (continued)

Homogyne alpina	.	+	.	.	+.5	+.5	+.3	.	.	III
Poa alpina	+	+	.	+	.	.	.	.	.	II
Luzula sudetica	+	+	+	.	.	.	.	.	.	II

Companion species with one occurrence: *Soldanella major* 7: +; *Crocus vernus* 1: +.5; *Pedicularis verticillata* 1: +; *Crepis paludosa* 3: +; *Caltha palustris* 3: +; *Polytrichum strictum* 3: +; *Saxifraga stellaris* 4: +; *Sphagnum compactum* 8: 1.3

Relevé: 1. Lacul Bucura (8.08.1968); 2. Tăul Ştirbului (16.07.1969); 3. V. Pietrele (8.08.1970); 4. V. Rea-Vf. Păpuşa (13.07.1971); 5. Lacul Zănoaga (8.07.1971); 6. Lacul glaciar Judele (9.07.1971); 7. Tăul Ascuns (9.07.1971); 8–9. Mt. Slăvei-Muchia Ascuţită (9.07.1971)

Table 6.25 Ass. *Carici dacicae-Drepanocladetum exannulati* Boşcaiu et al. 1972

Relevé No.	1	2	3	4	5	6	7	8	
Altitude (m a.s.l.)	2065	2060	2060	2050	2050	1990	1990	1950	
Aspect	–	–	–	–	–	–	–	–	
Slope (degrees)	0	0	0	0	0	0	0	0	
Herb cover (%)	100	100	100	100	100	100	100	100	
Sample area (sq. m)	25	25	25	25	25	25	25	25	K
Char. ass.									
Carex nigra subsp. *dacica*	4.5	4.5	4.5	4.5	4.5	4.5	4.5	3.5	V
Drepanocladus exannulatus	3.5	+	+	3.5	5.5	2.5	5.5	5.5	V
Drepanocladus fluitans	.	4.5	3.5	.	.	.	.	.	II
Caricion **et** ***Caricetalia fuscae***									
Carex curta	.	+	.	.	+	.	.	+	II
Juncus filiformis	.	.	.	.	1.4	+	+.3	.	II
Eriophorum scheuchzeri	.	.	.	+	.	.	.	+	II
Sphagnum teres	.	.	.	.	.	4.5	1.3	.	II
Calliergon stramineum	.	.	.	.	.	.	.	+.5	I
Sphagnum subsecundum	2.5	.	.	.	.	.	.	.	I
Sphagnum compactum	.	.	2.5	.	.	.	.	.	I
Plantago gentianoides	.	.	.	.	1.5	.	.	.	I
Oxycocco-Sphagnetea									
Sphagnum recurvum s.l.	.	1.5	.	1.3	.	.	.	1.4	II
Sphagnum amblyphyllum	.	.	.	1.3	.	.	.	.	I
Polytrichum strictum	.	.	.	.	1.5	.	.	.	I
Polytrichum ştefureacii	1.2	.	.	.	+.3	.	+.3	.	II
Polytrichum commune	.	.	.	.	.	.	.	+.2	I
Companion									
Nardus stricta	+	.	.	+	+.4	.	.	.	II
Philonotis seriata	.	.	.	1.2	+	.	.	.	I
Taraxacum panalpinum	.	.	.	+	.	.	.	.	I
Ligusticum mutellina	.	.	.	.	+.3	.	.	.	I

Relevé: 1–5. V. Judele (9.07.1971); 6–8. Mt. Slăvei-Muchia Ascuţită (9.07.1971)

Table 6.26 Ass. *Sphagno-Caricetum rostratae* Steffen 1931

Relevé No.	1	2	3	4	5	6	
Altitude (m a.s.l.)	1840	1840	1880	1880	1880	1880	
Aspect	W	W	–	–	–	–	
Slope (degrees)	3	5	0	0	0	0	
Herb cover (%)	100	100	100	100	100	100	
Sample area (sq. m)	25	25	25	25	25	25	K
Char. ass.							
Carex rostrata	5.5	4.5	4.5	4.5	3.5	4.5	V
Sphagnum compactum	.	.	+.5	2.4	5.5	1.3	IV
Sphagnum recurvum s.l.	1.5	1.5	.	.	.	.	II
Sphagnum russowii	.	.	+	.	.	.	I
Caricion **et** ***Caricetalia fuscae***							
Carex curta	2.5	1.5	.	.	.	+	III
Drepanocladus fluitans	.	.	5.5	5.5	.	5.5	III
Carex nigra subsp. *dacica*	+	.	.	.	.	1.3	II
Aulacomnium palustre	+	+	.	.	.	.	II
Sphagnum teres	4.5	4.5	.	.	.	.	II
Drepanocladus exannulatus	+	.	.	.	.	.	I
Carex echinata	.	.	.	+	.	.	I
Scheuchzerietalia palustris							
Carex limosa	.	.	+.4	.	+.3	.	II
Oxycocco-Sphagnetea							
Eriophorum vaginatum	1.5	2.5	+	+.2	1.3	1.3	V
Polytrichum commune	1.5	+	.	.	.	.	II
Carex pauciflora	.	.	.	.	2.5	.	I
Sphagnum magellanicum	.	.	+.2	.	.	.	I
Companion							
Luzula sudetica	+	+	.	.	+	.	III
Deschampsia caespitosa	+.3	.	.	.	+	.	II
Epilobium alsinifolium	+	.	.	.	.	.	I
Nardus stricta	.	.	.	.	1.3	.	I

Relevé: 1–2. Tăul Zănoguții (11.07.1968); 3–6. Lacul Zănoaga (8.07.1971)

development of vascular plants such as *Carex curta, Carex nigra* subsp. *dacica, Juncus filiformis,* and *Eriophorum scheuchzeri*. As the consolidation of the peat substrate continues, bryophyte species typical of the swampy grasslands, such as *Sphagnum teres, Sphagnum compactum, Sphagnum recurvum, Polytrichum strictum,* and *Polytrichum commune* direct the evolution of these coenoses toward terrestrial oligotrophic ones.

Ass. *Sphagno recurvi-Caricetum rostratae* Stefen 1931 (Table 6.26)

In this association are grouped the hydrophilic and eu-mesotrophic coenoses dominated by *Carex rostrata*, widely distributed in Europe. In the Retezat Mountain area, such phytocoenoses were identified in the small water bodies of the subalpine swamps near the Zănoaga and Zănoguța glacial lakes. Their floristic composition encompasses few species with different ecological requirements, making difficult the syntaxonomic collocation of the association. In addition, as the phytocoenoses are situated solely in the subalpine belt, the typical montane swamp (low moor) species are partially missing from their structure. We note that the species *Drepanocladus fluitans* has a high presence and coverage in the *Carex rostrata* coenoses from Retezat, similar to the communities described in the Hruby Jesenik Mountains in the Czech Republic under the name *Carici rostratae-Drepanocladetum fluitantis* Hadač et Vana 1967. The absence of oligotrophic species in the study phytocoenoses that are typical of the alliance *Sphagnion cuspidati*, led us to include the *Carex rostrata* coenoses from Retezat in the association *Sphagno recurvi-Caricetum rostratae* Steffen 1931. Over time, these coenoses successionally progress toward the oligotrophic association *Eriophoro vaginati-Sphagnetum recurvi* (Boşcaiu et al. 1972; Hájková et al. 2011; Soltés et al. 2001).

Class *Oxycoco-Sphagnetea* Br.-Bl. et Tx. 1943

In this class are grouped the oligotrophic and meso-oligotrophic swamp phytocoenoses, with a low number of species, dominated by some strongly acidophilic *Cyperaceae* and *Sphagnum* species. The main distribution area of oligotrophic swamps (peat bogs) is the Atlantic-Baltic region of Europe and the northern Eurasian region. The basic conditions for oligotrophic swamp formation and persistence are the abundant rainfall and soil deficiency in mineral nutrients, particularly base cations.

In the Romanian Carpathians, typical, convex peat bogs occur above 1000 m altitude in the Eastern Carpathians and the Apuseni Mountains (Pop 1960; Simon 1962). The oligotrophic swamps of the subalpine belt in the Carpathians, like the ones from the Retezat Mountains, only initiated their formation process at the beginning of Subatlantic period, as the climate became wetter and cooler (Boşcaiu 1971). Due to this fact, the peaty soil in these subalpine swamps has rich terrigenous content. The flora of these swamps (spring fen) (Quellmoor) is poor in boreal oligotrophic species.

Order *Sphagnetalia medii* Köstner et Flössner 1933
Natura 2000: priority habitat type 7110

The order includes the vegetation of convex oligotrophic swamps (raised bogs) and transitional mires of the montane and subalpine belts. The oligotrophic species typical of this order, common with those of the class, are well represented in the raised bogs of the montane belt (1000–1200 m) in the Romanian Carpathians. These species are *Andromeda polifolia, Carex pauciflora, Empetrum nigrum, Vaccinium oxycoccus, Vaccinium microcarpum, Vaccinium uliginosum, Drosera rotundifolia, Eriophorum vaginatum, Polytrichum strictum, Sphagnum magellanicum, Sphagnum capillifolium, Sphagnum recurvum* s.l. The transition mire phytocoenoses of the subalpine belt have few oligotrophic vascular species in their floristic structure.

Alliance *Sphagnion medii* Köstner et Flössner 1933

In this alliance are grouped the montane and subalpine herbaceous and shrubby oligotrophic swamp phytocoenoses from Central and Eastern Europe (Dierssen 1977, 1978; Hájková et al. 2011; Matuszkiewicz 2008; Soltés et al. 2001; Steiner 1993). Due to their phytogeographical position, the Retezat Mountains, like other massifs of the Southern Romanian Carpathians, are outside the climatic limits that support the proper development of genuinely shady oligotrophic formations (Beldie 1967; Boşcaiu 1971). The peat in these subalpine swamps has a pronounced mesotrophic character because it contains plenty of terrigenic material. Although the flora of these subalpine swamps has a pronounced acidophilic character (pH = 4.2), the most representative boreal oligotrophic species are missing from the massif (Simon 1962). Among the oligotrophic species present, we mention *Eriophorum vaginatum, Carex pauciflora, Sphagnum recurvum, Sphagnum subsecundum, Polytrichum strictum,* and *Polytrichum commune.*

Ass. *Eriophoro vaginati-Sphagnetum recurvi* Hueck 1925 (Table 6.27)

We include in this association all subalpine coenoses from the Retezat Mountains dominated by *Eriophorum vaginatum*, which are poorer in boreal oligotrophic species than those described from areas situated in the northern and northwestern Romania (Hájková et al. 2011; Malynovski and Kricsfalusy 2002; Soltés et al. 2001). Floristically, only some bryophyte species link them, namely *Sphagnum recurvum* s.l., *Sphagnum subsecundum, Sphagnum magellanicum, Sphagnum russowii, Polytrichum strictum,* and *Polytrichum commune*. The regional Carpathian-Balkan specificity of these phytocoenoses is given by *Pseudorchis frivaldii, Dactylorhiza cordigera, Plantago gentianoides,* and *Carex nigra* subsp. *dacica.*

Class *Molinio-Arrhenatheretea* Tx. 1937

This class groups hygrophilous and mesophilous grassland phytocoenoses of the temperate and temperate-continental regions of Europe. Such grasslands have, generally, a secondary character as they develop on deforested lands within the colline and montane belt. The soils of these grasslands vary greatly regarding their pedogenetic structure, texture, mineral content, and moisture, while the structure of grassland phytocoenoses changes according to these soil peculiarities. Such grasslands have an intrazonal distribution within the beech and spruce forests of the Retezat Mountains. Among the characteristic species of the class, the following are present within the mesophilous coenoses occurring in Retezat: *Festuca rubra, Festuca pratensis, Alopecurus pratensis, Rhinanthus minor, Plantago lanceolata, Stellaria graminea, Ranunculus acris, Rumex acetosa, Cerastium fontanum* subsp. *vulgare, Centaurea jacea, Anthoxanthum odoratum, Briza media, Silene flos-cuculi, Prunella vulgaris,* and *Poa pratensis.*

Table 6.27 Ass. *Eriophoro vaginati-Sphagnetum recurvi* Hueck 1925

Relevé No.	1	2	3	4	5	6	7	8	9	10	
Altitude (m a.s.l.)	1840	2000	2000	1880	1880	1870	1825	2050	2065	2060	
Aspect	–	W	W	–	–	–	–	S	S	–	
Slope (degrees)	0	3	5	0	0	0	0	3	3	0	
Herb cover (%)	100	100	100	100	100	90	90	100	100	100	
Sample area (sq. m)	20	20	20	20	20	20	20	20	20	20	K
Char. ass.											
Eriophorum vaginatum	3.5	4.5	4.5	3.5	4.5	2.4	2.5	3.5	3.5	3.5	V
Sphagnum recurvum	3.5	4.5	4.5	4.5	4.5	4.5	1.5	+	+	1.5	V
Sphagnum subsecundum	.	+.5	1.4	.	+.5	1.5	4.5	5.5	5.5	4.5	IV
Sphagnum russowii	.	.	.	1.3	.	.	.	.	.	.	I
Sphagnion **et** ***Sphagnetalia medii***											
Carex pauciflora	+	+	2.5	1.4	1.5	1.5	+	1.3	1.3	1.5	V
Polytrichum strictum	2.5	+	+	.	.	.	+	.	.	+	II
Sphagnum magellanicum	2.5	.	.	.	.	.	.	.	.	.	I
Sphagnum robustum	.	.	.	1.3	.	.	.	.	.	.	I
Polytrichum commune	.	.	1.3	.	+	.	.	.	.	.	I
Polytrichum gracile	.	.	.	+	.	.	.	.	.	.	I
Pohlia sphagnicola	+	.	.	.	.	.	.	.	.	.	I
Caricetalia fuscae **s.l.**											
Carex fusca subsp. *dacica*	1.5	1.5	+.5	+.3	+.4	+	+	+.3	+.4	1.4	V
Plantago gentianoides	.	.	+	+.3	+.4	.	+	+.3	+.3	+	IV
Carex curta	.	+	+	.	.	+	1.3	+.3	+.3	.	III
Dactylorhiza cordigera	.	.	.	.	+.5	+	.	+.5	1.5	1.4	III
Carex rostrata	+.5	+	.	.	.	.	.	+.3	2.5	.	II
Pseudorchis frivaldii	.	.	.	.	.	+	.	+.4	.	+	II

Juncus filiformis	.	.	.	.	.	1.3	.	.	.	+	I
Aulacomnium palustre	1.3	.	.	.	.	.	.	.	.	.	I
Drepanocladus exannulatus	.	.	.	+	.	.	+	.	.	.	I
Drepanocladus fluitans	.	.	.	+	.	.	.	+	.	.	I
Parnassia palustris	.	.	.	.	+.3	+.4	.	.	.	.	I
Companion											
Homogyne alpina	+	.	1.5	.	+	+.5	.	+.5	+	+	IV
Anthoxanthum alpinum	.	.	+.5	+.4	+.5	+	.	+.4	+.4	+	IV
Nardus stricta	.	.	1.5	2.5	2.5	.	.	+.4	+.5	1.5	III
Luzula sudetica	.	.	+	+	+	+	.	.	.	.	II
Soldanella major	.	.	+	.	1.5	.	.	+	+	.	II
Deschampsia caespitosa	.	.	.	+	.	+	.	+	+.4	.	II
Pseudorchis albida	.	.	+	.	+	.	.	.	.	+	I
Ligusticum mutellina	.	.	.	.	.	.	+	.	+.5	1.4	I
Vaccinium myrtillus	+.3	.	+	.	.	.	.	.	.	.	I
Potentilla ternata	.	.	+.3	+	.	.	.	.	.	.	I
Taraxacum panalpinum	.	.	.	.	+	.	.	.	.	+	I
Vaccinium vitis-idaea	+	.	.	.	.	.	.	.	.	.	I
Crocus vernus	.	.	+	.	.	.	.	.	.	.	I
Viola biflora	.	.	.	.	+	.	.	.	.	.	I
Crepis paludosa	.	.	.	.	+	.	.	.	.	.	I
Saxifraga stellaris	.	.	.	.	.	.	+	.	.	.	I
Scapania undulata	.	.	.	.	.	.	+	.	.	.	I
Senecio subalpinus	.	.	.	.	.	.	.	+	.	.	I

Relevé: 1–7. Lacul Zănoaga (25.08.1972); 8–10. Lacul Judele (26.08.1972)

Order *Arrhenatheretalia elatioris* Pawl. 1928

This order includes the fertilized grasslands of the colline and montane belts, used as meadows or pastures. They occur on flat areas or mild slopes with medium-deep eutricambosols, slightly acidic reaction and average moisture. The characteristic species of the order are: *Trifolium pratense, Trifolium repens, Cynosurus cristatus, Achillea millefolium, Taraxacum officinale, Plantago lanceolata, Leucanthemum vulgare, Lotus corniculatus, Knautia arvensis, Carum carvi, Veronica chamaedrys,* and *Rorippa pyrenaica.*

Alliance *Polygono-Trisetion* Br.-Bl. et Tx. ex Marschall 1947
Natura 2000: habitat type 6520

The alliance includes mesophilous, montane grasslands developed on acid brown soils, moderately wet and rich in nutrients. Mesotrophic species (*Trisetum flavescens, Alopecurus pratensis,* etc.) become less frequent when soil nutrient content decreases and *Festuca rubra* and *Agrostis capillaris* become dominant. Among the species mentioned as characteristic of the alliance (Dierschke 1981; Hegedüsová et al. 2020; Pott 1995), the following ones were identified in the investigated area: *Geranium sylvaticum, Astrantia major, Heracleum sphondylium, Leontodon hispidus* subsp. *danubialis, Trisetum flavescens*, and *Crocus vernus.*

Ass. *Violo declinatae-Agrostetum capillaris* Hegedüsová et al. 2020 (Table 6.28)

We group into this Carpathian association the mesophilous phytocoenoses, dominated by *Agrostis capillaris* and *Festuca rubra* in the Retezat submontane areas, developed on sunny, gentle slopes with moderately humid, eumesobasic, brown soils (eutricambosols). Overall, the *Agrostis capillaris/Festuca rubra*-dominated grasslands in the Carpathians display weak compositional variability, mainly induced by soil acidity and elevation (Gafta and Muncaciu 2016). The regional specificity of this association is given by the frequent species *Viola declinata* and *Campanula abietina*. The mesophilous character of the association is revealed by the characteristic species of the alliance *Polygono-Trisetion*, to which we assign this association, and by some species with high moisture requirements such as *Succisa pratensis, Lychnis flos-cuculi, Deschampsia caespitosa, Chaerophyllum hirsutum.* This association is floristically very similar to *Festuco rubrae-Agrostetum* Horvat 1951, described in the Illyrian subregion, but lacks the species *Phyteuma betonicifolium* and *Moenchia mantica,* considered as characteristic of the association. Also, the coenoses of this association resemble those from the northwestern Carpathians, described as *Anthoxantho-Agrostetum tenuis* Sillinger 1933 (Hájek 2007; Uhliarova et al. 2014), but differ from it by its characteristic species *Viola declinata* and *Campanula abietina.*

Order *Poo alpinae-Trisetetalia* Ellmauer et Mucina 1993
Natura 2000: habitat type 6520

We include in this order the mesophilous and mesotrophic *Poa alpina*-dominated grasslands of the subalpine belt from the Retezat Mountains. They are sporadically

Table 6.28 Ass. *Violo declinatae-Agrostetum capillaris* Hegedüsová et al. 2020

Relevé No.	1	2	3	4	5	6	
Altitude (m a.s.l.)	950	1000	1100	1050	1150	1100	
Aspect	W	SW	SW	–	SW	–	
Slope (degrees)	10	5	25	0	5	0	
Herb cover (%)	90	100	90	85	90	90	
Sample area (sq. m)	25	25	25	25	25	25	K
Char. ass.							
Agrostis capillaris	3.5	4.5	1.5	2.5	3.5	2.5	V
Viola declinata	+	1.5	+	.	+	.	IV
Campanula abietina	.	+	+	+	.	+	IV
Trifolium alpestre	.	+	+	.	.	.	II
Stachys officinalis	.	+	+	.	.	.	II
Polygono-Trisetion							
Alchemilla vulgaris	+	.	.	1.3	+	+	IV
Geranium sylvaticum	+	+	.	+	.	.	III
Heracleum spondylium	+	.	+	.	.	.	II
Leotodon hispidus subsp. *danubialis*	+	+	+	.	.	.	III
Viola tricolor subsp. *subalpina*	.	.	.	+	.	+	II
Crocus vernus	+	+	.	.	.	.	II
Astrantia major	.	+	.	.	.	.	I
Trisetum flavescens	.	+	.	.	.	.	I
Arrhenatheretalia							
Achillea millefolium	+	+	.	+	+	+.5	V
Trifolium pratense	1.5	+	+	2.5	2.5	+.5	V
Taraxacum officinale	+	+	.	1.3	1.5	+.4	V
Plantago lanceolata	.	1.5	+.5	.	+	+	IV
Veronica chamaedrys	+	.	.	+	1.5	+	IV
Leucanthemum vulgare	.	+	+	+	+	+	V
Trifolium repens	+	.	.	+	+	+	IV
Lotus corniculatus	.	.	+	.	+	+	III
Cynosurus cristatus	.	.	.	+.3	+	+	III
Carum carvi	+	.	.	.	+	+.3	III
Knautia arvensis	.	+	+	.	.	.	II
Rorippa pyrenaica	.	+	.	.	.	+	II
Molinio-Arrhenatheretea							
Festuca rubra	3.5	2.5	4.5	3.5	3.5	4.5	V
Rhinanthus minor	.	1.5	+	2.5	2.5	1.3	V
Stellaria graminea	1.5	2.5	+	1.5	1.3	+.3	V
Ranunculus acris	1.5	+	.	+	1.5	+	V
Rumex acetosa	1.5	.	+	+	+	+	V
Plantago lanceolata	.	1.5	+.5	.	+	+	IV
Cerastium fontanum subsp. *vulgare*	+	+	.	.	+	+	IV
Centaurea jacea	+	+	.	.	+	.	III

(continued)

Table 6.28 (continued)

Festuca pratensis	+	.	.	2.5	1.5	.	III
Anthoxanthum odoratum	.	+	1.5	+	.	.	III
Poa prtaensis	.	.	.	.	+	+	II
Briza media	.	+	1.5	.	.	.	II
Lychnis flos-cuculi	.	+	.	.	.	.	I
Prunella vulgaris	+	.	.	.	.	.	I
Alopecurus pratensis	.	.	.	.	+	.	I
Nardetalia							
Carex pallescens	+	.	+	+	.	.	III
Polygala vulgaris	.	.	+	.	+	+	III
Hieracium aurantiacum	+	.	.	.	.	+	II
Carlina acaulis	.	+	+	.	.	.	II
Euphrasia stricta	.	.	+	.	.	+	II
Thymus pulegioides	.	.	1.3	.	.	.	I
Scorzonera rosea	.	+	.	.	.	.	I
Potentilla erecta	.	.	+	.	.	.	I
Genista sagittalis	.	.	+	.	.	.	I
Companion							
Silene vulgaris	.	+	+	+	+	+	V
Campanula glomerata	.	+	+	+	+	.	IV
Aegopodium podagraria	+	+	.	+	.	.	III
Rumex acetosella	+	+	.	.	.	+	III
Luzula luzuloides	+	.	.	+	+	.	III
Plantago media	.	+	+	.	.	+	II
Ranunculus polyanthemos	.	.	.	+	+	.	II
Seseli libanotis	.	.	.	+	+	.	II

Companion species with one ocurrence: *Veratrum album* 1: +; *Cruciata glabra* 1: +.5; *Chaerophyllum hirsutum* 1: +; *Telekia speciosa* 1: +; *Milium effusum* 2: +; *Trifolium pannonicum* 2: +; *Campanula persicifolia* 2: +; *Orchis coriophora* 2: +; *Trifolium montanum* 3: +; *Pimpinella saxifraga* 3: +; *Thesium alpinum* 5: +; *Deschampsia caespitosa* 5: +; *Succisa pratensis* 5: +; *Senecio subalpinus* 5: +; *Potentilla argentea* 6: +

Relevé: 1–3. Piciorul Colțului (6.07.1989); 4–5. Vf. Piule (1.07.1986); 6. V. Lăpușnicu Mare (15.07.1970)

distributed on flat areas or gentle slopes with nutrient-enriched soils. They occasionally come in contact with phytocoenoses of the association *Potentillo ternatae-Festucetum supinae.* The main characteristic species of the order are: *Phleum alpinum, Poa supina, Rumex alpestris,* and *Hypericum maculatum* (Hegedüsová Vantarová 2014).

Alliance *Poion alpinae* Oberd. 1950

The subalpine, mesophilous grasslands that populate superficial, humus-rich, wet, podzolic soils were included in this alliance (Hegedüsová Vantarová 2014; Oberdorfer 1983). The main characteristic species of the alliance are: *Poa alpina,*

Ligusticum mutellina, Polygonum viviparum, Antennaria dioica, Gentiana verna, Taraxacum alpinum, and *Trifolium badium.*

Ass. *Alchemillo xanthochlorae-Poetum alpinae* Beldie 1967 corr. Coldea 2012 (Table 6.29)
Primordial name: ass. *Poa alpina-Alchemilla palmata* Beldie 1967 (Art. 43)

The phytocoenoses of this association occupy mainly the marginal areas of the massif (Mt. Zlata, Stănuleț, Stâna de Râu), where sheep grazing occurs intermittently in dry years. Such phytocoenoses are also found on small areas at the edge of some glacial lakes (Zănoaga, Bucura, Judele). They vegetate on superficial humic-silicatic soils, very rich in humus, with high trophicity, slightly acidic, and with favorable water regime. *Poa alpina* is the dominant species in these communities. Among the regional Carpathian species that prefigure its areal, we mention *Campanula serrata, Campanula abietina,* and *Ranunculus montanus* subsp. *pseudomontanus.* The species *Potentilla ternata, Anthoxantum alpinum, Geum montanum, Phleum alpinum,* and *Ligusticum mutellina* also have a high presence in the composition of this association. *Trifolium repens, Agrostis capillaris, Festuca nigrescens, Veronica serpyllifolia,* and *Leontodon autumnalis* connect these mesotrophic, subalpine phytocoenoses with the mesophilous montane grasslands of *Arrhenatheretalia.*

Class *Caricetea curvulae* Br.-Bl. 1948

We include in this class all alpine and subalpine primary grasslands from the Retezat Mountains that occur on humic-silicatic soils and lithosols with acidic reaction. Such grasslands are at climax stage, in concordance with the pedoclimatic factors. In their floristic composition are present artic-alpine species and some Carpathian-Balkan ones. Among the characteristic (diagnostic) species of the class, we mention: *Juncus trifidus, Hieracium alpinum, Campanula alpina, Pulsatilla alba, Luzula spicata, Festuca supina, Potentilla ternata,* and *Cetraria islandica* (Grabherr 1993).

Order *Caricetalia curvulae* Br.-Bl. et Jenny 1926

In this order are grouped the alpine and subalpine primary grasslands on acid substrates from the Alps and the Carpathians (Borza 1934; Boşcaiu 1971; Braun-Blanquet and Jenny 1926; Krajina 1933). Such alpine grasslands from the Balkans were classified in a distinct order, *Seslerietalia comosae* Simon 1958, differentiated floristically from the order *Caricetalia curvulae* by several Balkan species that are absent from Romania. Among the characteristic species of the order, we mention the following: *Campanula alpina, Primula minima, Avenula versicolor,* and *Agrostis rupestris.* The order includes two alliances.

Alliance *Caricion curvulae* Br.-Bl. et Jenny 1926
Natura 2000: habitat type 6150

We include in this alliance the *Carex curvula, Juncus trifidus,* and *Festuca supina* phytocoenoses of the alpine belt in the Retezat Mountains. They occur on superficial, skeleton-rich, humic-silicatic soils with acidic reaction. The following species are characteristic of the alliance: *Carex curvula, Oreochloa disticha, Festuca supina,*

Table 6.29 Ass. *Alchemillo xanthachlorae-Poetum alpinae* Beldie 1967 corr. hoc loco

Relevé No.	1	2	3	4	5	6	7	8	9	10	
Altitude (m a.s.l.)	2010	1950	2060	1800	1950	1700	1850	1825	1750	1600	
Aspect	–	S	–	SW	NW	–	W	SW	W	N	
Slope (degrees)	0	10	0	35	0	0	5	5	5	5	
Herb cover (%)	100	100	100	80	75	100	90	100	90	100	
Sample area (sq. m)	25	25	25	25	25	25	25	25	25	25	K
Char. ass.											
Poa alpina	4.5	5.5	5.5	3.5	3.5	4.5	4.5	5.5	4.5	4.5	V
Alchemilla xanthochlora	+	+	+	.	+	+	.	1.5	2.5	1.5	IV
Ranunculus pseudomontanus	+	.	.	+	+	+	.	.	+	.	III
Campanula abietina	+	.	+	+	.	+	+	.	.	+	III
Poion alpinae et ***Poo alpinae-Trisetetalia***											
Potentilla ternata	1.5	+	+	1.5	+	+	+	+.3	+.3	+	V
Phleum alpinum	+	.	+	.	.	+	+	.	.	+.3	III
Ligusticum mutellina	1.5	.	.	+	.	.	+.5	+.5	+.3	.	III
Hypericum maculatum	.	+	+	.	.	.	+	.	.	+	II
Polygonum viviparum	.	+	.	.	+	.	.	+	.	.	II
Rumex alpinus	.	.	+	.	.	.	+	.	.	+	II
Rumex alpestris	.	.	+	+	.	.	.	.	.	+	II
Poa supina	.	+	.	.	.	.	.	.	+	2.5	II
Viola declinata	.	.	.	.	.	+	.	+	+	.	II
Antennaria dioica	.	+	.	+.3	.	.	1.5	.	.	.	II
Taraxacum panalpinum	1.4	.	+	.	.	.	.	+	+	.	II
Campanula serrata	+	+	.	+	.	+	.	.	.	.	II
Gentiana verna	.	.	.	.	.	.	.	+	.	.	I
Trifolium badium	.	.	.	.	.	+	.	.	.	.	I

Molino-Arrhenatheretea											
Anthoxanthum odoratum	+	1.5	.	+	+	1.3	+	.	.	+	IV
Trifolium repens	1.3	.	.	.	2.5	+	+	+	+	+	IV
Festuca rubra agg.	+	.	+	+	.	+	.	.	.	+	III
Cerastium fontanum	+	.	.	+	.	.	.	+	.	.	II
Crocus vernus	.	+	.	.	.	+	.	.	.	+	II
Agrostis capillaris	.	.	.	.	.	+	.	.	+	+	II
Leontodon autumnalis subsp. *pratensis*	.	.	.	+	.	+	.	.	+	.	II
Caricetalia curvulae											
Geum montanum	+	+	+	+	.	+	+	+	+	+	V
Festuca supina	-	-	1.4	1.3	.	.	1.5	+	.	.	II
Luzula spicata	+	+	.	.	+	.	.	+	.	.	II
Avenula versicolor	+	.	.	2.5	.	.	2.5	.	.	.	II
Hieracium alpinum	+	.	+	.	.	.	.	.	+	+	II
Phyteuma confusum	.	.	.	+	.	.	+	.	.	.	I
Agrostis rupestris	.	.	.	+	.	.	1.5	.	.	.	I
Poa media	.	.	.	+	.	.	.	.	.	.	I
Companion											
Veronica serpyllifolia	+	.	+	.	.	.	+	+	+	.	III
Plantago gentianoides	+	+	+	.	.	+	.	.	+	.	III
Veratrum album	+	1.5	+	.	.	+	.	.	.	.	II
Achillea distans	+	.	.	.	.	+	.	.	.	.	II
Deschampsia caespitosa	+	.	1.5	.	.	.	.	+	.	.	II
Nardus stricta	.	1.5	.	.	.	+	.	.	.	.	I
Galium anisophyllon	.	.	.	.	1.3	.	.	+	.	.	I

Companion species with one occurrence: *Veronica alpina* 1: +; *Centaurea nervosa* 1: +; *Homogyne alpina* 2: +; *Carex atrata* 4: +; *Selaginella selaginoides* 5: +; *Acinos alpinus* 5: +; *Saxifraga adscendens* 5: +; *Trifolium pallescens* 5: +; *Veronica aphylla* 5: +; *Carex leporina* 6: +; *Trifolium montanum* 6: +; *Cardaminopsis halleri* subsp. *ovirensis* 7: +; *Alyssum repens* 8: +; *Myosotis alpestris* 9: +; *Festuca nigrescens* 10: +; *Plantago major* 10: +; *Urtica dioica* 10: +

Relevé: 1. Lacul Zănoaga (10.07.1970); 2. Zănoaga (10.07.1970); 3. Vf. Bucura (8.08.1970); 4. Fața Retezatului (15.07.1971); 5. Mt. Albele (9.07.1971); 6. V. Pelegii (12.07.1970); 7. Mt. Zlata (10.07.1970); 8. Mt. Albele-Mt. Piatra Iorgovanului (9.07.1970); 9. Mt. Stânuleț (9.07.1970); 10. Stâna de Râu (17.07.1971)

Phyteuma confusum, Senecio abrotanifolius subsp. *carpathicus, Senecio doronicum* subsp. *transylvanicus, Senecio incanus* subsp. *carniolicus*, *Minuartia recurva,* and *Leucanthemopsis alpina*. We specify that similar phytocoenoses from the Tatra Mountains, that lack *Carex curvula,* are grouped into the alliance *Juncion trifidi* Krajna 1933 (Dúbracová and Jarolimek 2007; Matuszkiewicz 2008).

Ass. *Primulo-Caricetum curvulae* Br.-Bl. 1926 em. Oberd. 1957 (Table 6.30)

The first phytocoenotic data on *Carex curvula* from Retezat were published by Borza (1934). The intermittent research we carried out over the years on the alpine grasslands from Retezat brought new additions to the floristic composition of these phytocoenoses and their distribution. *Carex curvula* is the dominant species in these communities. It is closely accompanied by *Primula minima* and *Phyteuma confusum,* achieving on average 70% coverage, similar to other phytocoenoses across the South-Eastern Carpathians described as *Caricetum curvulae* (Beldie 1967; Buia et al. 1962; Ghişa 1940; Puşcaru et al. 1956). The presence of *Loiseleuria procumbens* and *Vaccinium* dwarf scrubs in some coenoses (e.g., releve 9 in Table 6.30) is related to particular topography, i.e. upper slopes featuring relatively more xeric edaphic conditions (Puşcaş et al. 2005). In the composition of this association, some regional species are also sporadically present (*Senecio abrotanifolius* subsp. *carpathicus, Senecio doronicum* subsp. *transylvanicus, Potentilla ternata*), differentiating them from the *Carex curvula* coenoses from the Alps and Pyrenees (Braun-Blanquet 1948; Grabherr 1993; Oberdorfer 1978). The *Carex curvula* coenoses from Retezat occasionally come in contact with the ones from the alliance *Salicion herbaceae.*

Ass. *Phyteumo confusi-Juncetum trifidi* ass. nova hoc loco (Table 6.31)
Holotype: Table 6.31, rel. 6, *hoc loco*
Syntax syn: *Juncetum trifidi* auct. roman.

The coenoses dominated by *Juncus trifidus* in Retezat lie on high peaks and ridges continuously exposed to winds, with superficial, skeletal-rich, rocky soils, (lithosols), where snow is often blown off. Due to these eco-pedological conditions, the *Juncus trifidus* coenoses achieve low coverages. On the barren, rocky surfaces, some lichens species are present (*Cetraria islandica, Thamnolia vermicularis*). The constant presence of the Carpathian-Balkan species *Phyteuma confusum, Senecio doronicum* subsp. *transylvanicum* and *Potentilla ternata* in these phytocoenoses, led us to describe the new association regarded as a southeastern Carpathian synvicariant of the association *Juncetum trifidi* Szafer et al. 1923 em. Krajina 1933, described in the Tatra Mountains. Such phytocoenoses have been often described from the high mountain ranges of the Southern Romanian Carpathians under the name *Juncetum trifidi* (Beldie 1967; Borza 1934; Boşcaiu 1971; Buia et al. 1962; Puşcaru-Soroceanu et al. 1981). We specify that the saxicolous association *Senecioni rocheliani-Juncetum trifidi* Simon 1957, described at elevations between 2200 and 2550 m in the Pirin Mountains, floristically resembles this south-eastern Carpathian association.

Table 6.30 Ass. *Primulo minimae-Caricetum curvulae* Br.-Bl. In Br.-Bl et Jenny 1926 em. Oberd. 1957

Relevé No.	1	2	3	4	5	6	7	8	9	10	
Altitude (m a.s.l.)	2380	2300	2280	2260	2260	2280	2250	2280	2100	1980	
Aspect	W	N	–	–	W	S	–	S	–	NW	
Slope (degrees)	30	30	0	0	30	15	0	30	0	7	
Herb cover (%)	100	100	70	100	100	100	100	80	70	90	
Sample area (sq. m)	10	25	10	25	25	25	25	25	25	25	K
Char. ass.											
Carex curvula	4.5	5.5	3.5	4.5	4.5	4.5	5.5	4.5	4.5	5.5	V
Primula minima	2.5	1.5	2.5	+.5	3.5	+.3	+.4	+.5	1.5	+	V
Phyteuma confusum	.	.	+	+	+	.	+	+	+	+	IV
Caricion curvulae **et** ***Caricetalia curvulae***											
Campanula alpina	1.5	1.5	+.4	2.5	+	+	+	.	+.4	+	V
Potentilla ternata	+	+	.	+	+	+	+	+	+	+	V
Oreochloa disticha	+	+	.	+	+	+	+	+	+	+	V
Festuca supina	.	.	.	+	+	+	+	+	.	+	III
Avenula versicolor	.	.	.	+	.	.	.	+	+	+	II
Geum montanum	.	.	.	+	+	+	+	.	.	.	II
Leucanthemopsis alpina	.	.	+	.	+	+	+	.	.	.	II
Agrostis rupestris	.	.	+	+	.	.	.	.	.	.	I
Minuartia recurva	.	.	+	.	.	.	+	+	.	.	II
Caricetea curvulae											
Cetraria islandica	+	+	+	+	.	+	+	+	+	.	IV
Hieracium alpinum	.	+	+	+.5	.	.	+	+	+	+	IV
Juncus trifidus	.	+	.	+	+	+	+	.	.	.	III
Thamnolia vermicularis	+	+.5	.	+	.	+	+	+	.	.	III
Vaccinium gaultherioides	.	+	.	.	.	+	+	+	+	.	III

(continued)

Table 6.30 (continued)

Pulsatilla alba	.	+	.	.	.	+	.	.	+	.	II
Cetraria cucullata	.	+	.	+	.	+	.	.	.	.	II
Gentiana frigida	+	.	+	.	.	.	.	.	.	.	I
Luzula spicata	.	.	.	+	.	.	.	.	.	.	I
Rhododendro-Vaccinion											
Homogyne alpina	+	.	.	+.5	+	+	.	.	+	+	III
Vaccinium vitis-idaea	.	1.5	.	+	.	.	.	.	+	.	II
Loiseleuria procumbens	.	.	+	.	.	+	.	.	+	.	II
Antennaria dioica	+	.	.	.	.	.	.	.	+	+	II
Rhododendron myrtifolium	+	.	.	.	.	.	.	.	.	+	I
Vaccinium myrtillus	.	+	.	.	.	.	.	.	.	.	I
Salicio herbaceae											
Ligusticum mutellina	+	.	.	+	+	+	.	.	1.5	.	III
Festuca picturata	+.3	.	+	.	.	+	.	.	.	.	II
Soldanella pusilla	.	+	.	+	+	.	.	.	.	.	II
Salix herbacea	.	.	+.3	.	.	.	.	.	.	.	I

Companion species with 1 or 2 occurrence*: Polygonum viviparum* 1: +; *Pedicularis verticillata* 1, 3: +; *Senecio doronicum* subsp. *transylvanicus* 2, 6: +; *Senecio incanus* subsp. *carniolicus* 2, 6: +; *Sesleria coerulans* 3: +; *Gentiana acaulis* 4: +; *Gentiana punctata* 5, 6: +; *Polytrichum alpinum* 1: +; *Polytrichum sexangulare* 9: +; *Dicranum scoparium* 9: +; *Cetraria nivalis* 6: +; *Cladonia rangiferina* 6: +.; *Senecio abrotanifolius* subsp. *carpathicus* 7: +
Relevé: 1. Mt. Custura (12.07.1971); 2. Muchia Ascuțită (10.07.1970); 3. Mt. Peleaga (8.08.1970); 4. Vf. Judelui (10.07.1970); 5. Mt. Bucura (8.08.1970); 6. Vf. Zănoaga (12.07.1970); 7. Mt. Păpușa (10.07.1970); 8. Vf. Gruniu (12.08.1971); 9. Culmea Șeselor (5.07.1969); 10. V. Galeș-V. Rea (6.08.1970)

Table 6.31 Ass. *Phyteumo confusi-Juncetum trifidi* ass. nova hoc loco

Relevé No.	1	2	3	4	5	6	7	8	9	10	
Altitude (m a.s.l.)	2230	2220	2210	2100	2100	2046	2170	2110	2000	2100	
Aspect	SW	S	SE	W	SE	SW	SW	W	SE	W	
Slope (degrees)	15	5	15	10	20	35	35	10	25	15	
Herb cover (%)	70	70	70	80	70	60	90	80	80	80	
Sample area (sq. m)	25	25	25	25	25	25	10	25	10	10	K
Char. ass.											
Juncus trifidus	4.5	4.5	4.5	4.5	4.5	3.5	5.5	3.5	4.5	4.5	V
Phyteuma confusum	+	+	+	.	+.3	+.3	+	+	+.3	+.5	V
Senecio doronicum subsp. *transylvanicus*	.	.	+	.	+	+	+.5	+	+	.	III
Caricion curvulae et ***Caricetalia curvulae***											
Campanula alpina	.	+	+	+	+	+.5	+	.	+	+.3	IV
Avenula versicolor	.	+	.	+	1.5	+	+	.	1.3	+	IV
Oreochloa disticha	+	+	+	.	+.3	.	.	+	1.3	+	IV
Carex curvula	+	+	+	.	.	.	.	.	+	+.3	III
Festuca supina	+	+	+	.	+	.	+.3	.	1.3	.	III
Potentilla ternata	.	.	.	+	+.4	.	+	.	+	+	III
Agrostis rupestris	.	.	.	+	.	+	.	.	.	+	II
Geum montanum	.	.	.	1.4	.	.	.	.	.	.	I
Anthoxanthum alpinum	.	.	+.3	.	.	.	.	.	+	.	I
Luzula alpinopilosa subsp. *obscura*	.	.	.	.	.	.	+	+	.	.	I
Leucanthemopsis alpina	.	+	+	.	.	.	.	.	.	.	I
Caricetea curvulae											
Cetraria islandica	.	+	+	+	+	2.5	2.5	2.5	.	1.3	IV
Thamnolia vermicularis	+	+	+	+	+	.	.	+	1.3	1.2	IV
Hieracium alpinum	.	+	+	+	+	1.5	+	.	+	.	IV

(continued)

Table 6.31 (continued)

Cetraria nivalis	+	+	+	.	.	.	.	.	.	.	II
Pulsatilla alba	+	.	+	.	.	.	.	.	+	.	II
Primula minima	+	.	.	.	.	.	.	+	+	1.3	II
Alectoria ochroleuca	+	.	.	.	.	.	.	+	.	.	I
Rhododendro-Vaccinion											
Vaccinium vitis-idaea	.	.	.	.	+	.	+	.	+	+	II
Vaccinium gaultherioides	.	.	.	.	.	.	+	.	+	+	II
Homogyne alpina	.	.	.	1.5	+	.	.	.	.	.	I
Rhododendron myrtifolium	.	.	.	.	.	1.5	.	.	.	.	I
Companion											
Senecio incanus subsp. *carniolicus*	.	+	.	.	.	+.2	.	+	.	.	II
Ligusticum mutellina	.	.	.	1.5	.	+	+	.	.	+	II

Companion species with one occurrence: *Poa alpina* 5: +; *Soldanella pusilla* 4: +; *Poa media* 4: +; *Festuca nigrescens* 5: +; *Hieracium villosum* 5: +; *Solidago virgaurea subsp. minuta* 5: +; *Pedicularis verticillata* 4: +.3; *Taraxacum nigricans* 4: +; *Calamagrostis villosa* 8: +; *Doronicum carpaticum* 8: +; *Gentiana punctata* 8: +; *Veronica baumgartenii* 8: +; *Carex sempervirens* 7: +; *Luzula luzuloides* subsp. *cuprina* 7: +; *Anthennaria dioica* 7: +; *Cerastium alpinum* 9: +; *Carex atrata* 10: 7; *Sesleria bielzii* 10: +

Relevé: Mt. Păpuşa (8.08.1970); 2. Căldarea Lacului Zănoaga (10.07.1970); 3. Curmătura Judelui (10.07.1970); 4. Mt. Radeş (11.09.1970); 5. Sub Poarta Bucurei (12.07.1967); 6. Mt. Pietrele (8.08.1970); 7. Vf. Zlata (11.07.1970); 8. Vf. Radeş (13.09.1970); 9. Vf. Pietrele (7.08.1970); 10. Faţa Retezatului (12.09.1970)

Ass. *Potentillo ternatae-Festucetum supinae* Boşcaiu 1971 (Table 6.32)

The grasslands dominated by *Festuca supina* from the South-Eastern Carpathians, described earlier under different names by various geobotanists (Buia et al. 1962; Csűrös et al. 1956b; Gafta et al. 2012; Puşcaru-Soroceanu et al. 1981) were included by Boşcaiu (Boşcaiu 1971) in this regional southeastern Carpathian association. This is the southern synvicariant of the association *Junco trifidi-Festucetum supinae* Krajina 1933, described in the Tatra Mountains and the northern synvicariant the sub-association *Agrostio-Seslerietum comosae* subas. *festucetosum airoidis* (Bondev. 1959) Russakova 2000 reported in the Rila Mountains. The primary *Festuca supina* grasslands from Retezat are distributed frequently on high peaks and ridges between 2000 and 2250 m altitude (Zănoaga, Galeşu, Slăveiu), with superficial skeletal-rich soils (lithosols and humic-silicatic soils) and acidic reaction (pH = 4.8–5.1). At lower elevations, i.e. in the upper-montane and subalpine belts (1750–2000 m), *Agrostis rupestris* becomes (co)dominant in these phytocoenoses, although such a pattern was not observed at the scale of the whole Romanian Carpathians (Gafta et al. 2012). The characteristic species for the higher rank syntaxa are well represented in the association.

Alliance *Potentillo-Nardion* Simon 1957
Natura 2000: priority habitat type 6230

The subalpine and alpine grasslands of southeastern Europe, dominated by *Nardus stricta*, along with some regional Carpathian-Balkan species like *Potentilla ternata, Plantago gentianoides, Campanula abietina, Pseudorchis albida*, prompted Simon (1958) to describe this alliance in the Pirin Mountains and South-Eastern Carpathians and include it in the class *Nardo-Callunetea* Preisg 1949. Subsequently, the botanists who studied these high-altitude grasslands in the Carpathians (Boşcaiu 1971) and the Rila Mountains (Roussakova 2000), included the *Nardus stricta* grasslands in this class, based on the characteristic species of the class *Caricetea curvulae* present in such phytocoenoses.

Ass. *Centaureo nervosae-Nardetum strictae* Coldea 2012 (Table 6.33)
Syn.: *Nardetum strictae alpinum* Borza 1934 (Art. 34)

We grouped into this association the subalpine *Nardus stricta* grasslands developed on sunny, gentle slopes from Retezat, with semi-dry, humic-silicatic soils and acidic reaction. The floristic composition of this association includes many species characteristic of the alliance *Potentillo-Nardion* and also several that are typical of the order *Caricetalia* and class *Caricetea curvulae*. The presence of some microthermal species, such as *Soldanella pusilla, Gnaphalium supinum, Ligusticum mutellina, Festuca picturata,* and *Plantago gentianoides* reveals that *Nardus* grasslands from Retezat, although resembling floristically those from the Rila and Pirin Mountains, differ significantly from the ones described in Bulgaria as *Festuco-Nardetum strictae* Bonder 1959 corr. Roussakova 2000 and respectively, *Diantho-Nardetum strictae* (Bonder 1959) Roussakova 2000. The floristic composition of these grasslands comprises several Balkan species, such as *Gentiana pyrenaica, Crocus veluchensis,*

Table 6.32 Ass. *Potentillo ternatae-Festucetum supinae* Boşcaiu 1971

Relevé No.	1	2	3	4	5	6	7	8	9	10	
Altitude (m a.s.l.)	2210	2210	2070	2066	2000	2030	2010	1870	1750	1700	
Aspect	S	–	SE	–	SW	S	–	S	S	N	
Slope (degrees)	30	0	20	0	35	25	0	20	15	12	
Herb cover (%)	90	100	60	90	75	100	100	100	100	100	
Sample area (sq. m)	25	25	10	25	25	25	25	25	25	25	K
Char. ass.											
Potentilla ternata	2.5	2.5	+	1.3	+	1.5	+.3	+.3	+.2	2.5	V
Festuca supina	4.5	5.5	3.5	4.5	4.5	3.5	1.5	1.3	2.4	1.5	V
Caricion curvulae **et** ***Caricetalia curvulae***											
Agrostis rupestris	+	.	.	1.5	.	3.5	5.5	5.5	5.5	5.5	IV
Anthoxanthum alpinum	+.3	.	.	+	+	+.5	+.2	+	+	.	IV
Phyteuma confusum	.	+	+	+	+	+	+	+	.	.	III
Campanula alpina	.	+	+	1.5	+	+	+	.	.	.	III
Geum montanum	+.2	.	.	.	.	+.5	.	+	1.2	2.5	III
Avenula versicolor	.	.	+	.	+.3	1.5	+.2	+	.	.	III
Luzula spicata	+.2	.	.	+	.	.	.	+	.	.	II
Luzula alpinopilosa subsp. *obscura*	.	+	.	.	.	1.5	+.3	.	.	.	II
Veronica bellidioides	.	.	.	+	.	+	.	+	.	.	II
Carex curvula	.	+	.	.	+	.	.	.	.	.	I
Caricetea curvulae											
Hieracium alpinum	.	+	+	+	+	1.5	+.2	.	.	.	III
Vaccinium gaultherioides	.	+	+	+	+	.	+	+	.	.	III
Cetraria islandica	.	+	+	+	.	1.5	+	.	.	.	III
Pulsatilla alba	.	+	.	+	+	+	.	+	.	.	III
Primula minima	1.5	1.3	.	.	+	.	.	.	.	.	II

Juncus trifidus	.	.	+	1.3	1.3	.	.	.	.	.	II
Thamnolia vermicularis	.	+.5	.	+	.	.	.	.	.	.	I
Alectoria ochroleuca	.	.	.	+	+	.	.	.	.	.	I
Rhododendro-Vaccinion											
Vaccinium vitis-idaea	.	.	+	.	+	.	.	+	.	+	II
Homogyne alpina	1.5	.	.	.	.	+.4	.	.	.	.	I
Rhododendron myrtifolium	+.3	.	.	.	.	.	.	.	.	.	I
Vaccinium myrtillus	+.3	.	.	.	.	.	.	.	.	.	I
Companion											
Ligusticum mutellina	1.5	.	.	+	.	1.5	1.5	+	+	+	IV
Poa media	+.5	.	.	.	.	+	+	+	.	.	II
Carex atrata	.	+.5	.	+.5	.	.	.	+	.	.	II
Poa alpina	.	+	.	.	.	1.5	.	.	+	.	II
Deschampsia flexuosa	.	.	+	+	.	.	+	.	.	.	II
Antennaria dioica	.	.	.	.	.	+	.	+	.	+	II
Festuca nigrescens	.	.	.	.	.	.	.	+.3	1.5	1.2	II
Sesleria bielzii	+.4	.	.	.	.	.	.	.	.	.	I
Luzula sudetica	.	.	.	+	.	.	+	.	.	.	I
Carex sempervirens	+	.	+	.	.	.	.	.	.	.	I
Soldanella pusilla	+	.	.	.	.	.	+	.	.	.	I
Taraxacum panalpinum	.	+	.	.	.	+	.	.	.	.	I
Deschampsia caespitsa	.	.	.	.	.	+	.	.	.	+	I
Senecio doronicum subsp. *transylvanicus*	.	+	+	.	.	.	.	.	.	.	I

Companion species with one occurrence**:** *Sesleria coerulans* 1: +.4; *Nardus stricta* 4: 1.4; *Euphrasia stricta* 10: +; *Luzula campestris* 9: +; *Pseudorchis albida* 9: +; *Leucanthemopsis alpinum* 2: +.2; *Festuca picturata* 1: +; *Veronica baumgartenii* 2: +; *Epilobium angustifolium* 3: +; *Cerastium alpinum* 5: +; *Cerastium fontanum* 6: +; *Carex pyrenaica* 6: +; *Plantago gentianoides* 7: +; *Luzula campestris* 7: +; *Luzula multiflora* 6: +; *Phleum alpinum* 6: +; *Poa alpina* 6: +; *Alchemilla flabellata* 9: +; *Cerastium arvense* 9: +; *Galium anisophyllon* 9: +; *Thymus pulegioides* subsp. *montanus* 9: +; *Veronica serpyllifolia* 9: +; *Campanula abietina* 10: +; *Cardamine pratensis* 10: +; *Gnaphalium sylvaticum* 10: +; *Hieracium aurantiacum* 10: +; *Juniperus communis* subsp. *alpina* 10: +; *Leontodon autumnalis* 10: +; *Luzula luzuloides* 10: +; *Saxifraga rotundifolia* subsp. *heucherifolia* 10: +; *Trifolium repens* 10: +

Relevé: 1. Vf. Lazului (15.08.1970); 2. Vf. Zănoaga (10.07.1970); 3. Vf. Zlata (8.07.1970); 4. Şeaua Zănoguţii (8.08.1970); 5. Lacul Galeş (7.08.1970); 6. Mt. Slăveiul (13.07.1970); 7. Lacul Stânişoara (5.08.1970); 8. Mt. Zlata (10.07.1970); 9. Mt. Stănuleţ (9.07.1970); 10. Mt. Radeş (5.09.1969)

Table 6.33 Ass. *Centaureo nervosae-Nardetum strictae* Coldea 2012

Relevé No.	1	2	3	4	5	6	7	8	
Altitude (m a.s.l.)	1870	2000	2010	2045	2090	2000	1950	1900	
Aspect	–	SW	NW	W	–	SW	NE	NW	
Slope (degrees)	0	5	10	10	0	7	15	10	
Herb cover (%)	50	65	80	70	100	40	45	60	
Sample area (sq. m)	25	25	25	25	25	25	25	25	K
Char. ass.									
Nardus stricta	3.5	2.5	4.5	3.5	4.5	2.5	3.5	3.5	V
Centaurea nervosa	.	+	+	+	.	+	+	+	IV
Potentillo-Nardion									
Potentilla ternata	1.5	+	1.5	+.3	1.5	+	+	1.5	V
Geum montanum	+	+	1.4	+.5	+	+	.	+	V
Poa alpina	+	+	+	+.3	1.4	+	+	+	V
Luzula multiflora	.	.	.	+	.	.	+	.	II
Campanula abietina	.	+	.	.	.	.	+	.	II
Thymus praecox subsp. *polytrichus*	.	.	+	.	.	+	.	.	II
Hypochaeris uniflora	.	.	.	+	.	.	.	+	II
Pseudorchis albida	.	+	.	+	.	.	.	+	II
Plantago atrata	.	.	.	+	+	.	.	.	II
Luzula sudetica	+	.	.	+	+	.	.	.	II
Gentiana punctata	.	.	.	.	.	.	+	+	II
Gentiana acaulis	.	.	.	.	+	.	.	.	I
Caricetalia* et *Caricetea curvulae									
Anthoxanthum alpinum	1.5	+.3	+	1.5	1.5	+.3	1.5	1.5	V
Phyteuma confusum	.	+	1.1	+	+	+.5	+	+	V
Pulsatilla alba	.	.	+	+	1.5	+	+	.	IV
Cetraria islandica	+	.	+.4	.	+	.	+.5	+	IV
Agrostis rupestris	.	3.4	.	+	2.5	2.5	.	.	III
Veronica bellidioides	+	+	.	.	+	+	.	.	III
Festuca supina	+	+	.	.	.	+	.	.	II
Hieracium alpinum	.	.	+	.	.	.	+	+	II
Avenula versicolor	.	+	+	.	.	.	.	.	II
Carex atrata	+	.	+	.	.	.	.	.	II
Campanula alpina	.	.	+	.	.	.	.	+	II
Primula minima	.	.	+	.	.	.	.	.	I
Thamnolia vermicularis	+	.	.	.	.	.	.	.	I
Luzula spicata	.	+	.	.	.	.	.	.	I
Cladonia rangiferina	.	.	.	.	.	.	.	+	I
Salicion herbaceae* et *Salicetalia herbaceae									
Ligusticum mutellina	.	+	1.5	+	+	.	+.5	+	IV
Soldanella pusilla	+	.	+	+.3	.	.	.	+	III
Festuca picturata	.	.	.	+	.	.	+	+	II

(continued)

Table 6.33 (continued)

Gnaphalium supinum	.	.	.	.	+	.	.	.	I
Junipero-Bruckenthalion									
Homogyne alpina	+	+	+	3.5	+	.	1.5	1.5	V
Vaccinium myrtillus	.	.	.	+	.	.	+	+	II
Juniperus communis subsp. *alpina*	.	+	.	.	.	+	.	.	II
Vaccinium vitis-idaea	.	+	.	.	.	+	.	.	II
Rhododendron myrtifolium	.	.	.	.	.	.	+	+	II
Companion									
Antennaria dioica	.	+	.	.	.	+.5	.	.	II
Deschampsia flexuosa	.	.	.	+	.	.	+	.	II
Taraxacum panalpinum	.	.	+	.	.	+	.	.	II
Pedicularis verticillata	.	.	.	+	.	.	.	+	II
Loiseleuria procumbens	.	.	.	.	.	+	.	+	II

Companion species with one occurrence: *Luzula alpinopilosa* subsp. *obscura* 8: +; *Phleum alpinum* 8: +; *Trifolium repens* 8: +; *Cardaminopsis halleri* subsp. *ovirensis* 5: +; *Cerastium cerastioides* 5: +; *Cerastium fontanum* 5: +; *Luzula campestris* 6: +; *Veratrum album* 6: +
Relevé: 1. Vf. Huta (8.07.1970); 2. Mt. Zlata (10.07.1970); 3. Vf. Radeş (13.09.1970); 4. Vf. Zănoaga (10.07.1970); 5. Vf. Zănoaga (8.07.1970); 6. Vf. Zlata (7.07.1970); 7. Mt. Şesele (10.07.1969); 8. Mt. Şesele (13.09.1969)

Jasione bulgarica, Dianthus microlepis, Sesleria comosa, Carex bulgarica, and *Geum coccineum*, conferring to these associations distinct phytogeographic peculiarities compared to those from Retezat (Photo 6.2).

Ass. *Carici nigrae-Nardetum strictae* (Borza 1934) Kliment 2007 (Table 6.34)
Syn: *Hygronardetum strictae* Borza 1934 (Art. 34)

We included in this association the *Nardus stricta* coenoses developed on flat terrain near some glacial lakes and subalpine springs in Retezat, with wet, acidic, nutrient-poor, humic-silicatic soils. *Nardus stricta* is the dominant species of the association, achieving on average 65% coverage. Along with it, local and general ecological differential species such as *Eriophorum vaginatum, Carex nigra* subsp. *dacica, Carex pyrenaica,* and *Plantago gentianoides* are also frequent, strengthening the association's meso-hygrophytic character, similarly to the one described in Tatra (Kliment 2007). Considering the large number of subalpine species characteristic of the alliance *Potentillo-Nardion* and order *Caricetalia curvulae*, we assigned this association, along with the other primary grasslands, in the *Caricetea curvulae* class.

Ass. *Poetum mediae* Csűrös et al. 1956 (Table 6.35)
Lectotype: Csűrös et al. 1956, Table 1, rel. 2., lectotype *hoc loco*
Syn.: *Poeto mediae-Nardetum strictae* Resmeriță 1987 (Art. 29)

The first phytocoenoses dominated by *Poa media*, a Carpathian-Balkan species, were described in the northern and eastern areas of the Retezat Mountains (Csűrös et al. 1956b). Such coenoses were later found also in the southeastern and

Photo 6.2 *Centaurea nervosa* Willd. (photo M. Ciobanu)

southwestern parts of Retezat (Resmeriță 1987). Since the floristic composition of these coenoses comprises a large number of characteristic species of the alliance *Potentillo-Nardion* and order *Caricetalia curvulae,* we included this association in the class *Caricetea curvulae.* We specify that coenoses having *Poa media* as dominant species were subsequently described in Bulgaria (Rila Mountains) under several names (*Carici curvulae-Poetum mediae* Bender 1959, *Poetum mediae* Gantcher 1963), but were only considered as communities (Gesellschaft) by Roussakova (Roussakova 2000). The *Poa media* grasslands in Retezat develop on flat areas and gentle slopes, with wet, nutrient-rich, slightly acidic (pH = 5), humico-silicatic soils. In some places, they come in contact with the chionophilic coenoses of the association *Nardo-Gnaphalietum supini* Bartsch 1940, allowing penetration of some species of the alliance *Salicion herbaceae* in the composition of the *Poa media* coenoses.

Class *Loiseleurio-Vaccinietea* Eggler ex Schubert 1960

This vegetation class has been described to differentiate syntaxonomically the dwarf shrublands widely distributed in the arctic tundra from the boreo-nemoral forests developed in the high mountains of Europe (Eggler 1952). A detailed

Table 6.34 Ass. *Carici nigrae-Nardetum strictae* (Borza 1934) Kliment 2007

Relevé No.	1	2	3	4	5	6	7	8	9	10	
Altitude (m a.s.l.)	1700	1820	1860	1850	1910	1990	1925	2070	2080	2125	
Aspect		SW	–	–	SW	–	–	S	S	–	
Slope (degrees)	0	3	0	0	3	0	0	3	3	0	
Herb cover (%)	100	100	100	100	100	100	100	100	100	90	
Sample area (sq. m)	25	10	25	25	25	25	10	10	10	25	K
Char. ass.											
Nardus stricta	4.5	4.5	5.5	5.5	4.5	5.5	5.5	4.5	4.5	4.5	V
Eriophorum vaginatum	.	1.5	1.5	+	+.3	1.3	+	+	.	1.4	IV
Plantago gentianoides	+	+	.	+	.	+	.	1.5	+	+	IV
Carex nigra subsp. *dacica*	.	.	.	.	+	+.3	+	.	1.5	1.5	III
Carex pyrenaica	.	.	.	.	.	.	+	.	1.5	+	II
Potentillo-Nardion											
Potentilla ternata	.	+	+	+.5	.	+	+	+	+	+	IV
Ligusticum mutellina	.	+	+	.	+.5	1.4	.	+	+	+	IV
Geum montanum	+	.	.	.	.	+	+	+	+	.	III
Carex leporina	+	.	.	+	.	.	+	+	.	.	II
Festuca nigrescens	+	+	+	.	.	.	.	.	.	.	II
Luzula multiflora	.	.	+	1.5	.	.	.	.	.	.	I
Pseudorchis albida	.	.	.	.	.	.	.	.	+	.	I
Gentiana punctata	.	.	.	.	+	.	.	.	.	.	I
Luzula sudetica	+	.	+	.	.	.	.	.	.	.	I
Solidago virgaurea subsp. *minuta*	.	.	.	.	+	.	.	.	.	.	I
Caricetalia curvulae											
Anthoxanthum alpinum	.	+	+	+	+	+	+	+	+	.	IV
Luzula spicata	.	.	.	+	.	.	+	.	.	.	I

(continued)

Table 6.34 (continued)

Phyteuma confusum	.	.	.	.	.	.	+	.	.	.	I
Cetraria islandica	.	.	.	.	.	.	.	.	+	.	I
Agrostis rupestris	.	.	.	+	.	.	.	.	.	.	I
Avenula versicolor	.	.	.	.	+	.	.	.	.	.	I
Campanula alpina	.	.	.	.	+	.	+	.	.	.	I
Companion											
Deschampsia caespitosa	+	+	+	+	.	+	+	.	+	.	IV
Homogyne alpina	.	.	+	.	+	.	+	+	.	+	III
Sphagnum girgensohnii	+.3	1.5	1.3	1.5	.	+	.	+	.	.	III
Carex echinata	+	.	+	+	.	.	.	.	.	.	II
Soldanella pusilla	.	.	.	.	.	.	+	+	.	+	II
Veratrum album	.	+	.	.	.	+	.	+	.	.	II
Saxifraga stellaris	+	+	.	.	.	.	.	.	.	.	I
Carex curta	.	.	+	+	.	.	.	.	.	.	I
Soldanella major	.	.	.	.	+	+	.	.	.	.	I
Vaccinium myrtillus	.	.	.	.	+	.	.	+	.	.	I

Companion species with one occurrence: *Potentilla erecta* 1: +; *Poa media* 4: +; *Caltha palustris* 1: +; *Prunella vulgaris* 1: +; *Trifolium repens* 1: +; *Phleum alpinum* 3: +; *Deschampsia flexuosa* 4: +; *Cerastium caespitosum* 4: +; *Poa annua* 4: +; *Crocus vernus* 9: +; *Alchemilla hybrida* 10: +; *Veronica officinalis* 5: +; *Campanula abietina* 3: +

Relevé: 1. Gura Bucurii (17.07.1967); 2. Lacul Gemenele (16.07.1967); 3. Stânişoara (6.08.1970); 4. Vf. Lolaia (6.08.1970); 5. Căldarea Gemenele (16.07.1967); 6. Vf. Gemenele (16.07.1967); 7. Căldarea Şeselor (18.07.1967); 8–9. V. Judelui (10.07.1970); 10. V. Judelui (8.07.1971)

Table 6.35 Ass. *Poetum mediae* Csűrös 1956

Relevé No.	1	2	3	4	5	6	7	8	9	10	
Altitude (m a.s.l.)	2000	2000	1900	1800	1900	2000	1950	2100	1900	1850	
Aspect	–	–	–	E	NE	S	W	W	N	–	
Slope (degrees)	0	0	0	3	3	5	5	30	3	0	
Herb cover (%)	100	100	100	100	100	100	100	100	100	100	
Sample area (sq. m)	25	25	25	25	25	25	25	25	25	25	K
Char. ass.											
Poa media	3.5	2.5	3.5	3.5	3.5	4.5	4.5	3.5	4.5	4.5	V
Cerastium alpinum	+	+	1.1	+	+	.	+	.	.	.	IV
Potentillo-Nardion											
Nardus stricta	3.5	4.5	3.5	3.5	3.5	1.5	2.5	+	3.5	1.5	V
Geum montanum	+.5	+.5	2.5	1.5	1.5	+	+.5	.	+.5	.	V
Potentilla ternata	2.4	2.5	1.5	+.5	2.5	1.5	1.5	+	+.5	1.5	V
Campanula abietina	+	+	.	+	.	+	+	.	+	.	III
Pseudorchis albida	+	+	.	.	+	+	+	.	.	+	III
Plantago atrata	+	+	.	.	+	.	+	.	.	.	II
Plantago gentianoides	.	.	.	.	+	+	+	.	.	.	II
Ranunculus pseudomontanus	+	+	.	+	.	.	.	.	+	.	II
Alchemilla flabellata	.	.	.	+	+	.	+	.	.	.	II
Festuca nigrescens	+	+	.	1.3	.	.	.	.	.	.	II
Caricetalia et ***Caricetea curvulae***											
Avenula versicolor	+	+	+	.	+	+	+	.	+	+	IV
Phyteuma confusum	+	+	.	.	+	.	.	.	.	+	II
Festuca supina	.	.	.	.	.	+	.	+	.	.	I
Veronica bellidioides	.	.	.	.	.	+	.	+	.	.	I
Luzula spicata	.	.	.	.	.	+	.	.	.	+	I

(continued)

Table 6.35 (continued)

Hieracium alpinum	.	.	.	.	.	+	.	.	.	+	I
Gnaphalium supinum	.	.	.	.	.	+	.	.	.	.	I
Campanula alpina	.	.	.	.	.	+	.	.	.	.	I
Agrostis rupestris	.	.	.	.	.	+	.	.	.	.	I
Primula minima	.	.	.	.	.	+	.	.	.	.	I
Salicion herbaceae **et** ***Salicetalia herbaceae***											
Ligusticum mutellina	1.5	+.5	+	2.5	2.5	+	1.5	+	+.5	+	IV
Gnaphalium supinum	+	+	.	.	+	.	.	+	+	+	III
Cerastium cerastioides	.	.	.	+	+	.	.	.	.	.	I
Sedum alpestre	.	.	.	.	.	+	.	+	.	.	I
Leucanthemopsis alpina	.	.	.	.	.	+	.	.	.	+	I
Soldanella pusilla	.	.	.	.	.	+	.	.	.	.	I
Rhododendro-Vaccinion											
Vaccinium gaultherioides	.	.	.	.	.	+	.	1.5	.	.	I
Cetraria islandica	.	.	.	.	.	+	.	+	.	.	I
Rhododendron myrtifolium	.	.	.	.	.	+	.	.	.	.	I
Juniperus communis subsp. *alpina*	.	.	.	.	.	+	.	.	.	.	I
Vaccinium vitis-idaea	.	.	.	.	.	+	.	.	.	.	I
Homogyne alpina	.	.	.	.	.	+	.	.	.	.	I
Companion											
Poa supina	+	+	.	.	.	+	+	.	+	.	III
Ranunculus pseudomontanus	+	+	.	+	.	.	.	.	+	.	II
Thymus pulegioides subsp. *montanus*	.	.	.	+	+	.	+	.	+	.	II
Poa alpina	.	.	.	+	+	.	+	.	+	.	II
Phleum alpinum	.	.	.	+	+	.	+	.	+	.	II
Festuca nigrescens	+	+	.	1.3	.	.	.	.	.	.	II
Veronica alpina	.	.	.	+	+	.	+	.	.	.	II

Alchemilla flabellata	.	.	.	+	+	.	+	.	.	.	II
Agrostis alpina	.	.	.	+	+	.	.	.	.	.	I
Trifolium repens	.	.	.	.	+	+	.	.	.	.	I
Taraxacum panalpinum	.	.	.	.	.	+	.	+	.	.	I

Companion species with one occurrence: *Deschampsia caespitosa* 6: +; *Festuca picturata* 6: +; *Veronica officinalis* 6: +; *Antennaria dioica* 6: +
Relevé: 1–5. Mt. Drăgşanu (7.07.1972); 6. Vf. Pietrele (7.07.1972); 7. Mt. Drăgşanu (7.07.1972); 8. Mt. Păpuşa (8.08.1970); 9. Mt. Drăgşanu (8.08.1972); 10. Cioaca Radeşului (8.08.1970)

characterisation of this class from ecological, floristic, syntaxonomic, and nomenclatural points of view has been carried out by Grabherr (1993). The class is well individualised from ecological and physiognomical points of view, however there are few charcateristic species (Koči and Chytrý 2007; Péterfi 1993; Roussakova 2000; Šibik et al. 2007a). In the structure of the phytocoenoses included in this class, we can find many species belonging to the *Caricetalia curvulae* and *Piceetalia abietis* orders. Based on these arguments, the dwarf alpine shrublands were included either in the *Caricetalia curvulae* (Boşcaiu 1971) or as a distinct order *Loiseleurio-Vaccinietalia* in the class *Vaccinio Piceetea* (Matuszkiewicz 2008; Seibert 1992). This class includes a single order, namely *Rhododendro-Vaccinietalia*.

Order *Rododendro-Vaccinietalia* Br.-Bl. in Br.-Bl. et Jenny 1926
Syn.: *Loiseleurio-Vaccinetalia* Eggler 52 (Art. 29).
Natura 2000: habitat type 4060

The order encompasses arctic and boreo-nemoral acidic dwarf shrublands from the mountainous regions of Europe. The characteristic species are similar to those of the class, of which in the region of the Retezat Mountains, we found *Loiseleuria procumbens, Vaccinium gaultherioides, Vaccinium vitis-idaea* and *Cladonia rangiferina*. The order includes two alliances, well-individualised in therm of climatic conditions and origin (Grabherr 1993; Mucina et al. 1993).

Alliance *Loiseleurio-Vaccinion* Br.-Bl. in Br.-Bl. et Jenny 1926

Here we group the primary vegetation consisting of the dwarf alpine shrubs, occurring in extreme climatic conditions from the windswept crests of high mountain ranges, superficial skeletal soils that have an acidic reaction (pH = 4.5–4.8). The characteristic species of the alliance that can also be found in the phytocoenoses from the Retezat Mountains are: *Loiseleuria procumbens, Vaccinium gaultherioides, Phyteuma confusum* and *Primula minima*.

Ass. *Cetrario-Loiseleurietum* Br.-Bl. in Br.-Bl. et Janny 1926 (Table 6.36)

The plant communities dominated by *Loiseleuria* (*Kalmia*) *procumbens* can be found in the Retezat Mountains in the extreme habitats found on wind-exposed ridges from high altitudes. These phytocoenoses develop on lithosols, that are frozen for a long period throughout the cold months, mainly due to the lack of snow coverage, while being dry during the warmer months due to high insolation and strong winds. They usually develop together with the phytocoenoses of *Caricion curvulae* alliance, leading to a rich pool of common species shared by the two types of vegetation. Adapted to the dry and cold climate conditions in the high altitude environment, the phytocoenoses dominated by *Loiseleuria* have favoured the development of lychen synusia dominated by *Cetraria islandica*, *Cetraria nivalis,* and *Alectoria ochroleuca*. Floristically, the phytocoenoses from the Retezat Mountains resemble the ones described in the Alps (Aichinger 1933). Being developed in sites that are highly exposed to wind and sun, these plant communities have a crucial role in the protection of skeletal soils against erosion.

Table 6.36 Ass. *Cetrario-Loiseleurietum procumbentis* Br.-Bl. in Br.-Bl. et Jenny 1926

Relevé No.	1	2	3	4	5	6	7	8	9	10	
Altitude (m a.s.l.)	2370	2320	2240	2220	2140	2160	2110	2070	2200	2000	
Aspect	–	W	–	–	SW	W	W	NE	N	NW	
Slope (degrees)	0	45	0	0	10	20	5	15	5	10	
Herb cover (%)	100	100	100	90	100	90	100	95	100	70	
Sample area (sq. m)	25	25	25	25	25	25	25	25	25	25	K
Char. ass.											
Loiseleuria procumbens	4.5	4.5	4.5	5.5	4.5	4.5	4.5	4.5	4.5	3.5	V
Loiseleurio-Vaccinion											
Cetraria islandica	2.5	2.5	2.5	2.5	+	+	+	2.5	+	.	V
Vaccinium gaultherioides	+	1.3	+	+	+	.	.	+	1.5	2.5	IV
Cetraria nivalis	1.3	+	+	.	+	.	+	.	.	.	III
Alectoria ochroleuca	.	+.3	+	+	1.5	+.5	+.5	.	.	.	III
Thamnolia vermicularis	+	1.5	.	+	.	.	+.5	.	.	.	II
Rhododendro-Vaccinietalia											
Rhododendron myrtifolium	.	.	+	.	2.3	1.3	1.3	+.3	1.5	+	IV
Vaccinium vitis-idaea	.	.	.	.	+	+	+.3	+	.	.	II
Vaccinium myrtillus	.	.	.	.	.	+	+	.	.	+	II
Homogyne alpina	.	.	.	.	.	+	.	.	.	+	I
Juniperus communis subsp. *alpina*	.	.	.	.	.	.	+	+	.	.	I
Caricetalia curvulae s.l.											
Carex curvula	+.4	1.5	1.5	+	1.5	+	2.5	.	+	1.3	V
Phyteuma confusum	+	+	+	+	.	+	+	+	.	.	IV
Oreochioa disticha	+	.	+	+	+	.	+	.	+	+.3	IV
Campanula alpina	.	1.5	1.5	1.3	+	+	+	+.5	+	.	IV
Juncus trifidus	.	+	+	+	+	+	.	1.5	+	1.4	IV

(continued)

Table 6.36 (continued)

Primula minima	+	+	.	.	+.5	+	.	+	.	.	III
Festuca supina	.	+	.	+	.	+	+	+	.	.	III
Pulsatilla alba	+	+	.	.	+	.	.	.	.	.	II
Hieracium alpinum	.	.	.	+	+	+	.	.	+	.	II
Senecio abrotanifolius subsp. *carpathicus*	+	.	.	.	.	.	.	.	.	.	I
Avenula versicolor	.	.	.	.	.	+	.	.	.	+	I
Potentilla ternata	.	.	.	.	.	.	.	+	.	.	I
Agrostis rupestris	.	.	.	.	.	.	.	.	.	1.5	I
Companion											
Leucanthemopsis alpina	.	+	+	+	+	.	.	.	.	.	II
Ligusticum mutellina	.	.	.	.	.	+	.	+	.	+	II

Companion species with one or two occurrences: *Carex atrata* 4: +; *Senecio incanus* subsp. *carniolicus* 4, 5: +; *Senecio doronicum* subsp. *transylvanicus* 3, 5: +; *Saxifraga aizoides* 1: +; *Salix herbacea* 1, 9: +; *Sesleria coerulans* 2, 6: +; *Veronica bellidioides* 3, 5: +; *Veronica baumgartenii* 3, 5: +; *Anthemis carpatica* 8: +; *Carex sempervirens* 8: +.5; *Silene acaulis* 9: +; *Huperzia selago* 9: +; *Cladonia rangiferina* 2, 9: +; *Polytrichum alpinum* 3, 6: +

Relevé: 1. Mt. Custura (18.07.1971); 2. Muchia Ascuțită (10.07.1970); 3. Curmătura Judelui (10.07.1970); 4. Căldarea Lacului Zănoaga (10.07.1970); 5. Vf. Zănoaga (10.07.1970); 6. Mt. Peleaga (11.07.1970); 7. V. Rea (17.07.1971); 8. Vf. Zlata (7.07.1970); 9. Stâna de Râu-Căldare (17.07.1971); 10. Vf. Piule (6.08.1970)

Ass. *Cetrario-Vaccinietum gaultherioidis* (Hadač 1956) Hadač ex Šibik et al. 2007 (Table 6.37)

We include in this plant association the arctic-alpine phytocoenoses dominated by *Vaccinium gaultherioides* from the Retezat Mountains. Similarly to the plant communities from the Țarcu and Godeanu Mountains (Boșcaiu 1971), the species *Empetrum nigrum* s.l. is missing from the phytocoenoses from Retezat. These communities have a more pronounced microthermal character compared to those developed by *Loiseleuria procumbens*, despite their similar habitat (superficial lithosols, strongly wind-battered sites, crystalline bedrock, alpine belt). The phytocoenoses formed by *Vaccinium gaultherioides* represent primary vegetation and have developed mainly in the alpine belt, with rare and small transgressions into the subalpine belt, especially following the deforestation of *Juniperus communis* subsp. *alpina* (Buia et al. 1962; Pușcaru-Soroceanu et al. 1981). The floristic composition of the phytocoenoses developed in the Retezat Mountains closely resembles the one described from the Western Carpathians (Šibik et al. 2007b), differing only through the absence of few regional species occurring in the Tatra Mountains, like *Soldanella carpatica, Carex firma, Sesleria tatrae, Sesleria albicans* and *Campanula tatrae*.

Alliance *Rhododendrion myrtifolii* de Faucault ex Theurillat et Mucina 2016

This alliance, initially described under the name *Rhododendrion kotschyi* by Foucault 1991 in order to differentiate the shrub plant communities with *Rhododendron ferrugineum* from the Alps from the ones developed in the Carpathians, has recently been typified in Mucina (2016). The Romanian phytocoenologists that studies the *Rhododendron myrtifolium* plant communities from various mountain ranges of the South-Eastern Carpathians included them in the alliance *Rhododendro-Vaccinion* Br.-Bl. in Br.-Bl. et Jenny 1926 (Beldie 1967; Boșcaiu 1971; Coldea 1990) based on the argument that the few existing Carpathian species do not support its inclusion in a separate, regional alliance. In addition, *Rhododendron myrtyfolium* has a large altitudinal amplitude, occurring from the upper montane to the alpine belt (Coldea 2015). In the structure of these phytocoenoses, we can find transgressive species from spruce forests, *Pinus mugo* and *Juniperus communis* subsp. *alpina* shrublands or from acidophilic alpine grasslands included in the *Caricion curvulae* alliance. The authors who validated the alliance mentioned along *Rhododendron myrtifolium* the following diagnostic species: *Soldanella major* and *Potentilla ternata*, species that are widely distributed in the Southern Carpathians (Mucina 2016). Considering that these dwarf shrublands dominated by *Rhododendron myrtifolium* have a well-individualised physiognomy, a well-delimited distribution in the region and a particular ecology, occurring in the lower alpine belt of the South-Eastern Carpathians, we consider this alliance valid and fully justified through solid scientific arguments.

Ass. *Rhododendro myrtifolii-Vaccinietum* Coldea et al. 1981 (Table 6.38)
Syn.: *Rhododendro-Vaccinietum austro-carpaticum* Borza 1959 (Art. 34)

Table 6.37 Ass. *Cetrario-Vaccinietum gaultherioidis* Hadač 1956 ex. Sibik et al. 2007

Relevé No.	1	2	3	4	5	6	7	8	9	10	
Altitude (m a.s.l.)	2090	2190	2070	2150	2040	2050	1990	1910	1880	1900	
Aspect	NW	S	E	S	N	SW	NE	NW	NW	NW	
Slope (degrees)	15	30	10	10	30	15	3	15	15	15	
Herb cover (%)	100	80	80	100	70	100	70	70	75	75	
Sample area (sq. m)	25	25	25	4	25	10	25	25	25	25	K
Char. ass.											
Vaccinium gaultherioides	4.5	4.5	4.5	5.5	4.5	4.5	4.5	4.5	4.5	4.5	V
Cetraria nivalis	+	.	.	.	+	.	+	.	.	.	II
Cetrario-Loiseleurion											
Cetraria islandica	2.5	1.5	+	+	+	3.5	+	.	+	.	IV
Thamnolia vermicularis	.	.	+	.	+	.	+	.	.	.	II
Loiseleuria procumbens	.	.	.	.	.	.	.	.	+	+	I
Alectoria ochroleuca	.	.	.	.	.	.	+	.	.	.	I
Rhododendro-Vaccinietalia											
Rhododendron myrtifolium	+	1.5	.	+	+	+	+	+	1.3	+	V
Vaccinium myrtillus	+	+.5	+	+	1.5	.	.	.	+	+	IV
Homogyne alpina	+.5	+.3	.	+.4	+	+	.	.	+	.	III
Juniperus communis subsp. *alpina*	.	+	1.3	.	.	+	.	.	+	+	III
Vaccinium vitis-idaea	.	.	+	.	+.5	+	.	+	.	.	II
Bruckenthalia spiculifolia	+	.	.	+	.	.	.	.	.	.	I
Caricetalia curvulae s.l.											
Avenula versicolor	+.5	+	1.5	+	+	.	+	.	+	.	IV
Campanula alpina	1.5	+	.	+	+	.	+	.	+	1.5	IV
Potentilla ternata	.	+	+	+	+	.	.	+	+	+	IV
Phyteuma confusum	.	.	+	+	.	.	+	+.5	+	+	III

Anthoxanthum alpinum	+	+.5	+	+	.	.	.	.	.	.	II
Hieracium alpinum	.	1.5	.	.	.	.	+	.	+	+	II
Pulsatilla alba	+	+.3	.	.	.	.	.	.	+	+	II
Festuca supina	+	.	+	.	+	.	+.2	.	.	.	II
Carex curvula	.	+	.	1.5	.	.	+	.	.	.	II
Juncus trifidus	.	.	1.3	.	+	+	.	.	.	.	II
Geum montanum	.	.	.	+	.	.	.	.	+	+	II
Cladonia rangiferina	1.3	.	+	+	.	.	.	.	.	.	II
Primula minima	+	.	+	.	.	.	.	.	.	.	I
Oreochloa disticha	.	.	.	.	+	.	+	.	.	.	I
Agrostis rupestris	1.5	.	.	.	.	.	.	.	.	.	I
Companion											
Ligusticum mutellina	+	+.5	+	+	.	.	.	.	+	.	III
Poa alpina	+	+	+	.	.	.	.	.	.	.	II
Soldanella pusilla	+	.	.	.	+	.	.	.	+	.	II

Companion species with one or two occurrences: *Luzula sudetica* 1, 3: +; *Hypochaeris uniflora* 3: +; *Gentiana punctata* 9, 10: +; *Calamagrostis villosa* 2: +; *Gentiana acaulis* 2: +; *Senecio doronicum* subsp. *transylvanicus* 2: +; *Antennaria dioica* 3: +; *Luzula luzuloides* 3: +; *Pedicularis verticillata* 3: +; *Veronica bellidioides* 3: +; *Huperzia selago* 5: +; *Carex sempervirens* 6: +; *Bartsia alpina* 8: +; *Gentiana verna* 8: +; *Onobrychis montana* subsp. *transsilvanica* 8: +; *Polygonum viviparum* 8: +; *Ranunculus oreophilus* 8: +; *Selaginella selaginoides* 8: +; *Sedum atratum* 8: +; *Poa media* 5: +; *Dicranum scoparium* 1: +; *Pleurozium schreberi* 1: +; *Polytrichum juniperinum* 9, 10: +

Relevé: 1. Mt. Huta-Mt. Judele (10.07.1970); 2. Lacul Zănoaga (8.07.1970); 3. Ridge Zănoguței; (8.07.1970); 4. Ridge Bucurei (15.07.1967); 5. Lacul Bucura (11.08.1970); 6. Vf. Zlata (11.08.1970); 7. Vf. Șeselor (16.07.1967); 8. Mt. Soarbele (26.06.1971); 9. Lacul Șeselor (12.07.1967); 10. Tăul Negru (14.07.1967)

Table 6.38 Ass. *Rhododendro myrtifolii-Vaccinietum* (Borza 1959) Coldea et al. 1981

Relevé No.	1	2	3	4	5	6	7	8	9	10	
Altitude (m a.s.l.)	2020	2150	2080	2100	2150	2090	2100	1900	2020	2150	
Aspect	NE	S	E	SE	NW	NW	NW	W	N	S	
Slope (degrees)	40	25	50	15	20	15	50	50	40	30	
Herb cover (%)	90	95	80	90	90	90	90	90	90	75	
Sample area (sq. m)	25	25	25	25	25	25	25	25	25	25	K
Char. ass.											
Rhododendron myrtifolium	4.5	3.5	3.5	3.5	4.5	4.5	4.5	4.5	4.5	3.5	V
Vaccinium gaultherioides	+.3	+.3	1.3	3.5	2.5	2.5	+	+.3	+	+	V
Rhododendrion **et** ***Rhododendro-Vaccinietalia***											
Vaccinium myrtillus	+	1.5	3.5	+.4	2.5	1.5	1.5	1.3	2.5	1.4	V
Homogyne alpina	1.5	+.5	+	+.5	+.5	1.5	+.5	+	+.5	+.5	V
Phyteuma confusum	+	+.4	.	+.4	+	.	.	+	.	+.5	III
Calamagrostis villosa	.	.	.	+.3	+	+	.	+	1.5	.	III
Soldanella major	.	+	.	+	+	+	.	.	.	.	II
Potentilla ternata	+	+	.	.	.	.	.	.	.	+.3	II
Huperzia selago	+	.	.	.	+	.	+	.	.	.	II
Vaccinium vitis-idaea	.	+	+	.	.	.	.	+	.	.	II
Campanula serrata	.	+	.	.	.	.	.	.	.	.	I
Centaurea nervosa	.	.	.	+	.	.	.	.	.	.	I
Bruckenthalia spiculifolia	.	.	.	.	.	.	.	+	.	.	I
Caricetalia curvulae s.l.											
Campanula alpina	+	+.3	1.5	.	+	.	+	+	1.5	+	IV
Pulsatilla alba	.	+	+	+	.	+	+	.	.	+	III
Juncus trifidus	.	+	+	+	.	.	+	+	+	.	III
Agrostis rupestris	.	+.5	.	+	.	1.5	2.5	+	.	+.4	III

Primula minima	+.5	+	+	+	.	.	.	.	+	.	III
Carex curvula	1.5	2.5	.	+.5	+	.	.	.	.	1.5	III
Hieracium alpinum	.	+	+	+	+.4	.	.	.	+.4	+	III
Avenula versicolor	.	+	+.4	+	.	.	.	.	.	+	II
Oreochloa disticha	.	.	.	+	+	.	+	.	.	.	II
Sesleria bielzii	.	+	.	.	.	.	.	.	.	+	I
Cetrario-Loiseleurion											
Cetraria islandica	1.5	2.5	2.5	3.5	1.5	1.5	1.4	2.5	1.5	.	V
Loiseleuria procumbens	2.5	+.5	+.3	+.3	1.3	2.5	.	1.3	+	+.3	V
Companion											
Pinus mugo	+	+	+	2.2	.	2.3	2.3	1.3	1.3	+	V
Ligusticum mutellina	+.4	+.4	+	+.5	+	1.5	+	+	+.5	+.3	V
Gentiana punctata	.	+	.	+	+	+	+	+	.	+	IV
Soldanella pusilla	+.4	.	.	+	+	+.5	+	.	+	.	III
Cladonia rangiferina	.	1.3	+	.	+.3	.	1.3	+	.	.	III
Dicranum scoparium	1.4	.	.	.	4.5	1.5	+	.	+	.	III
Geum montanum	+	+	.	+	+	.	.	.	.	.	II
Luzula alpinopilosa subsp. *obscura*	+	.	+	.	.	.	.	+	.	.	II
Deschampsia flexuosa	.	+	.	+	+	.	.	.	.	.	II
Gentiana acaulis	.	+	.	+.3	.	+	.	.	.	.	II
Nardus stricta	.	+	.	.	.	1.5	.	.	.	1.5	II
Anthoxanthum alpinum	.	.	.	+	.	.	.	.	+	+	II
Senecio incanus subsp. *carniolicus*	.	.	.	.	.	.	1.4	+	+	.	II
Juniperus communis subsp. *alpina*	.	+	.	2.2	+	.	.	1.3	.	.	II
Pleurozium schreberi	1.3	.	1.4	.	1.3	.	.	.	.	.	II
Polytrichum juniperinum	+	.	+	.	.	+	.	.	.	.	II

(continued)

Table 6.38 (continued)

Thamnolia vermicularis	.	.	.	+	.	.	+	+	.	.	II
Hylocomium splendens	.	.	.	.	+	.	+	.	+	.	II

Companion species with one occurrence: *Veratrum album* subsp. *lobelianum* 1: +; *Gnaphalium supinum* 2: +; *Poa alpina* 3: +.3; *Veronica baumgartenii* 7: +; *Leucanthemopsis alpina* 7: +; *Senecio doronicum* subsp. *transylvanicus* 8, 9: +; *Alectoria ochroleuca* 8: +; *Diplophyllum albicans* 1: +, 6: 1.3; *Grimia hartmanni* 1: +; *Rhytidiadelphus triquetrus* 1: +.3; *Polytrichum formosum* 5: +; *Dicranum montanus* 6: +; *Drepanocladus uncinatus* 6: +; *Rhacomitrium sudeticum* 7: +; *Dicranum montanum* 6: +; *Polytrichum piliferum* 8: +

Relevé: 1. Căldarea Bucurii (6.08.1969); 2. Curmătura Bucurii (16.07.1967); 3. Tăul Zănoguții (10.07.1970); 4. Vf. Șesele (16.07.1970); 5. Adăpătoarea Caprelor (7.07.1969); 6. Tăul Secat (8.07.1969); 7. Tăul Negru (5.07.1969); 8. Mt. Zlata (9.07.1970); 9. Tăul Cârligului (8.07.1969); 10. Fața Retezatului (16.07.1967)

We have included in this regional Carpathian plant association the phytocoenoses with *Rhododendron myrtifolium* from the Retezat Mountains, where the species covers large areas (thousands of hectares) on the semi-shaded slopes from the Retezat, Şesele, Zănoaga, Radeşu and Zlata peaks and in the glacial circuses Bucura, Zănoaga or Peleaga. These phytocoenoses, occurring on the screes of the high mountain ranges, represent primary vegetation. They prefer the sites sheltered from the wind and with long term snow coverage. Other frequently occurring shrub species (some with up to 10% coverage) in these plant communities are: *Vaccinium gaultherioides, Vaccinium myrtillus, Loiseleuria procumbens,* and *Pinus mugo*. Among herbaceous species, more frequently encountered are *Homogyne alpina, Ligusticum mutellina* and *Campanula alpina*, similar to the plant communities described in the Ţarcu and Godeanu Mountains (Boşcaiu 1971).

Ass. *Campanulo abietinae-Vaccinietum myrtilli* Boşcaiu ex Coldea 1991 (Table 6.39)

The subalpine plant communities dominated by *Vaccinium myrtillus* from the Retezat Mountains are included in this association that was thoroughly described both ecologically and syntaxonomically by Boşcaiu (1971) in the Ţarcu and Godeanu Mountains. Within its floristic composition we can find both regional Carpathian-Balkan species (*Campanula abietina, Senecio doronicum* subsp. *transylvanicum, Phyteuma confusum*), as well as more widespread species characteristic of the alliance, order and the class *Loiseleurio-Vaccinietea*. Ecologically and floristically it resembles the *Avenastro versicoloris-Vaccinietum myrtilli* Krajina 1933, a syntaxon described in the Tatra Mountains (Šibik et al. 2007a). At lower altitudes, in the mountain ranges from the South-Eastern Carpathians (Dihoru 1975), the phytocoenoses with *Vaccinium myrtillus* are poorer in characteristic species and incorporate in their floristic composition more species of the order *Nardetalia*, rendering the assignation of this association to a higher-ranked syntaxon more difficult. In general, such phytocoenoses occuring in the upper montane belt are secondary vegetation and develop as disclimax stages over a long period. The phytocoenoses with *Vaccinium myrtillus* have a major role in the protection of soil from erosion and in the regeneration of *Pinus mugo* in the subalpine belt.

Class *Elyno-Seslerietea* Br-Bl. 1948

This class includes mainly natural grassland phytocoenoses developed on calcareous bedrock from the montane to alpine belt. The characteristic soils for these sites are poorly developed and superficial, mostly rendzinic leptosols with frequent rocky outcrops. The structure of these phytocoenoses is dominated by graminoids belonging to *Festuca* and *Carex* genera, while the characteristic taxa of the class are, among others: *Gentiana verna, Helianthemum alpestre, Anthyllis vulneraria* subsp. *alpestris, Biscutella laevigata, Galium anisophyllum, Dryas octopetala* or *Thesium alpinum* (Koči 2007a).

Table 6.39 Ass. *Campanulo abietinae-Vaccinietum myrtilli* Boşcaiu ex Coldea 1991

Relevé No.	1	2	3	4	5	6	
Altitude (m a.s.l.)	1980	1850	1900	1850	1800	1900	
Aspect	SW	E	SE	SW	W	W	
Slope (degrees)	35	40	35	25	25	35	
Herb cover (%)	80	60	85	90	75	95	
Sample area (sq. m)	25	50	50	50	50	50	K
Char. ass.							
Campanula abietina	+	+	+	+	+	+	V
Phyteuma confusum	.	+	.	+.3	+	+	IV
Senecio doronicum subsp. *transylvanicus*	.	+	.	+	+	+.2	IV
Rhododendrion myrtifolii							
Rhododendron myrtifolium	1.5	1.5	1.5	1.2	+.3	1.2	V
Soldanella major	+	+	+	.	+	+	V
Potentilla ternata	+	.	+	+.3	.	+	IV
Rhododendro-Vaccinietalia							
Vaccinium myrtillus	4.5	3.5	4.5	4.5	4.5	4.5	V
Homogyne alpina	+	+.3	+.3	1.3	+	1.3	V
Vaccinium gautherioides	.	1.3	.	1.3	+	1.2	IV
Luzula luzuloides subsp. *cuprina*	+	.	+	+	+	.	IV
Solidago virgaurea subsp. *minuta*	+	+	+	.	+	.	IV
Centaurea nervosa	+	.	+	.	.	+	III
Vacinium vitis-idaea	+	+	.	.	.	+	III
Huperzia selago	.	+	.	.	+	+	III
Loiseleuria procumbens	.	.	.	+	+	.	II
Gentiana acaulis	+	.	+	.	.	.	II
Leontodon hispidus subsp. *montanum*	+	.	.	.	.	.	I
Caricetalia curvulae							
Avenula versicolor	1.5	+	1.5	+	+	+.3	V
Campanula alpina	+	+	+	.	+	+	V
Pulsatilla alba	1.2	+	+	+	+	+	V
Ligusticum mutellina	.	1.2	+	+	+	1.3	V
Hieracium alpinum	+	.	+	+	.	+	IV
Oreochloa disticha	.	+	.	.	+	+	III
Agrostis rupestris	.	.	.	1.2	+	+	III
Juncus trifidus	.	.	.	.	+	1.3	II
Senecio incanus ssp. *carniolicus*	.	+	.	.	.	+	II
Primula minima	.	+	.	.	.	+	II
Companion							
Calamagrostis vilosa	.	1.2	1.3	1.3	1.5	1.3	V
Anthoxanthum alpinum	+	+	1.2	+	.	+	V
Gentiana punctata	.	+	.	+	+	+.2	IV
Pinus mugo	.	+	.	+	.	+	IV
Festuca picturata	.	+	+	.	+	+	IV

(continued)

Table 6.39 (continued)

Cetraria islandica	.	1.3	1.1	.	+	2.3	IV
Cladonia rangiferina	+	.	.	+	+	+.3	IV
Veratrum album subsp. *lobelianum*	.	+	.	.	+	+	III
Geum montanum	+	.	+	.	+	.	III
Deschampsia flexuosa	.	+	.	.	+	+	III
Pseudorchis albida	+	.	+	+	.	.	III
Dicranum scoparium	.	1.1	+	.	2.3	.	III
Pleurozium schreberi	+	.	.	+	1.3	.	III
Juniperus communis subsp. *alpina*	.	.	+	.	+	.	II
Hypochaeris uniflora	+	.	+.3	.	.	.	II
Gnaphalium supinum	.	+	.	.	.	+	II
Pedicularis verticillata	+	.	+	.	.	.	II
Veronica baumgartenii	.	+	.	+	.	.	II
Nardus stricta	.	.	.	1.5	.	.	I
Carex atrata	.	.	.	.	+	.	I

Relevé: 1. Căldarea Bucura (20.07.1967); 2. Lacu Gemenea (18.07.1967); 3. Căldarea Judele (19.07.1967); 4. Fața Retezatului (17.07.1967); 5. Tăul Știrbului (15.07.1984), 6. Căldarea Șesele (15.07.1968)

Order *Seslerietalia caerulaeae* Br.-Bl. in Br.-Bl. et Jenny 1926

Natura 2000: habitat type 6170

This order encompasses basophilic primary grasslands developed on sunny slopes and calcareous soils from the subalpine and alpine belts from the Alps and the Carpathians. The typical soils are rendzines and proto-rendzines rich in bases and with a neutral to basic reaction (pH = 6.7–7.3). Among the main species characterising the studied phytocoenoses from the Retezat Mountains included in this order (Koči 2007a; Oberdorfer 1978) are: *Carex sempervirens, Ranunculus oreophilus, Myosotis alpestris, Leontopodium alpinum, Phyteuma orbiculare, Scabiosa lucida, Oxytropis montana, Carex rupestris*, *Cerastium arvense* subsp. *molle, Carex capillaris, Saxifraga moschata, Hedysarum hedysaroides, Festuca versicolor, Hieracium villosum, Polygala alpestris, Polygonum viviparum,* and *Minuartia verna*. This order includes the following alliances: *Festuco saxatilis-Seslerion bielzii* and *Seslerion rigidae*, well-individualised ecologically and floristically.

Alliance *Festuco saxatilis-Seslerion bielzii* (Pawl. et Walas 1949) Coldea 1984

We list here heliophilous alpine and subalpine plant communities from the sunny calcareous ridges of Retezat Mountains (including Piatra Iorgovanului, Piule, Albele and Scorota peaks). Among the characteristic plant taxa for these phytocoenoses are several Carpathian endemics, which act as regional diagnostic taxa differentiating them from the Western Carpathian plant communities from the *Seslerion tatrae* Pawl. 1935 (Matuszkiewicz 2008) and *Agrostion alpinae* Janik et al. 1980 (Koči 2007a) alliances. Other species typical of the *Festuco-Seslerion bielzii* alliance are:

Sesleria bielzii, Sesleria rigida subsp. *haynaldiana, Androsace villosa* subsp. *arachnoidea, Thymus pulcherrimus, Onobrychis montana* subsp. *transsilvanica, Festuca stricta* subsp. *rumelica, Erysimum witmanii* subsp. *transsilvanicum, Alyssum repens, Carduus kerneri* subsp. *lobulatiformis.*

Ass. *Seslerio haynaldianae-Festucetum veriscoloris* Beldie 1967 (Table 6.40)

The phytocoenoses included in this association are distributed mainly on the sun-exposed ridges of Piule and Piatra Iorgovanului calcareous massifs from the south-eastern part of the Retezat Mountains. The superficial rendzinic leptosol from these sites is characterised by a basic reaction (pH = 7–7.4). The dominant species, with an average cover of 40% in the investigated sites, are *Festuca versicolor* and *Sesleria rigida* subsp. *haynaldiana*. Along these, several other frequently occurring species have been found to develop well in these phytocoenoses, among which the Carpathian endemics *Onobrychis montana* subsp. *transsilvanica, Festuca stricta* subsp. *rumelica* and the Alpine-Carpathian *Alyssum repens* are characteristic of the alliance, while *Carex sempervirens, Helianthemum alpestre, Ranunculus oreophilus* and *Helianthemum nummularium* subsp. *tomentosum* are characteristic of the order and class. The phytocoenoses from the Retezat Mountains are floristically differentiated from the ones occurring in the Western Carpathians (Koči 2007a) through the group of species characterising the alliances *Seslerion tatrae* Pawl. 1935, *Caricion firmae* Gams 1936 and *Agrostion alpinae* Janik et al. 1980.

Ass. *Seslerion hyanaldianae-Caricetum sempervirentis* Puşcariu et al. 1956 (Table 6.41)
– *saxifragetosum marginatae* subass. nova hoc loco
Holotype: Table 6.41, rel. 5 hoc loco

The heliophilous phytocoenoses of this association frequently inhabit the sunny and abrupt calcareous slopes from the subalpine and alpine belts of the Retezat Mountains (Stânuleți, Albele, Piatra Iorgovanului and Piule peaks). The soils of these sites are superficial rendzines, saturated in bases and with a high content of calcium carbonate. The dominant species, with an average cover of 50% are *Carex sempervirens* and *Sesleria rigida* subsp. *haynaldiana*. Along these, characteristic species of the alliance, order and the class are well represented as well. Compared to the plant communities described in the Bucegi Mountains (Beldie 1967; Puşcaru et al. 1956), the ones from the Retezat Mountains are differentiated by the presence of several regionally distributed species, like the Carpathian-Balkan *Saxifraga marginata* and *Edraianthus graminifolius*. Based on the presence of these taxa, we hereby describe a new subassociation—*saxifragetosum marginatae*, characterised by ecological particularities, typical for meridional and Balkan phytocoenoses. We also want to highlight that this syntaxon resembles floristically the association *Edraiantho-Seslerietum juncifoliae* Horvat 1974 described in the Biokov Mountains (Croația) and included in the *Seslerion juncifoliae* Horvat 1930 alliance from *Elyno-Seslerietea* Br.-Bl. 1948 (Photos 6.3 and 6.4).

Ass. *Poo molinerii-Festucetum pachyphyllae* Boşcaiu et al. 1978 (Table 6.42)

Table 6.40 Ass. *Seslerio haynaldianae-Festucetum versicoloris* Beldie 1967

Relevé No.	1	2	3	4	5	6	7	
Altitude (m a.s.l.)	1640	1660	1680	1810	1830	1850	1560	
Aspect	S	S	S	S	S	S	S	
Slope (degrees)	45	45	45	45	50	20	70	
Herb cover (%)	80	80	70	80	60	70	80	
Sample area (sq. m)	25	25	25	25	25	25	25	K
Char. ass.								
Festuca vesicolor	2.5	3.5	2.5	4.5	3.5	3.5	3.5	V
Helianthemum nummularium subsp. *obscurum*	1.3	1.5	+	+	+	.	+	V
Cerastium transsilvanicum	+	+	.	+	+	+	.	IV
Festuco-Seslerion bielzii								
Onobrychis montana subsp. *transsilvanica*	1.5	+	+	+.5	1.5	1.5	+.4	V
Alyssum repens	+	+	+	+	+	.	+	V
Acinos alpinus subsp. *alpinus*	.	1.5	+	+	+	+.3	+	V
Festuca stricta subsp. *rumelica*	2.5	2.5	.	1.5	3.5	1.5	1.5	V
Thymus pulcherrimus	+	.	.	.	+	+.3	+	III
Dianthus tenuifolius	.	+	+	+.4	.	.	+	III
Carduus kerneri subsp. *lobulatiformis*	+	+	+	+	.	.	.	III
Androsace villosa subsp. *arachnoidea*	.	.	.	+	.	+	+	III
Bupleurum diversifolium	.	.	+	.	.	.	.	I
Silene nutans subsp. *dubia*	.	.	.	+	.	.	.	I
Trisetum fuscum	.	.	.	.	.	.	+	I
Seslerion rigidae								
Asperula capitata	+	+.5	+	+.5	+.5	+.3	+	V
Dianthus petraeus	.	+	+	1.5	1.5	1.5	+	V
Edraianthus graminifolius	.	+	+	+	+	.	+	IV
Draba lasiocarpa	.	+	.	.	.	.	+	II
Primula veris subsp. *columnae*		.	+	.	.	.	+	II
Seslerietalia* et *Elyno-Seslerietea								
Sesleria rigida subsp. *haynaldiana*	2.5	2.5	1.5	1.5	1.5	1.5	3.5	V
Carex sempervirens	+	2.5	3.5	+.4	+.4	+.3	1.5	V
Ranunculus oreophilus	+	+	1.5	+.5	+	+.5	+	V
Helianthemum alpestre	2.5	+.5	.	1.5	1.5	1.5	+	V
Cerastium arvense subsp. *molle*	+	+.5	+	+	+	+	+	V
Polygala alpestris	+	+	+	+	+	.	+	V
Myosotis alpestris	+	+.5	+.5	+	+	.	+	V
Anthyllis vulneraria subsp. *alpestris*	+	+	+	.	.	.	+	III
Biscutella laevigata	+	+	+	.	.	.	+	III
Leontopodium alpinum	.	.	.	+	+.2	.	+	III
Poa molinerii	+	.	.	.	+	+	.	III

(continued)

Table 6.40 (continued)

Saxifraga adscendens	+	+	.	.	.	+	.	III
Carex capillaris	.	.	.	+	1.5	1.5	.	III
Minuartia verna	.	+	.	+.3	.	.	.	II
Phyteuma orbiculare	.	.	.	+	.	+	.	II
Daphne cneorum	.	.	.	.	.	+	+.4	II
Thesium alpinum	.	+	.	+	.	.	.	II
Pedicularis baumgartenii	+	+	.	.	.	.	.	II
Polygonum viviparum	+	.	+	.	.	.	.	II
Pedicularis verticillata	+	+	.	.	.	.	.	II
Oxytropis pyrenaica	.	.	.	.	.	+	+	II
Nigritella nigra	.	.	.	+	.	.	+	II
Gentiana verna	.	.	+	.	.	.	.	I
Scabiosa lucida	.	.	+	.	.	.	.	I
Galium anisophyllon	.	.	.	.	.	+	.	I
Companion								
Trifolium alpestre	+	+	+	.	.	.	+	III
Hippocrepis comosa	+.3	+.4	+	.	.	.	.	III
Campanula glomerata	+	+	+	.	.	.	.	III
Galium album s.l.	.	+	+	.	.	.	+	III
Saxifraga paniculata	.	.	.	+	+	.	+	III

Companion species with one occurrence**:** *Trifolium pallescens* 1: +; *Bartsia alpina* 3: +; *Botrychium lunaria* 4: +; *Pedicularis comosa* 3: +; *Taraxacum panalpinum* 1: +; *Orchis sambucina* 1: +; *Ornithogalum gussonei* 1: +; *Potentilla thuringiaca* 1: +; *Luzula multiflora* 2: +; *Seseli libanotis* 2: +; *Avenula praeusta* 3: +; *Gymnadenia conopsea* 7: +; *Trinia glauca* 3: +; *Lotus corniculatus* 7: +; *Hieracium villosum* 5: +; *Agrostis alpina* 5: +; *Aconitum anthora* 7: +; *Saxifraga marginata* 7: +; *Arabis hirsuta* 7: +; *Origanum vulgare* 7: +

Relevé: 1–3. Mt. Piatra Iorgovanului (26.06.1971); 4–6. Mt. Piule-Vf. Scorotei (6.07.1972); 7. V. Scorotei (25.06.1971)

The phytocoenoses dominated by *Festuca stricta* subsp. *rumelica* (*F. pachyphylla*) are often encountered in the subalpine belt of the calcareous Retezat (Piule, Piatra Iorgovanului, Albele and Stănuleți), where they inhabit the sun-exposed, moderately steep slopes and rendzinic, superficial soils with frequent rock outcrops. A frequently occurring mesoxerophilous species with preference for saxicolous sunny habitats encountered in these phytocoenoses is *Poa molinerii*, which can locally develop up to 15–20% coverage. The other species that are found in the composition of these plant communities usually have a lower coverage, despite their high frequency of occurrence. The regional discriminant species, specific for the Southern Carpathians are *Festuca stricta* subsp. *rumelica* and several species characteristic of the alliance, like: *Sesleria rigida* subsp. *haynaldiana, Cardus kerneri* subsp. *lobulatiformis*, *Minuartia setacea* subsp. *banatica* and *Silene nutans* subsp. *dubia.*

Ass. *Salici kitaibelianae-Dryadetum octopetalae* ass. nova hoc loco (Table 6.43)
Holotype: Table 6.43, rel. 4, hoc loco

Table 6.41 Ass. *Seslerio haynaldianae-Caricetum sempervirentis* Pușcaru et al. 1956 *saxifragetosum marginatae* subass. nova

Relevé No.	1	2	3	4	5	6	7	8	9	10	
Altitude (m a.s.l.)	1900	1960	1960	1850	1640	1860	2060	1880	1850	1860	
Aspect	SE	S	S	E	SE	SW	SE	N	NE	W	
Slope (degrees)	60	40	60	35	50	30	30	30	60	60	
Herb cover (%)	75	75	75	75	90	75	75	75	75	80	
Sample area (sq. m)	25	25	25	25	25	25	25	25	25	25	K
Diff. subass.											
Saxifraga marginata	2.5	1.3	1.4	1.5	2.4	1.3	1.5	1.4	2.5	3.5	V
Edraianthus graminifolius	+.5	.	+.3	+.5	+.4	1.4	+.3	+	1.5	.	IV
Festuco-Seslerion bielzii											
Onobrychis montana subsp. *transsilvanica*	.	+.4	.	+	2.5	+	+	+	+	.	IV
Alyssum repens	+	+	+	.	+	+	.	+	.	.	III
Thymus pulcherrimus	+	+	+.4	.	+	.	+	+	.	.	III
Acinos alpinus subsp. *alpinus*	+	+	+.4	.	+	.	.	+	.	.	III
Eritrichum nanum subsp. *jankae*	.	.	.	+	.	.	.	+	.	.	I
Dianthus tenuifolius	.	.	.	.	+	.	.	.	+	.	I
Silene nutans subsp. *dubia*	.	+	.	.	.	.	.	.	.	.	I
Seslerion rigidae											
Dianthus petraeus	.	+.4	.	+.3	+	+	+	+	+	+	IV
Draba lasiocarpa	+	+	+	.	+	.	.	+	.	.	II
Asperula capitata	.	+	.	.	+.4	.	+	+	.	.	III
Festuca stricta subsp. *rumelica*	.	1.5	+	.	+.3	.	.	.	.	.	II
Primula veris subsp. *columnae*	.	.	.	.	+	.	.	.	.	.	I
Seslerietalia et ***Elyno-Seslerietea***											
Sesleria rigida subsp. *haynaldiana*	3.5	4.5	3.5	3.5	3.5	2.5	2.5	3.5	2.5	3.5	V
Carex sempervirens	2.5	1.4	2.5	2.5	2.5	3.5	3.5	1.5	1.5	1.4	V

(continued)

Table 6.41 (continued)

Oxytropis carpatica	.	+	.	+.5	+	+	+	+	+.4	.	IV
Helianthemum alpestre	1.4	1.3	1.5	2.5	1.5	2.5	2.5	.	+.4	+	IV
Ranunculus oreophilus	+	1.5	+.5	+.5	.	+.5	+	+.5	+.5	+	V
Polygonum viviparum	.	+	+.4	+.4	+.4	+	.	+.5	+.4	+	IV
Poa molinerii	+	+	+	+.4	.	+	+.3	+	.	.	IV
Gentiana verna	+	+	+.3	+	.	+	.	+	.	+	IV
Myosotis alpestris	+	+	.	+	+	.	.	.	+	+	III
Polygala alpestris	.	+	+.3	.	+.3	+	.	.	+	.	III
Androsace villosa subsp. *arachnoidea*	.	+	.	+.5	.	1.3	+.3	+	.	.	III
Biscutella laevigata	.	+	.	.	+	.	.	+	+	+	III
Dryas octopetala	.	+	.	.	+	.	.	2.5	3.5	2.5	III
Galium anysophyllon	+	+	+.4	.	.	+	.	.	+	.	III
Veronica aphylla	+	.	+	.	+	.	.	+	.	.	II
Cerastium arvense subsp. *molle*	.	+.3	.	.	+	+	+	.	.	.	II
Saxifraga moschata	+	+	+	.	.	.	.	.	.	.	II
Festuca versicolor	.	.	.	+	+	.	.	+	.	.	II
Bartsia alpina	.	+	.	.	+	.	.	.	.	+	II
Scabiosa lucida	+	.	.	.	.	+	.	.	.	.	I
Daphne cneorum	.	+.3	.	+	.	.	.	.	.	.	I
Bupleurum falcatum subsp. *cernuum*	.	.	.	+	+	+	.	+	.	.	II
Leontopodium alpinum	.	.	.	.	.	.	+.3	+	+	.	II
Erysmum witmannii subsp. *transsilvanicum*	.	.	.	.	+	+	.	.	.	.	I
Sedum atrataum	.	.	.	.	+	.	.	+	.	.	I
Carduus kerneri subsp. *lobulatiformis*	.	+	.	.	+	.	.	.	.	.	I
Hieracium villosum	.	.	+	+	.	.	.	.	.	.	I
Anthyllis vulneraria subsp. *alpestris*	.	.	+	.	.	.	+	.	.	.	I
Hedysarum hedysaroides	.	.	.	.	.	+	.	+	.	.	I

Salix retusa	.	.	.	.	.	.	.	1.3	.	+	I
Asplenietea											
Saxifraga paniculata	+	+	2.5	+	+	+	+	.	+	.	IV
Asplenium viride	.	.	+.3	.	+	.	.	.	.	.	I
Androsace lactea	.	.	.	.	.	.	.	.	+	+	I
Cystopteris fragilis	.	.	+	.	.	.	.	.	.	.	I
Asplenium ruta-muraria	.	.	.	.	+	.	.	.	.	.	I
Kernera saxatilis	.	.	.	.	+	.	.	.	.	.	I
Companion											
Lotus corniculatus	+	+	.	.	.	.	.	.	.	.	I
Luzula multiflora	.	+	+	.	.	.	.	.	.	.	I
Sedum atratum	.	+	.	+	.	.	.	.	.	.	I
Helianthemum nummularium subsp. *tomentosum*	.	+	.	.	+	.	.	.	.	.	I
Hippocrepis comosa	.	+	.	.	+	.	.	.	.	.	I
Potentilla thuringiaca	.	+	.	.	+	.	.	.	.	.	I
Phyteuma confusum	.	+	.	.	.	.	+	.	.	.	I
Botrychium lunaria	.	+	.	.	.	.	.	+	.	.	I
Viola alpina	.	+	.	.	.	.	.	.	+	.	I
Seseli libanotis	.	.	.	+	+	.	.	.	.	.	I
Gymnadenia conopsea	.	.	.	.	+	.	.	.	+	.	I

Companion species with one occurrence: *Saxifraga adscendens* 2: +; *Trifolium pallescens* 5: *Galium album* s.l. 2: +; *Saxifraga androsacea* 1: +; *Trifolium alpestre* 2: +; *Campanula alpina* 3: +; *Pedicularis comosa* 5: +; *Avenula versicolor* 5: +; *Soldanella major* 8: +; *Rhododendron myrtifolium* 8: +; *Selaginella selaginoides* 8: +; *Saxifraga aizoides* 8: +; *Vaccinium gaultherioides* 8: 1.3; *Pedicularis verticillata* 9: +; *Minuartia verna* 5: +; *Euphrasia salisburgensis* 5: +; *Thesium alpinum* 3: +; *Ditrichum flexicaule* 4: +; *Bryum caespiticeum* 4: +

Relevé: 1. Curmătura Soarbele-Stănuleți (23.06.1971); 2–3. Mt. Stănuleți (23.06.1971); 4. Mt. Piatra Iorgovanului-Mt. Albele (24.06.1971); 5–7. Mt. Piatra Iorgovanului (26.06.1971); 8. Mt. Piule (6.07.1972); 9. Mt. Albele (24.06.1971); 10. Vf. Piule (25.06.1971)

Photo 6.3 *Edraianthus graminifolium* (L.) A.DC. Willd. (photo M. Ciobanu)

Photo 6.4 *Centaurea triumfetti* All. (photo M. Ciobanu)

We include here the phytocoenoses dominated by the alpine dwarf shrub species *Dryas octopetala* and *Salix retusa* subsp. *kitaibeliana* from the alpine and subalpine belts of the calcareous Retezat Mountains (Piule, Custura, Piatra Iorgovanului, and Albele). These heathlands usually develop on north-facing, steep slopes and

Table 6.42 Ass. *Poo molinerii-Festucetum pachyphyllae* Boşcaiu et al. 1978

Relevé No.	1	2	3	4	5	6	7	8	
Altitude (m a.s.l.)	1930	1910	1900	1850	1845	1800	1830	1750	
Aspect	SE	S	–	E	SW	S	S	S	
Slope (degrees)	30	35	0	15	10	45	35	45	
Herb cover (%)	50	60	80	90	80	60	55	70	
Sample area (sq. m)	25	25	25	25	25	25	100	100	K
Char. ass.									
Festuca stricta subsp. *rumelica*	3.5	3.5	1.5	4.5	4.5	3.5	3.5	3.5	V
Poa molinerii	1.4	2.4	4.5	2.5	1.5	2.5	1.3	2.3	V
Festuco-Seslerion bielzii									
Thymus pulcherrimus	1.4	1.4	+	+	.	+.4	1.4	1.2	V
Sesleria rigida subsp. *haynaldiana*	+	1.3	.	+	.	+	1.4	1.3	IV
Alyssum repens	.	+	+	.	+	1.4	+	+	IV
Carduus kerneri subsp. *lobulatiformis*	+	+	.	.	.	.	+	+	III
Minuartia setacea subsp. *banatica*	.	.	+	.	.	+	.	.	II
Cerastium lerchenfeldianum	.	.	.	.	.	+.5	.	.	I
Seslerion rigidae									
Draba lasiocarpa	+	.	.	.	+	.	.	.	II
Dianthus petraeus	.	.	.	.	+	+	.	.	II
Asperula capitata	.	.	.	.	.	+	.	+	II
Edraianthus graminifolius	.	.	.	+	.	.	.	.	I
Seslerietalia et ***Elyno-Seslerietea***									
Acinos alpinus subsp. *alpinus*	.	+	+	+	+	1.5	+	1.5	V
Myosotis alpestris	+	+	+	+	+	+.3	+	+	V
Cerastium arvense subsp. *molle*	+	1.4	1.4	+	+	.	.	+	IV
Saxifraga adscendens	+	+	.	+	+	.	+	+	IV
Biscutella laevigata	+	+	.	+	.	.	+	+	IV
Ranunculus oreophilus	.	.	+	+.5	+.5	.	+	1.2	IV
Saxifraga moschata	+	.	+	.	+	.	+	.	III
Helianthemum alpestre	.	.	.	1.5	1.3	+	.	1.5	III
Polygala alpestris	+	+	.	.	.	.	+	+	III
Trifolium pallescens	+	.	+	.	+	.	.	.	II
Galium anisophyllum	.	.	+	+	+	.	.	.	II
Carex sempervirens	.	.	.	+.3	+	.	.	1.2	II
Scabiosa lucida	.	.	+	.	.	.	.	+	II
Polygonum viviparum	.	.	.	+	.	.	.	+	II
Veronica aphylla	.	.	+	.	.	.	.	.	I
Dryas octopetala	.	.	.	+	.	.	.	.	I

(continued)

Table 6.42 (continued)

Oxytropis pyrenaica	.	.	.	+	.	.	.	.	I
Potentilla crantzii	.	.	.	.	.	+	.	.	I
Companion									
Sedum atratum	+	+	+	.	.	.	+	.	III
Taraxacum panalpinum	.	+	+	+	.	.	.	+	III
Silene nutans subsp. *dubia*	+	+	.	.	.	.	+	.	II
Senecio rupestris	.	+	.	.	.	+	+	.	II
Saxifraga paniculata	+	.	.	.	.	.	.	.	I
Botrychium lunaria	+.2	.	.	.	.	.	.	.	I
Lotus corniculatus	+	.	.	.	.	.	.	.	I

Relevé: 1–2. Mt. Stănulete (23.06.1971); 3. Mt. Piatra Iorgovanului (23.06.1971); 4–5. Mt. Albele (24.06.1971); 6–8. Mt. Piule (25.06.1971)

windswept crests. The rendzinic leptosol, also characteristic of these plant communities, has a patchy distribution and is usually developed superficially. The dominant species attain an average cover of 50%, while among the species with a relatively high cover (10–25%) we mention *Carex sempervirens, Carex rupestris, Sesleria rigida* subsp. *haynaldiana, Polygonum viviparum, Ranunculus oreophilus,* and *Bartsia alpina*. Other taxa with arctic-alpine distribution, typical of the xero-cryophilous communities, can be sporadically encountered here as well: *Silene acaulis, Carex atrata, Oxytropis pyrenaica,* and *Oxytropis carpatica*. These species confer these phytocoenoses a closer resemblance to those of the alliance *Oxytropido-Elynion* Br.-Bl. 1949 (Šibik et al. 2007a). However, the absence of *Elyna myosurioides* and the predominance of diagnostic species of the order *Seslerietalia* and the class *Elyno-Seslerietea* have determined us to assign these communities to a new association from the Retezat Mountains. We finally highlight the partial floristic resemblance with the ass. *Dryado octopetalae-Caricetum firmae* Sillinger 1933 described in the Western Carpathians (Tatra) (Šibik et al. 2007a) (Photos 6.5 and 6.6).

Alliance *Seslerion rigidae* Zólyomi 1939

The calcareous grasslands developed on rocks from the South-Eastern Carpathians are included in this alliance, which is characterised by the presence of superficial rendzinic lithosols, with small volume and a neutral to moderately basic pH reaction. The characteristic species of the alliance are *Sesleria rigida, Athamanta turbith* subsp. *hungarica, Thymus comosus, Asperula capitata, Dianthus petraeus, Aconitum anthora, Bromus riparius,* and *Linum uninerve* (Boşcaiu 1971; Nechita 2003).

Ass. *Festucetum xanthinae* Boşcaiu 1971 (Table 6.44)

We included here the phytocoenoses dominated by *Festuca xanthina* found in the southern part of calcareous Retezat Mountains (Scorota and Jiul de Vest valleys), where they occur on abrupt cliffs and at the base of rocky slopes, with superficial

Table 6.43 Ass. *Salici kitaibelianae-Dryadetum octopetalae* ass. nova hoc loco

Relevé No.	1	2	3	4	5	6	7	8	9	10	
Altitude (m a.s.l.)	1850	1860	1800	1830	1890	1950	1900	1900	2150	2100	
Aspect	N	NE	NE	NE	NE	N	N	NW	NW	N	
Slope (degrees)	55	25	45	50	30	10	40	45	50	40	
Herb cover (%)	65	55	80	75	75	75	65	80	75	60	
Sample area (sq. m)	25	25	25	25	25	25	25	25	25	25	K
Char. ass.											
Salix kitaibeliana	+	2.5	1.4	1.3	1.3	3.5	3.5	3.5	3.5	3.5	V
Dryas octopetala	3.5	3.5	3.5	3.5	3.5	3.5	3.4	3.5	3.5	1.2	V
Festuco-Seslerion bielzii											
Sesleria rigida subsp. *haynaldiana*	3.5	+	3.5	2.5	3.5	1.3	+	.	.	.	IV
Oxytropis carpatica	+	+	+	+.3	+	+	+	.	.	.	IV
Androsace villosa subsp. *arachnoidea*	+	+	+	+	+	.	.	.	.	.	III
Carex capillaris	.	.	.	+	.	+	+	.	.	.	II
Sesleria bielzii	.	.	.	.	.	.	.	+	+	+	II
Onobrychis montana subsp. *transsilvanica*	.	.	.	+	+	.	.	.	.	.	I
Oxytropis pyrenaica	.	.	.	.	.	+	+	.	.	.	I
Carex atrata	.	.	.	.	.	.	.	.	+	+	I
Seslerietalia et ***Elyno-Seslerietea***											
Carex sempervirens	1.3	2.5	1.5	1.3	1.5	1.5	1.5	+	+	.	V
Polygonum viviparum	+	1.3	+	+.3	+.4	1.5	+.3	+	+	+.5	V
Ranunculus oreophilus	.	+	+	1.3	+.4	+	+.3	+	+	.	V
Bartsia alpina	.	+.3	+	.	.	+	+	+	+	+.3	IV
Saxifraga marginata	.	+	2.5	2.5	1.4	.	+	.	.	.	III
Myosotis alpestris	.	.	+	+	.	.	+	+	+	+	III
Biscutella laevigata	+	+	+	+	.	+	+	.	.	.	III

(continued)

Table 6.43 (continued)

Veronica aphylla	.	+	.	.	+	.	.	+	+	+	III
Edraianthus graminifolius	+	.	+.3	1.5	+	.	+	.	.	.	III
Helianthemum alpestre	.	.	.	+.3	+.3	+	+	1.3	.	.	III
Hedysarum hedysaroides	.	.	.	.	+	.	+	.	+.3	.	II
Pedicularis verticillata	.	.	.	+	.	.	.	+	+	+	II
Carex rupestris	.	+	.	.	+	.	.	+	.	.	II
Gentiana verna	.	+	.	.	.	+	+	+.3	.	.	II
Eritrichium nanum subsp. *janke*	+	.	.	+	+	.	.	.	.	.	II
Alyssum repens	.	+	.	.	.	+	+	.	.	.	II
Cerastium alpinum subsp. *lanatum*	+	.	.	.	.	.	.	.	+	.	I
Silene acaulis	.	.	.	.	.	.	.	+	1.3	.	I
Gentiana nivalis	.	+	.	.	.	.	+	.	.	.	I
Dianthus petraeus	.	.	.	+	.	.	+	.	.	.	I
Carex rupestirs	+	.	.	.	.	+	.	.	.	.	I
Draba lasiocarpa	.	.	.	.	+	.	+	.	.	.	I
Aster alpinus	.	.	.	.	.	+	+	.	.	.	I
Artemisia eriantha	.	.	.	.	.	.	.	+.3	+.3	.	I
Thymus pulcherrimus	.	.	.	.	+	.	.	.	.	.	I
Anemone narcissiflora	.	.	.	.	.	.	.	.	.	+	I
Companion											
Saxifraga paniculata	+	+	+	+	+	.	+	+	+	.	IV
Viola alpina	+	.	+	.	+	+	.	.	.	.	II
Soldanella major	.	.	.	.	+	.	+	.	+	+	II
Salix alpina	+	.	.	.	+	.	.	+	.	.	II
Vaccinium vitis-idaea	.	+	.	.	.	+	.	.	+	.	II
Huperzia selago	.	.	.	+	.	.	.	+	+	.	II
Saxifraga aizoides	.	.	.	.	1.3	+	+	.	.	.	II

Rhododendron myrtifolium	.	.	.	.	.	.	.	1.4	1.5	2.5	II
Oreochioa disticha	.	.	.	.	.	.	.	1.3	1.4	+	II
Campanula alpina	.	.	.	.	.	.	.	+	+	+	II
Homogyne alpina	.	.	.	.	.	.	.	+	+	+	II
Coeloglossum viride	+	.	.	.	.	.	.	.	.	+	I
Selaginella selaginoides	.	+	.	.	+	.	.	.	.	.	I
Primula minima	.	.	.	.	.	.	.	1.3	+	.	I
Vaccinium gaultherioides	.	.	.	.	.	.	.	1.3	.	1.5	I

Companion species with one ocurrence: *Galium anisophyllum* 2: +; *Asplenium viride* 2: +; *Gymnadenia conopsea* 3: +; *Polygala alpestris* 3: +; *Poa molinerii* 4: +; *Geum reptans* 8: +; *Doronicum columnae* 9: +; *Anthoxanthum alpinum* 10: +; *Hieracium alpinum* 10: +; *Ligusticum mutellina* 10: +; *Pulsatilla alba* 10: +; *Avenula versicolor* 10: +; *Gentiana punctata* 10: +; *Ditrichum flexicaule* 2: +; *Tortella tortuosa* 2: +; *Mnium orthorhynchum* 2: +; *Tritomaria quinquedentata* 2: +; *Scapania aequilobata* 2: +

Relevé: 1–4. Mt. Albele (24.06.1971); 5–6. Mt. Piatra Iorgovanului (25.06.1971); 7. Mt. Piule (25.06.1971); 8–10. Mt. Custura (15.07.1970)

Photo 6.5 *Salix kitaibeliana* Willd. (photo M. Ciobanu)

rendzinic soils and a low edaphic volume. Together with the dominant species *Festuca xanthina*, which attains an average cover of 35%, we can find other saxicolous species with Carpathian-Balkan distribution that confer on this association a regional character: *Atamanta turbith* subsp. *hungarica, Dianthus petraeus, Asperula capitata, Thymus comosus*, and *Silene saxifraga*. We mention that a closely-resembling association was described in eastern Serbia, namely *Festuceum xanthino-variae* Jovanović-Dunjić 1955, included in the alliance *Seslerio-Festucion xanthinae* Horvat 1974.

Class *Mulgedio-Aconitetea* Hadač et Klika in Klika 1948

The rich hydrographic network of the Retezat Mountains and particularly the permanent lakes (over 50) (Pişota 1971) in the subalpine and alpine belt, favored the formation of numerous valleys and streams on both northern and southern slopes. The abundant rainfall in the summer season and the water flows resulting from spring snow thawing enriched the riverbed of subalpine streams with alluvia and favored growth and development of lush vegetation, with a distinct physiognomic appearance and high nitrophilic requirements. Such phytocoenoses make up the type of vegetation called "Altherbosa" (Hochstandenflur), present in all European

Photo 6.6 *Dryas octopetala* (photo M. Ciobanu)

mountain ranges. The Central European orophytes predominate in the floristic composition of these phytocoenoses, indicating their alpine origin. Many Carpathian endemic species such as *Heracleum palmatum, Phyteuma vagneri, Leucanthemum rotundifolium, Poa granitica* subsp. *disparilis, Doronicum carpaticum,* and rarely *Luzula alpinopilosa* subsp. o*bscura,* and *Rhododendron myrtifolium* occur in these phytocoenoses. Among the main characteristic species of the class we mention: *Aconitum napellus* subsp. *tauricum, Geranium sylvaticum, Rumex alpestris, Senecio subalpinus, Poa chaixi, Viola biflora*, and *Veratrum album* subsp. *lobelianum.* We grouped this type of phytocoenoses from Retezat in three distinct orders, namely: *Calamagrostietalia villosae, Adenostyletalia alliariae,* and *Petasito-Chaerophylletalia.*

Order *Calamagrostietalia villosae* Pawl. et al. 1928

In this order we group tall grass phytocoenoses that usually vegetate on the slopes of some glacial cirques, close to permanent lakes (e.g. Negru, Gemenele, Zănoaga, Slăvei, Spurcat, Judele). The structure of these coenoses is dominated by acidophilic and mesophilic species of *Poaceae*. The main characteristic species of the order are: *Crepis conyzifolia, Gentiana punctata, Luzula luzuloides* subsp. *cuprina, Solidago virgaurea* subsp. *minuta,* and *Campanula serrata* (Koči 2007b).

Alliance *Calamgrostion villosae* Pawlowski et al. 1928

The tall grass coenoses on siliceous substrates, dominated by *Calamagrostis villosa* and *Festuca picturata,* from the subalpine and alpine belts of the Retezat are included in this alliance. They vegetate on acidic, superficial, skeleton-rich lithosols,

Table 6.44 Ass. *Festucetum xanthinae* Boşcaiu 1971

Relevé No.	1	2	3	4	5	
Altitude (m a.s.l.)	1400	1430	1180	1170	1160	
Aspect	E	E	S	S	S	
Slope (degrees)	30	30	60	60	40	
Herb cover (%)	80	80	60	40	40	
Sample area (sq. m)	25	25	25	16	16	K
Char. ass.						
Festuca xanthina	3.5	3.5	3.5	3.5	3.5	V
Seslerion rigidae						
Dianthus petraeus	1.3	+.3	.	1.3	1.3	IV
Poa nemoralis	3.5	3.5	3.5	.	.	III
Seseli libanotis	.	+	.	+	+	III
Sesleria rigida	+	1.3	.	+	.	III
Silene nutans subsp. *dubia*	+.3	+	.	.	+	III
Athamantha turbith subsp. *hungarica*	+	.	+.4	.	+	III
Asperula capitata	+	+	.	+	.	III
Primula veris subsp. *columnae*	+.4	+.5	.	+	.	III
Aconitum anthora	+	+	+	.	.	III
Linum uninerve	+	+	.	.	.	II
Festuca stricta subsp. *rumelica*	1.5	+.3	.	.	.	II
Bromus riparius	2.5	2.4	.	.	.	II
Silene saxifraga	.	.	.	+	+	II
Thymus comosus	+	.	+	.	.	II
Festuco-Seslerion bielzii						
Acinos alpinus subsp. *alpinus*	+.3	+	.	.	.	II
Alyssum repens	+	+	.	.	.	II
Onobrychis montana subsp. *transsilvanica*	+.3	+	.	.	.	II
Dianthus tenuifolius	+.3	+	.	.	.	II
Erysimum witmannii subsp. *transsilvanicum*	+	.	.	.	.	I
Seslerietalia						
Helianthemum nummularium subsp. *tomentosum*	+	+	.	.	.	II
Cerastium arvense subsp. *molle*	+.3	.	.	.	.	I
Kernera saxatilis	.	+	.	.	.	I
Minuartia setacea subsp. *banatica*	.	.	+	.	.	I
Companion						
Origanum vulgare	+.4	+	+	.	.	III
Galium album s.l.	+	.	+	.	.	II
Sedum hispanicum	+	.	+	.	.	II
Linum catharticum	+	+	.	.	.	II
Asplenium trichomanes	.	.	+	.	+	II
Aspleium ruta-muraria	.	.	+	.	+	II

(continued)

Table 6.44 (continued)

Saxifraga paniculata	.	.	+	.	+	II
Campanula rapunculoides	.	.	.	+	+	II

Companion species with one or two occurrences: Galium album 3: +; *Pimpinella saxifraga* 1: +; *Campanula persicifolia* 1: +; *Cardaminopsis arenosa* 1, 5: +; *Linum catharticum* 1, 2: +; *Parnassia palustris* 1: +; 5: +; *Campanula glomerata* 3: +; *Digitalis grandiflora* 3, 5: +; *Cystopteris fragilis* 3, 5: +; *Doronicum columnae* 3: +

Relevé: 1–2. V. Scorotei (8.07.1972); 3–5. V. Jiului-Câmpușel (8.07.1972)

with variable moisture during the growing season. The characteristic species of the alliance according to Jarolimek and Šibik (2008) are: *Calamagrostis villosa, Festuca picturata, Gnaphalium norvegicum, Hieracium alpinum,* and *Hypochaeris uniflora.*

Ass. *Hyperico grisebachii-Calamagrostetum villosae* Pawl. et Walas 1949 (Table 6.45)
Syn.: *Calamagrostidetum villosae retezatense* Borza 1934 (Art. 34)

In this association we group the coenoses dominated by *Calamagrostis villosa* vegetating on the sunny slopes of some glacial cirques from Retezat (Judele, Gemenele, Zănoaga, Slăveiul) with superficial lithosols, rich in semi-fixed gravel and with lower moisture in summer. The species *Calamagrostis villosa* achieves on average 60% coverage in these herbaceous coenoses. Along with this species there is present, in some places, the regional Carpathian-Balkan species *Hypericum richeri* subsp. *grisebachii,* designated as characteristic of association. The Southeastern Carpathian specificity of this syntaxon is also strengthened by the presence of several other regional species such as: *Phyteuma vagneri, Campanula transsilvanica, Potentilla ternata,* and *Campanula abietina,* which occur sporadically in these phytocoenoses (Borza 1934) (Photos 6.7 and 6.8).

Ass. *Festuco picturatae-Calamagrostetum villosae* Pawł. et al. 1928 corr. Kliment et al. 2004 (Table 6.46)
Syn.: *Festucetum pictae* Borza 1934 (Art. 31, 32)

The coenoses dominated by *Festuca picturata* in the glacial cirques from Retezat (Zănoaga, Gemenele, Judele), vegetate on slopes that are shadier and milder, and where the snow lasts longer than in the case of the *Hyperico-Calamagrostetum* association. The soils in these locations are poorly evolved lithosols, with a transition horizon between A horizon of 5–8 cm and R horizon, made up of rock fragments. In the composition of the association are present the characteristic species of alliance and order and also those for the class *Mulgedio-Aconitetea*, similar to those described in the Tatra Mountains (Hegedüsová Vantarová 2014; Koči 2007b). The specific microclimatic conditions in the locations from Retezat, where the *Festuca picturata* coenoses vegetate, also facilitate the occurrence of some other chionophilic species in their composition (*Soldanella pusilla, Veronica alpina, Sedum alpestre, Gnaphalium supinum*), giving the association distinct ecological particularities compared to coenoses of the association *Hyperico-Calamagrostetum* (Borza 1934).

Table 6.45 Ass. *Hyperico grisebachii-Calamagrostietum villosae* Pawł. et Walas 1949

Relevé No.	1	2	3	4	5	6	7	8	9	
Altitude (m a.s.l.)	1950	1900	1900	1930	2000	1970	2020	1900	1980	
Aspect	E	NE	–	E	NE	E	NE	S	S	
Slope (degrees)	35	20	0	45	10	30	45	15	45	
Herb cover (%)	95	90	90	100	100	100	90	80	90	
Sample area (sq. m)	25	25	25	25	25	25	25	25	25	K
Char. ass.										
Calamagrostis villosa	5.5	4.5	4.5	4.5	3.5	3.5	4.5	4.5	4.5	V
Hypericum richerii subsp. *grisebachii*	1.1	+	.	+	.	.	+	.	+	III
Calamagrostion* et *Calamagrostidetalia										
Gentiana punctata	+	+.3	+	+	.	+	+	.	.	IV
Festuca picturata	+	.	+	+.4	2.5	1.4	1.3	.	.	IV
Solidago virgaurea subsp. *minuta*	+	+	.	+	+	.	+	.	.	III
Laserpitium crapfii	+	+	+	.	.	.	.	.	.	II
Gnaphalium norvegicum	+	.	.	.	.	+	.	.	.	II
Campanula transsilvanica	.	.	.	+	.	+	.	.	.	II
Hypochaeris uniflora	+	+	.	.	.	.	.	.	.	II
Crepis conyzifolia	.	.	.	.	.	.	+	.	.	I
Phyteuma vagneri	+	.	.	.	+	.	.	.	.	I
Mulgedio-Aconitetea										
Adenostyles alliariae	.	+.3	1.2	1.3	+	+	+	1.3	.	IV
Veratrum album subsp. *lobelianum*	+	+.3	+	+.4	+	+.4	+	.	.	IV
Silene vulgaris	+	+	.	+.3	+	+	+	.	.	IV
Ranunculus platanifolius	.	+	.	+.2	.	+.3	+	.	.	III
Rumex alpestris	+	.	.	.	+	+	+	.	.	II
Senecio nemorensis	+	.	.	+.3	.	+	.	.	.	II
Aconitum napellus subsp. *tauricum*	.	.	+.3	.	+	.	.	.	.	II
Geranium sylvaticum	+	.	+	.	.	+.2	.	.	.	II
Achillea distans	.	.	.	.	.	.	.	+	+	II
Hypericum maculatum	+	.	.	.	.	+.4	.	.	.	II
Milium effusum	.	+	.	+	.	.	.	.	.	II
Doronicum austriacum	.	+	.	.	.	.	.	+	.	II
	.	.	.	+	.	+	.	.	.	II

(continued)

Table 6.45 (continued)

Deschampsia caespitosa										
Angelica archangelica	.	.	.	.	+	.	.	.	.	I
Companion										
Luzula alpinopilosa subsp. *obscura*	+	+	+.3	.	+	+	+	+	1.5	V
Ligusticum mutellina	+	.	1.5	.	+	+	+	.	+	IV
Vaccinium myrtillus	+	.	.	+	.	+	2.4	+	1.3	IV
Geum montanum	+	.	.	.	+	+	+	.	+	III
Homogyne alpina	+	+	.	.	.	+	+	+.5	.	III
Potentilla ternata	+	.	.	.	+	+	+	.	+	III
Rhododendron myrtifolium	+	+	.	.	.	+	1.3	.	.	III
Campanula abietina	.	.	.	+	+	.	+	.	1.5	III
Pulsatilla alba	+	.	.	.	.	.	+	.	+	II
Stellaria nemorum	.	.	+.4	+	+	.	.	.	.	II
Centaurea nervosa	+	.	.	.	.	+	.	.	+	II
Pseudorchis albida	.	.	.	.	.	+	.	+	+	II
Deschampsia flexuosa	.	.	.	.	.	.	+	1.3	1.5	II

Companion species with one or two occurrences: *Anthoxanthum alpinum* 3, 6: +; *Doronicum carpaticum* 3: +; *Saxifraga rotundifolia* subsp. *heucherifolia* 3: +; *Cardaminopsis halleri* subsp. *ovirensis* 3: +; *Senecio doronicum* subsp. *transylvanicus* 4: +, 7: +.3; *Myosotis sylvatica* 5: +; *Carex atrata* 6: +; *Pedicularis verticillata* 6: +; *Melampyrum sylvaticum* 6: +; *Soldanella major* 6: +; *Plantago gentianoides* 6: +; *Senecio subalpinus* 6: +; *Poa chaixii* 7: +; *Senecio fuchsii* 7: +, 8: 1.3; *Valeriana tripteris* 8: +; *Chamaerion angustifolium* 8, 9: +; *Scorzonera rosea* 8, 9: +; *Agrostis rupestris* 9: +; *Antennaria dioica* 9: +; *Vaccinium vitis-idaea* 9: +; *Gymnadenia conopsea* 9: +; *Crocus vernus* 9: +

Relevé: 1. Tăul Negru-Vf. Judele (5.07.1969); 2. Tăul Negru-Şesele (5.07.1969); 3. Tăul Şteviei (6.07.1969); 4. Gemenele (15.07.1967); 5. Căldarea Şeselor (14.07.1967); 6. Vf. Judele

(continued)

Order *Adenostyletalia alliariae* Br.-Bl. 1930

Natura 2000: habitat type 6430

The tall grass coenoses grouped in this order frequently vegetate at the edges of streams and valleys in the subalpine and alpine belts of the Retezat, on moist alluvial soils, rich in nutrients and gravel brought by floods. Such phytocoenoses are also found near glacial lakes, especially in lower areas that are flooded in spring. Different plant species participate in building up these phytocoenoses, assembling an impressive multicolored image during the flowering period. Among the characteristic species designated for this order we mention: *Rumex alpestris, Epilobium alpestre, Saxifraga rotundifolia,* and *Tozzia alpina* (Oberdorfer 2001). The particular ecological conditions of this order and its core of characteristic species, which are different from those of the order *Calamagrostietalia villosae*, determined us to keep

Photo 6.7 *Campanula transsilvanica* (photo B. Hurdu)

it as a separate syntaxon, similar to Kliment et al. 2007 (Hegedüsová Vantarová 2014).

Alliance *Adenostylion alliariae* Br.-Bl. 1926

We include in this alliance all species-rich hygrophilic communities of tall forbs populating the streams in the subalpine belt of the Retezat, with alluvial, shallow soils, rich in minerals and, sporadically, rich in gravel. The characteristic species designated for this alliance are: *Adenostyles alliariae, Doronicum austriacum, Athyrium distentifolium, Ranunculus platanifolius,* and *Heracleum palmatum* (Borza 1934; Hegedüsová Vantarová 2014).

Ass. *Aconitetum taurici* Borza ex. Coldea 2017 (Table 6.47)
Syn.: *Aconitetum taurici retezatense* Borza 1934 (Art. 34)

The *Aconitum napellus* subsp. *tauricum* coenoses are frequently neighbouring subalpine springs and streams of the Crystalline Retezat, populating their narrow

Photo 6.8 *Phyteuma wagneri* A. Kern. (photo M. Ciobanu)

banks, with alluvial soils, permanently wet and sporadically gleyed to a small extent. The species *Aconitum napellus* subsp. *tauricum* and *Saxifraga rotundifolia* subsp. *heucherifolia*, which we consider characteristic of the association, achieve on average 30% coverage. Due to their phytogeographical specificity, these species strengthen the regional character of this association and position it as a synvicariant syntaxon of *Aconitetum firmi* Pawł. et al. 1928 described in the Tatra Mountains (Borza 1934). Since both characteristic species of *Adenostylion* and *Adenostyletalia* are present in the structure of this association, we include it in these syntaxa and not in the order *Calamagrostietalia villosae* (Hegedüsová Vantarová 2014). In some places, at higher altitudes, the coenoses of this association come in contact with the fontinal coenoses of the class *Montio-Cardaminetea*.

Ass. *Adenostylo-Doronicetum* Horvat 1956 (Table 6.48)

The *Adenostyles alliariae* var. *kerneri* coenoses grouped in this Balkan-Carpathian association are widespread in the glacial cirques, where they populate scree and boulder aggregates continuously watered by subalpine streams (Gemenele, Bucura, Zănoaga). In the structure of the association, along the Central-European alpine elements, there are also present some regional Carpathian species such as:

Table 6.46 Ass. *Festuco picturatae-Calamagrostietum villosae* Pawł. et al. 1928 corr. Kliment et al. 2004

Relevé No.	1	2	3	4	5	6	7	
Altitude (m a.s.l.)	2050	2050	2020	1940	1930	1950	1920	
Aspect	NW	N	N	SE	E	NE	W	
Slope (degrees)	40	60	5	20	30	10	5	
Herb cover (%)	90	95	85	65	65	70	85	
Sample area (sq. m)	15	15	20	25	25	25	25	K
Char. ass.								
Festuca picturata	3.4	5.5	4.4	3.5	2.4	3.5	4.5	V
Gentiana punctata	+.2	+	+	.	+	+	1.3	V
Ligusticum mutellina	1.1	2.2	1.2	+	+.5	1.1	2.5	V
Phyteuma vagneri	.	1.2	.	.	+	.	.	II
Calamagrostion villosae et Calamagrostietalia								
Geum montanum	1.2	1.1	.	.	+	+	1.3	IV
Calamagrostis villosa	2.3	.	.	2.5	3.5	3.5	+	IV
Centaurea nervosa	.	.	+	+	1.3	+	+	IV
Solidago virgaurea subsp. *minuta*	.	.	.	+	+	+	.	III
Hieracium alpinum	+	.	+	.	.	.	.	II
Deschampsia flexuosa	1.2	.	.	.	.	.	.	I
Hypochaeris uniflora	+	.	.	.	.	.	.	I
Luzula luzuloides subsp. *cuprina*	.	.	.	2.5	.	.	.	I
Campanula transsilvanica	.	.	.	.	+	.	.	I
Mulgedio-Aconitetea								
Veratrum album subsp. *lobelianum*	1.1	1.2	.	+	+	+	+	V
Adenostyles alliariae	+.2	+	.	.	+	1.2	.	III
Rumex alpestris	+	.	.	+	+	+	.	III
Deschampsia caespitosa	.	.	+	.	1.2	1.2	+	III
Geranium sylvaticum	.	.	.	+	+	+	+	III
Luzula alpinopilosa subsp. *obscura*	.	2.2	.	.	+	+	+	III
Doronicum austriacum	.	.	.	+	.	+	.	II
Angelica archangelica	.	.	.	+	.	+	.	II
Ranunculus platanifolius	.	.	.	.	+	+	.	II
Allium victorialis	.	.	.	+	.	.	.	I
Stelllaria nemorum	.	.	.	.	.	+	.	I
Aconitum napellus subsp. *tauricum*	.	.	.	.	.	+	.	I
Poa chaixii	.	.	.	.	.	+	.	I
Caricion curvulae								
Potentilla ternata	1.2	.	2.2	+.3	+	+	+	V
Poa media	2.3	.	2.2	+	+	.	+	IV
Carex atrata	+	.	.	.	+	+	.	III
Avenula versicolor	.	.	2.3	+	.	.	+	III
Campanula alpina	.	.	1.1	+	.	.	+	III
Phyteuma confusum	+	.	+	.	.	.	.	II

(continued)

Table 6.46 (continued)

Agrostis rupestris	2.3	.	.	.	.	.	.	I
Festuca supina	1.2	.	.	.	.	.	.	I
Companion								
Homogyne alpina	2.2	1.2	1	+.2	+	.	+.3	V
Anthoxanthum alpinum	+	+	.	+.2	+.3	+	+	V
Campanula abietina	1.2	+	.	+	.	+	.	III
Pedicularis verticillata	.	.	1.2	+	+	.	+	III
Vaccinium myrtillus	+	.	.	+	.	+	.	III
Luzula sudetica	.	.	+	+	.	.	+.2	III
Soldanella major	.	.	.	+	+	+	.	III
Silene vulgaris	+	.	.	+	.	.	.	II
Campanula serrata	+	.	.	.	+	.	.	II
Taraxacum nigrescens	+	.	.	.	.	+	.	II
Veronica alpina	.	+	.	.	.	+	.	II
Soldanella pusilla	.	1.1	.	.	.	.	+	II
Pseudorchis albida	.	.	+	.	.	.	+	II
Nardus stricta	.	.	2.2	.	.	.	+	II
Gentiana acaulis	+	.	.	.	.	.	+	II
Sedum alpestre	.	+	.	.	.	.	+	II
Gnaphallum supinum	+	.	.	.	.	.	.	I
Hypericum maculatum	.	.	.	.	+	.	.	I
Senecio subalpinus	.	.	.	.	+	.	.	I
Phleum alpinum	.	.	.	.	+	.	.	I

Relevé: 1. Zănoaga; 2. Tăul Negru; 3. Şesele; 4. Căldarea Judele (1967); 5. Căldarea Gemenea (1967); 6. Căldarea Şesele (1967); 7. Tăul Ştirbului (1969)

Heracleum palmatum, Pulmonaria rubra, Leucanthemum rotundifolium, Aconitum napellus subsp. *tauricum, Doronicum carpaticum,* which floristically differentiates it from the *Adenostyles alliariae* coenoses of the Tatra Mountains (Kliment 2007; Matuszkiewicz 2008). Based on these geographic differential species, Borza (1959) and then Boşcaiu (1971) described the sub-association "*dacicum*" as specific to the Southern Romanian Carpathians.

Ass. *Carduo personatae-Heracleetum palmati* Beldie 1967 (Table 6.49)

The hygrophilic and sciophilous coenoses of this endemic association of the South-eastern Carpathians are frequently distributed in Retezat along torrents and streams with wet boulders from the subalpine belt (Ştirbului, Judele, Bucura and Valea Rea, Galeşu and Radeşu), on skeletal soils, rich in organic matter. The characteristic species of this association, *Heracleum palmatum* and *Carduus personata*, achieve on average 35% coverage. In addition to these, some hygrophilic species characteristic of the tall-herb communities of the subalpine belt are also present with a significant coverage, such as: *Stellaria nemorum, Adenostyles alliariae, Rumex alpinum, Aconitum napellus* subp. *tauricum, Deschampsia caespitosa,* and *Angelica archangelica.* The conservative character of these hygrophilic phytocoenoses is

Table 6.47 Ass. *Aconitetum taurici* Borza 1934

Relevé No.	1	2	3	4	5	6	7	
Altitude (m a.s.l.)	1980	1950	2010	2010	1820	1050	2100	
Aspect	SW	SW	SW	S	S	SW	SW	
Slope (degrees)	10	10	20	40	30	30	15	
Herb cover (%)	65	65	50	60	30	40	60	
Sample area (sq. m)	25	25	25	25	25	25	100	K
Char. ass.								
Aconitum napellus subsp. *tauricum*	3.5	3.5	3.5	2.3	2.5	2.5	2.4	V
Doronicum carpaticum	+	+.5	+	.	.	.	.	III
Adenostylion et ***Adenostyletalia***								
Veratrum album subsp. *lobelianum*	+	+	+	+	+	1.3	+	V
Saxifraga rotundifolia subsp. *heucherifolia*	+	+	+	.	1.4	+.3	1.2	V
Stellaria nemorum	1.5	1.5	+.3	.	.	.	2.5	III
Rumex alpestris	.	.	+	.	+	+	+	III
Adenostyles alliariae	.	+	+	.	.	.	1.3	III
Chaerophyllum hirsutum	.	.	+	2.4	.	.	+	III
Rumex alpinus	.	.	.	1.3	+	.	+	III
Mulgedio-Aconitetea								
Deschampsia caespitosa	+	1.3	.	1.3	+	.	2.5	IV
Calamagrostis villosa	3.5	3.5	.	.	.	.	+	III
Doronicum austriacum	.	.	+.3	+	.	.	+	III
Athyrium distentifolium	.	.	+	.	+	.	+	III
Viola biflora	.	.	.	.	+.4	2.4	1.2	III
Angelica archangelica	.	.	2.5	1.4	.	.	.	II
Senecio subalpinus	.	.	.	+	.	.	+	II
Montio-Cardaminetea (incl. ***Montio-Cardaminetalia***)								
Caltha palustris	.	+	+	+	.	.	+	III
Saxifraga stellaris	.	.	+	+	1.3	+	.	III
Chrysosplenium alternifolium	+	.	+	.	.	.	.	II
Philonotis seriata	.	.	.	.	.	1.3	2.3	II
Companion								
Luzula alpinopilosa subsp. *obscura*	+	+	1.4	2.5	+	1.3	+	V
Ligusticum mutellina	+	+	+	.	.	.	+	III
Rhodiola rosea	+	+	.	2.4	.	.	+	III

Companion species with one or two occurrences: *Festuca picturata* 1, 3: +; *Cardaminopsis halleri* subsp. *ovirensis* 1, 3: +; *Gentiana punctata* 1, 3: +; *Epilobium angustifolium* 4: +; *Senecio fuchsii* 4: +; *Taraxacum panalpinum* 6: +; *Leucanthemum rotundifolium* 7: +; *Carduus personata* 7: +; *Geum montanum* 7: +; *Platago gentianoides* 7: +

Relevé: 1. Tăul Şteviei (6.06.1969); 2. Tăul Caprelor (7.06.1969); 3. Tăul Negru (14.07.1967); 4. V. Judelui (13.07.1970); 5. V. Pelegii (8.08.1970); 6. Stâna Pelegii (12.07.1971); 7. Căldarea Bucurii (12.09.1970)

Table 6.48 Ass. *Adenostylo-Doronicetum austriacae* Horvat 1956

Relevé No.	1	2	3	4	5	6	7	8	9	
Altitude (m a.s.l.)	1950	1950	2060	2020	1600	1970	2010	2070	1980	
Aspect	E	NE	S	E	W	–	S	SE	NE	
Slope (degrees)	30	45	30	45	45	0	20	20	25	
Herb cover (%)	90	95	70	95	70	90	85	85	80	
Sample area (sq. m)	25	25	25	25	25	25	25	25	25	K
Char. ass.										
Adenostyles alliariae	4.5	4.5	3.5	3.5	2.5	4.5	3.5	3.5	4.5	V
Doronicum austriacum	+	+	.	+	+.4	1.3	.	+	+	IV
Doronicum carpaticum	.	.	1.3	+	.	.	.	.	+.3	II
Adenostylion **et** ***Adenostiletalia***										
Rumex alpinus	.	.	+	+	.	2.5	1.5	+	.	V
Rumex alpestris	+	+.4	+	1.5	.	+	+	2.5	+	IV
Veratrum album subsp. *lobelianum*	+	+	1.4	1.3	.	.	.	+	+	IV
Saxifraga rotundifolia subsp. *heucherifolia*	+	.	+	.	.	.	1.4	2.5	+	III
Achillea distans	.	.	.	+	.	+	.	.	.	II
Chaerophyllum hirsutum	.	.	.	.	+	.	.	+	.	II
Heracleum palmatum	.	.	.	.	.	+	.	+	.	II
Stellaria nemorum	.	.	.	.	.	+	.	+	.	II
Senecio nemorensis	.	3.5	.	+	.	.	.	.	.	II
Leucanthemum rotundifolium	.	.	.	+	.	.	.	.	.	II
Carduus personata	.	.	.	.	+	.	.	.	.	I
Tozzia alpina subsp. *carpathica*	.	.	.	.	.	.	+	.	.	I
Pulmonaria rubra	.	.	.	.	.	.	+			I
Mulgedio-Aconitetea										
Aconitum napellus subsp. *tauricum*	+.3	.	1.3	+	+	+.3	+.3	1.5	+.3	V
Deschampsia caespitosa	.	+	+	+	+	+	+	+	.	IV
Angelica archangelica	+.2	.	.	+	+	1.3	.	+	+	IV
Cicerbita alpina	+	+	.	+	.	.	.	+	+	III
Ranunculus platanifolius	+.4	.	.	1.5	+.3	+	.	.	+	III
Geranium sylvaticum	.	.	+	+	.	+	+.3	+	.	III

(continued)

Table 6.48 (continued)

Athyrium distentifolium	+.4	.	.	3.5	1.3	.	1.3	.	.	III
Senecio subalpinus	+	.	.	+	.	.	+	.	.	II
Milium effusum	+	.	.	+	.	.	.	.	.	II
Viola biflora	.	+	.	.	.	.	.	.	.	I
Companion										
Silene vulgaris	+	.	+	+	.	+	+	+	.	IV
Senecio ovatus	.	.	.	1.5	2.5	+	+	.	.	III
Gentiana punctata	+	.	.	+	.	+	.	+	+.3	III
Calamagrostis villosa	+	3.5	.	1.5	1.4	.	.	.	.	III
Geum montanum	+	+	.	.	.	.	+	.	+.3	III
Festuca picturata	+	.	1.3	1.5	.	.	.	.	+	III
Luzula alpinopilosa subsp. *obscura*	+	.	.	+	+	.	.	1.5	+	III
Myosotis sylvatica	+	+	.	.	.	.	.	.	+	II
Ligusticum mutellina	+	+	.	.	.	.	.	.	+	II
Poa alpina	+	.	+	.	.	.	.	.	+	II
Campanula abietina	.	.	+	+	.	+	.	.	.	II
Dryopteris filix-mas	.	.	+	+	.	.	+	.	.	II
Caltha palustris	.	.	+	.	.	+	.	+	.	II

Companion species with one or two occurrences: *Vaccinium myrtillus* 1: +; *Rhododendron myrtifolium* 1: +; *Pedicularis exaltata* 1: +; *Solidago virgaurea* subsp. *minuta* 1, 4: +; *Senecio doronicum* ssp. *transylvanicus* 1: +; *Hypericum maculatum* 4, 7: +; *Centaurea nervosa* 4: +; *Taraxacum nigricans* 1: +; *Rubus idaeus* 5: +; *Chrysosplenium alternifolium* 6: 1.5, 8: +; *Plantago gentianoides* 7: +; *Veronica bellidioides* 9: +

Relevé: 1. Lacul Gemenea (1.07.1967); 2. Tăul Secat (8.06.1969); 3. Lacul Bucura (7.08.1970); 4. Lacul Ana (12.08. 1970); 5. V. Pelegii (6.08.1970); 6. Tăul Spurcat (11.10.1970); 7. Zănoaga-V. Judele (12.06.1970); 8. Şesele (10.07.1970); 9. Lacul Ana (5.07.1969)

emphasized by the occurrence of some Carpathian-Balkan species, such as *Leucanthemum rotundifolium, Aconitum toxicum, Doronicum carpaticum, Senecio doronicum* subsp. *transylvanicus,* and *Campanula abietina*, similar to those described from the Ţarcu-Godeanu Mountains (Boşcaiu 1971).

Ass. *Phleo alpini-Deschampsietum caespitosae* (Krajina 1933) Coldea 1983 (Table 6.50)

Syn.: *Deschampsietum caespitosae transsilvanicum* Borza 1934 (Art. 34)

The coenoses dominated by *Deschampsia caespitosa* are frequently distributed on the margins of glacial lakes from the Retezat (Bucura, Galeşu, Lia, Tău Negru) and the meadows of the rivers in some glacial valleys (Judele, Galeşu, Valea Rea), where they vegetate on flat terrain or gentle slopes, with alluvial, deep, moist soils, rich in humus. The characteristic species designated for this association, *Phleum alpinum* and *Plantago gentianoides,* define the subalpine-alpine specificity of these phytocoenoses (Borza 1934). Since the hygrophilic species characteristic of

Table 6.49 Ass. *Carduo personatae-Heracleetum palmati* (Borza 1934) Beldie 1967

Relevé No.	1	2	3	4	5	6	7	8	9	
Altitude (m a.s.l.)	1870	1930	2100	1850	1910	1940	2050	1790	1940	
Aspect	S	W	SW	N	NW	S	S	SW	S	
Slope (degrees)	20	5	15	10	20	30	40	40	15	
Herb cover (%)	95	90	95	90	80	90	75	70	65	
Sample area (sq. m)	25	25	25	25	25	25	25	25	25	K
Char. ass.										
Heracleum palmatum	3.5	2.5	3.5	3.5	3.5	4.5	3.5	3.4	3.5	V
Carduus personata	+.3	+	1.5	.	+.2	+	+	+	+	V
Adenostylion* et *Adenostyletalia										
Rumex alpinus	2.5	3.5	+	2.5	+	1.4	3.5	2.4	1.3	V
Adenostyles alliariae var. *kerneri*	2.5	+	1.3	1.5	1.4	1.3	1.3	1.3	.	V
Stellaria nemorum	3.5	+	2.5	2.5	.	2.5	.	+	.	IV
Ranunculus platanifolius	+	.	+	+	.	+	+	+	.	IV
Saxifraga rotundifolia subsp. *heucherifolia*	+.3	+	.	+	+.4	.	.	.	1.3	III
Milium effusum	+	+	.	+	.	+	+	.	.	III
Doronicum austriacum	+	.	.	.	.	+	+	+	.	III
Leucanthemum rotundifolium	+	.	.	+	.	.	.	+	+.3	III
Athyrium distentifolium	.	.	.	.	+	.	+	+	+	III
Chaerophyllum hirsutum	.	.	.	+	.	.	.	+	.	II
Senecio nemorensis	1.5	.	+	+	.	.	.	.	.	II
Tozzia alpina subsp. *carpathica*	.	+	.	.	.	.	.	.	.	I
Achillea distans	.	.	.	.			+	.	.	I
Aconitum toxicum	.	.	+	.	.	.	.	.	.	I
Mulgedio-Aconitetea										
Aconitum napellus subsp. *tauricum*	+	+.4	2.5	+.5	1.3	+.3	+	+	+	V
Deschampsia caespitosa	+	+	1.5	+	1.3	.	+	+	+.3	V
Geranium sylvaticum	+	+	+	.	1.3	+	+	+	+	V
Rumex alpestris	1.3	+	+	.	+	+	+	+	+	V
Veratrum album subsp. *lobelianum*	+	+	+	.	1.3	+	+	+	+	V
Angelica archangelica	+	.	1.4	+	.	+	+	1.3	.	IV

(continued)

Table 6.49 (continued)

Senecio subalpinus	+	.	.	.	+	.	.	.	.	II
Viola biflora	.	.	.	.	+	.	.	.	+.3	II
Cicerbita alpina	.	.	.	.	.	+	.	+	.	II
Companion										
Silene vulgaris	+	.	+	.	.	+	1.3	.	.	III
Epilobium angustifolium	.	.	+	.	.	+	+	+	.	III
Luzula alpinopilosa subsp. *obscura*	.	.	.	+	1.4	.	.	+	+	III
Saxifraga stellaris	+	.	.	.	+	.	.	.	+	III
Gentiana punctata	.	+	.	.	.	+	+	.	.	II
Caltha palustris	.	.	.	.	+	.	.	1.4	+	II
Senecio fuchsii	.	.	.	.	.	+	1.3	+	.	II

Companion species with one or two occurrences: *Valeriana sambucifolia* 1: +; *Pinus mugo* 1: +; *Oxalis acetosella* 1: +; *Soldanella major* 1: +; *Calamagrostis arundinacea* 1, 3: +; *Ligusticum mutellina* 2: +; *Festuca nigrescens* 3: +; *Doronicum columnae* 2: +, 5: 1.4; *Thalictrum aquilegifolium* 4: +; *Hypericum maculatum* 1, 7: +; *Doronicum carpaticum* 4, 8: +; *Cardaminopsis halleri* subsp. *ovirensis* 5: +; *Festuca picturata* 7: 1.4; *Solidago virgaurea* subsp. *minuta* 7: +; *Geum montanum* 7: +; *Senecio doronicum* subsp. *transylvanicus* 7: +; *Gnaphalium norvegicum* 7: +; *Campanula abietina* 7, 8: +; *Rubus idaeus* 8: +; *Calamagrostis villosa* 8: 1.3; *Plantago gentianoides* 9: +; *Allium schoenoprasum* subsp. *sibiricum* 9: +; *Carex pyrenaica* 9: +
Relevé: 1. Pârîul Ştirbului (10.09.1970); 2. Căldarea Vf. Judele (16.07.1967); 3. Pârîul Judele (12.09.1970); 4. V. Rea (8.08.1970); 5. V. Galeşului (13.07.1971); 6. Tăul Spurcat (11.10.1970); 7. Lacul Ana (11.08.1970); 8. V. Bucurii (7.08.1970); 9. Tăul Răsucit (11.07.1970)

Adenostylion and *Adenostyletalia* syntaxa are better represented in the association's structure than those of *Calamagrostietalia villosae,* we assigned the association to the former syntaxa (Coldea 1983).

Ass. *Aconito taurici-Rumicetum alpini* ass. nova hoc loco (Table 6.51)
Holotype: Table 6.51, rel. 4 *hoc loco*
Syn.: *Senecio subalpini-Rumicetum alpini* Horvat 1949 (Art. 3b)

The *Rumex alpinus*-dominated coenoses from the Retezat Mountains, established near the springs and subalpine streams, as a result of sheep grazing in the last two centuries, are included within this new regional association specific to the South-eastern Carpathians. From the ecological and floristic points of view, this syntaxon is a synvicariant of the association *Aconito firmi-Rumicetum alpini*, described in the Tatra Mountains (Unar et al. 1985). However, the phytocoenoses from Retezat have in their floristic structure many hygrophilic species characteristic of the *Adenostylion* and *Adenostyletalia* syntaxa, for which reason we assigned it to these syntaxonomic units. The optimal ecological conditions for the *Rumex alpinus* coenoses are within the perimeter of glacial cirques (e.g. Gemenele, Judele, Şesele, Galeşu), on gentle slopes with moist soils, rich in nitrites. Given the high altitudes where the *Rumex alpinus* coenoses from Retezat are located, and particularly due to the protection actions set after the establishment of the "Retezat National Park" (Borza 1934), such

Table 6.50 Ass. *Phleo alpini-Deschampsietum caespitosae* (Krajina 1933) Coldea 1983

Relevé No.	1	2	3	4	5	6	7	8	9	10	
Altitude (m a.s.l.)	2000	2040	2100	2000	2050	1920	1810	2120	1950	1810	
Aspect	–	–	–	SW	S	–	S	W	–	–	
Slope (degrees)	0	0	0	20	10	0	15	10	0	0	
Herb cover (%)	80	90	100	90	90	100	100	100	100	100	
Sample area (sq. m)	25	25	25	25	25	25	25	25	25	25	K
Char. ass.											
Deschampsia caespitosa	3.5	4.5	5.5	4.5	4.5	5.5	5.5	5.5	5.5	5.5	V
Plantago gentianoides	.	+	+.5	.	1.5	+	+	.	+.3	.	III
Phleum alpinum	+	+	.	+	.	.	+	+	.	+	III
Adenostylion* et *Adenostyletalia											
Saxifraga rotundifolia subsp. *heucherifolia*	.	.	+	.	.	.	.	1.3	.	.	II
Rumex alpinus	.	.	.	2.5	.	.	+.3	.	.	+	II
Adenostyles alliariae var. *kerneri*	+	.	+	.	.	.	.	.	.	.	I
Chaerophyllum hirsutum	.	.	.	.	.	.	1.3	.	.	+	I
Heracleum palmatum	.	.	.	.	.	.	+	.	.	+	I
Doronicum austriacum	.	.	.	.	.	.	.	.	.	+	I
Stellaria nemorum	.	.	.	+	.	.	.	+	.	.	I
Aconitum toxicum	.	.	.	.	+	.	.	.	.	.	I
Leucanthemum rotundifolium	.	.	.	.	.	.	+	.	.	.	I
Alopecurus laguriformis	.	.	.	.	.	.	.	+.1	.	.	I
Mulgedio-Aconitetea											
Veratrum album subsp. *lobelianum*	+	+	+	.	.	+	+.4	+	+	1.5	IV
Aconitum napellus subsp. *tauricum*	.	.	.	+	.	.	+	1.3	1.3	1.3	III
Rumex alpestris	+	.	.	+	+	.	+	.	.	.	II
Senecio subalpinus	.	.	+	.	.	+	+	.	.	+	II

(continued)

Table 6.50 (continued)

Angelica archangelica	.	.	+	.	.	+	.	.	.	.	I
Hypericum maculatum	.	.	.	+	+	.	.	.	.	.	I
Companion											
Ligusticum mutellina	1.4	1.5	1.5	.	1.5	+	.	+	+	+	IV
Geum montanum	+	+	.	+	.	.	+	.	.	.	II
Potentilla ternata	+	+	.	+	.	.	+	.	.	.	II
Luzula alpinopilosa subsp. *obscura*	+	.	+	+	.	.	.	.	+	.	II
Campanula abietina	+	.	.	+	.	.	+	+	.	.	II
Poa alpina	.	1.3	.	.	.	+	+	1.5	.	.	II

Companion species with one or two occurrences: *Carex atrata* 1: +; *Pedicularis verticillata* 1: +; *Pulsatilla alba* 1: +; *Phyteuma vagneri* 1: +; *Festuca picturata* 1: 2.5, 8: +; *Anthoxanthum alpinum* 3, 10: +; *Centaurea nervosa* 1: 2.5; *Carex nigra* subsp. *dacica* 2, 6: +; *Juncus trifidus* 2: +; *Soldanella major* 3: +; *Viola dacica* 4: +; *Rubus idaeus* 4: +; *Urtica dioica* 4: +; *Ranunculus repens* 4: +; *Milium effusum* 4: +; *Alchemilla alpestris* 5: +, 6: +.3; *Nardus stricta* 5: 1.4; *Carex curta* 6: +; *Crocus vernus* 6: +; *Caltha palustris* 7: +; *Cerastium holosteoides* 7: +, 9: 1.3; *Cardaminopsis halleri* subsp. *ovirensis* 8: +; *Taraxacum alpinum* 8: +.4, 9: +; *Ranunculus nemorosus* 9: +; *Philonotis seriata* 6: 2.5, 9: 1.3

Relevé: 1. Căldarea Șeselor-Tăului Negru (14.07.1967); 2. Lacul Lia (6.07.1969); 3. Izvorul Șeselor (8.07.1969); 4. Fața Retezatului (9.07.1969); 5. V. Judele (13.07.1970); 6. Plaiul Zănoaga (13.07.1970); 7. V. Bucurii (7.08.1970); 8. Lacul Bucura (9.08.1970); 9. Lacul Galeș (13.07.1971); 10. V. Galeșului (13.07.1971)

Table 6.51 Ass. *Aconito tatarici-Rumicetum alpini* ass. nova. hoc loco

Relevé No.	1	2	3	4	5	6	7	8	9	
Altitude (m a.s.l.)	2000	1800	1900	1930	1925	1850	1800	1860	1925	
Aspect	W	S	SW	S	W	NW	S	NW	SW	
Slope (degrees)	10	15	25	15	15	20	5	15	10	
Herb cover (%)	90	85	75	75	90	80	80	90	80	
Sample area (sq. m)	25	25	25	25	25	25	25	25	25	K
Char. ass.										
Rumex alpinus	4.5	4.5	4.5	4.5	4.5	4.5	5.5	5.5	4.5	V
Adenostylion **et** ***Adenostyletalia***										
Aconitum napellus subsp. *tauricum*	+.3	+	+	+	+	+	+	+	+.2	V
Stellaria nemorum	+	1.4	.	+.4	+.4	2.5	2.5	1.5	+	V
Adenostyles alliariae var. *kerneri*	.	1.3	+	+	1.3	+	+	+	1.3	V
Heracleum palmatum	.	+	+.3	+.3	1.3	+	.	.	+	IV
Saxifraga rotundifolia subsp. *heucherifolia*	.	.	+	+	.	+	.	+	+	III
Doronicum austriacum	.	+	.	+	+	+	.	.	.	III
Ranunculus platanifolius	.	+	+	+	+	.	.	.	+	III
Athyrium distentifolium	.	+	+	.	.	.	.	+	+	III
Senecio nemorensis	.	1.4	.	.	1.5	.	+	.	.	II
Chaerophyllum hirsutum	.	+.3	.	.	.	+	+	.	.	II
Milium effusum	.	+.2	.	.	1.5	.	.	.	+	II
Tozzia alpina subsp. *carpathica*	.	+	.	.	.	+	.	.	+	II
Valeriana sambucifolia	.	.	.	.	.	+	.	.	.	I
Leucanthemum rotundifolium	.	.	.	.	.	.	+	.	.	I
Mulgedio-Aconitetea										
Deschampsia caespitosa	1.3	+	+.3	+	+	+	+	+	.	V
Veratrum album subsp. *lobelianum*	+	+	+.4	+	+.3	.	+	1.3	+.3	V
Rumex alpestris	+	+.3	+	+	+	.	+	.	+	IV
Geranium sylvaticum	+.2	+.3	.	.	+	+	+	.	.	III
Angelica archangelica	2.4	.	.	+	+	+	.	.	+	III
Poa chaixii	+	.	.	+	.	.	.	.	.	II

(continued)

Table 6.51 (continued)

Senecio subalpinus	.	+	.	.	.	.	+	.	.	II
Epilobium alpestre	.	+	.	+	.	.	.	.	.	II
Cicerbita alpina	.	.	.	.	+	.	.	.	.	I
Companion										
Ligusticum mutellina	+	.	+	+	.	+	.	+	.	III
Campanula abietina	+	+	.	+	.	.	+	.	.	III
Geum montanum	+	.	.	+	.	.	+	.	+	III
Luzula alpinopilosa subsp. *obscura*	+	.	.	.	.	+	.	.	+	II
Gentiana punctata	.	.	+	+	.	.	.	.	+	II
Silene vulgaris	.	.	.	+	+	.	.	.	+	II
Potentilla ternata	.	.	.	+	.	.	+	.	+	II

Companion species with one or two occurrences: *Cardaminopsis halleri* subsp. *ovirensis* 1, 9: +; *Plantago gentianoides* 1: +; *Homogyne alpina* 1, 4: +; *Anthoxanthum alpinum* 1, 4: +; *Juniperus communis* subsp. *alpina* 2: +; *Lilium martagon* 2: +; *Solidago virgaurea* subsp. *minuta* 2, 4: +; *Festuca picturata* 3, 6: +; *Pedicularis exaltata* 3: +; *Pinus mugo* 3: +; *Rhodeola rosea* 3: +; *Hypericum maculatum* 2: +; *Phyteuma vagneri* 4: +; *Saxifraga stellaris* 4, 8: +; *Trollius europaeus* s.l. 5: +; *Chrysosplenium alternifolium* 7: +; *Senecio ovatus* 8: +; *Alchemilla alpestris* 7: +; *Calamagrostis villosa* 3, 6: +; *Myosotis sylvatica* 7, 9: +; *Epilobium alpestre* 2, 4: +
Relevé: 1. Căldarea Şeselor (14.07.1967); 2. Gemenea (15.07.1967); 3. Căldarea Judele (15.07.1967); 4. Pârîul Gemenea-Vf. Judele (15.07.1967); 5. Tăul Negru-Gemenea (15.07.1967); 6. Tăurile Cârligului (8.07.1969); 7. V. Galeş-Lacul Brazilor (7.08.1970); 8. Muntele Zlata (9.07.1970); 9. Tăul Negru (14.07.1967)

coenoses acquired their naturality and lost the ubiquitous species from their floristic structure. This process did not occur in the coenoses described in other mountain ranges of the Southeastern Carpathians, where grazing has continued (Alexiu 1998; Buia et al. 1962; Mihăilescu 2001; Puşcaru et al. 1956).

Ass. *Salici silesiacae-Alnetum viridis* Colič et al. 1962 (Table 6.52)

The scrub coenoses of this association preserved their structural authenticity and distribution area in the perimeter of the scientific reserve of the Retezat National Park (Gemenele glacial cirque, Faţa Retezatului) and the Custura Retezatului area (Stâna de Râu). They vegetate on siliceous rocks, along torrents and ravines on the steep and slightly sunny slopes inside the massif, frequently exposed to avalanches. Sometimes they colonize the slopes of glacial cirques near the lakes. The *Alnus viridis* coenoses from Retezat are most often found at the upper limit of spruce forests, in contact with the *Pinus mugo* and *Pinus cembra* coenoses, on rocky and wet substrates where juniper shrubs cannot settle. Such phytocoenoses have been interpreted as a final evolution stage of hygrophilic coenoses of the alliance *Adenostylion* and considered as an edaphogenic paraclimax stage (Boşcaiu 1971). Based on these syndynamic considerations and the predominance of the species typical of tall-forb communities specific for the class *Mulgedio-Aconitetea*, we include the association *Salici-Alnetum viridis* in this class. We mention that the

Table 6.52 Ass. *Salici-Alnetum viridis* Colič et al. 1962

Relevé No.	1	2	3	4	5	6	
Altitude (m a.s.l.)	1900	1950	1950	1900	1630	1650	
Aspect	E	SE	E	E	E	NE	
Slope (degrees)	35	45	25	30	50	45	
Herb cover (%)	75	85	65	50	65	50	
Sample area (sq. m)	50	25	25	25	25	25	K
Char. ass.							
Alnus viridis	4.5	5.5	3.5	3.5	5.5	5.5	V
Salix silesiaca	+	.	+	.	.	+	III
Adenostylion **et** ***Adenostyletalia***							
Senecio nemorensis	+.4	1.5	+	+.3	2.5	+.4	V
Adenostyles alliariae var. *kerneri*	1.3	+	2.5	+.5	.	.	IV
Saxifraga rotundifolia	.	.	.	+	+	+	III
Chaerophyllum hirsutum	.	.	.	.	+	+	II
Leucanthemum rotundifolium	.	.	.	.	+	+	II
Doronicum austriacum	+	.	.	.	.	.	I
Carduus personata	.	.	.	.	+	.	I
Hypericum tetrapterum	.	.	.	.	.	+	I
Mulgedio-Aconitetea							
Athyrium distentifolium	+	+	+	2.3	.	+.4	V
Veratrum album subsp. *lobelianum*	+.3	+	+.3	1.4	.	+	V
Calamagrostis villosa	+.2	1.5	+	+	.	.	IV
Aconitum napellus subsp. *tauricum*	.	.	+	+	+.4	+	IV
Ranunculus platanifolius	+	+	1.4	.	.	.	III
Valeriana tripteris	.	+	.	.	1.3	+	III
Angelica archangelica	.	+	.	.	1.5	.	II
Deschampsia caespitosa	.	+	.	.	.	2.5	II
Rumex alpestris	.	.	.	.	+	1.5	II
Doronicum columnae	.	.	+	.	+	.	I
Thalictrum aquilegifolium	.	.	.	.	+.3	+	I
Geranium sylvaticum	.	.	.	.	.	+	I
Companion							
Gentiana punctata	+.3	+	+.3	+.3	+	.	V
Dryopteris dilatata	+.3	+.3	1.4	+	.	+	V
Rhododendron myrtifolium	+	+	+	+.3	.	1.5	V
Pinus mugo	.	+	+.3	+.2	+	+.3	V
Silene vulgaris	.	+	+	+	+	+	V
Soldanella major	+	+	+	.	+	.	IV
Vaccinium myrtillus	+	+.4	.	1.2	.	+	IV
Pinus cembra	+.2	+	.	.	+	+	IV
Rhodiola rosea	.	+	+	+	+	.	IV
Campanula abietina	.	+	+	.	+	+	IV
Anthoxanthum alpinum	+	+	.	+	.	.	III

(continued)

Table 6.52 (continued)

Juniperus communis subsp. *alpina*	+	+	.	.	+	.	III
Solidago virgaurea subsp. *minuta*	+	+	.	.	.	+	III
Rubus idaeus	.	+	.	.	+	+	III

Companion species with one or two occurrences: *Pseudorchis albida* 1: +; *Homogyne alpina* 2: +; *Hypochaeris uniflora* 2: +; *Ligusticum mutellina* 2: +; *Picea abies* 1: +; *Senecio doronicum* subsp. *transylvanicus* 2, 3: +; *Festuca picturata* 4, 5: +; *Potentilla ternata* 4: +; *Poa alpina* 5: +; *Geum montanum* 6: +; *Alchemilla alpestris* 5, 6: +; *Senecio subalpinus* 6: +; *Myosotis sylvatica* 6: +
Relevé: 1–4. Lacul Gemenea (15.07.1967); 5–6. Stâna de Râu (14–15.07.1971)

spore-pollen analyses from Tăul Zănoguței (1850 m) determined the pre-boreal age of the *Alnus viridis* coenoses from Retezat (Pop et al. 1970).

Order *Petasito-Chaerophylletalia* Morariu 1967
Natura 2000: habitat type 6430

This order groups nitrophilous tall-herb communities, populating the eroded riversides along the valleys and streams of the submontane and montane belts in the Carpathians. Initially, the order was placed in the class *Epilobietea angustifolii* (Morariu 1967), and later some authors placed it in the class *Galio-Urticetea* (Borhidi 1996; Coldea 2012). We consider that assigning this order to the class *Mulgedio-Aconitetea* is more judicious (Kliment et al. 2007).

Alliance *Petasition officinalis* Sillinger 1933

The alliance encompasses nitrophilous and hygrophilic tall-forb communities in the submontane and montane areas of the Carpathians and the Alps. The floristic structure of these phytocoenoses in the intramontane valleys, with discontinuous forest vegetation, is enriched with nemoral species typical of the alliance *Alnion incanae*. The main characteristic species of the alliance are: *Chaerophyllum hirsutum, Mentha longifolia, Impatiens noli-tangere, Stellaria nemorum, Geranium phaeum, Crepis paludosa, Myosotis scorpioides, Caltha palustris,* and *Urtica dioica* (Kliment et al. 2007).

Ass. *Telekio-Petasitetum hybridi* Morariu ex Resmeriță et Rațiu 1974 (Table 6.53)
Lectotype: Resmeriță et Rațiu 1974, Table 3, rel. 5. *hoc loco*

The representative coenoses for this association were identified between 800 and 1100 m altitude, on the intramontane streams in the northwestern (Zlătuia), western (Radeșu Mic) and southern (Lăpușnicu Mare) parts of the Retezat massif. They vegetate in the sinuous parts of the streams, where various alluvia are deposited over the rocky substrate. The dominant species of the association is *Petasites hybridus,* achieving on average 40% coverage. In addition to it, the regional species *Telekia speciosa* together with *Petasites kablikianus* and *Pulmonaria rubra*, give the association a specific southeastern European character. The association includes a high

Table 6.53 Ass. *Telekio-Petasitetum hybridi* Morariu ex Resmeriţă et Raţiu 1974

Relevé No.	1	2	3	4	5	
Altitude (m a.s.l.)	980	900	950	1100	820	
Aspect	W	W	W	S	W	
Slope (degrees)	15	3	5	3	10	
Herb cover (%)	100	100	100	50	50	
Sample area (sq. m)	25	25	25	25	25	K
Char. Ass.						
Telekia speciosa	1.5	+	3.5	+	1.2	V
Petasites hybridus	4.5	5.5	2.5	5.5	3.5	V
Petasition et ***Petasito-Chaerophylletalia***						
Stellaria nemorum	+	+	1.3	1.5	+	V
Impatiens noli-tangere	+	+.3	+	.	+	IV
Cirsium oleraceum	1.3	+	+	+	.	IV
Rumex alpinus	+	.	+	+	.	III
Poa trivialis	+	.	+	.	+	III
Angelica sylvestris	+	.	.	+	+	III
Lamium maculatum	.	+	+	.	+	III
Aegopodium podagraria	.	+.3	+	.	+	III
Mentha longifolia	+	+	.	.	.	II
Stachys palustris	+	.	+	.	.	II
Petasites kablikianus	.	.	.	+	+	II
Pulmonaria rubra	.	+	.	+	.	II
Geranium phaeum	.	.	+.3	.	+	II
Chaerophyllum hirsutum	.	.	.	.	+	I
Mulgedio-Aconitetea						
Milium effusum	+	.	+	+	+	IV
Cicerbita alpina	+	.	.	+	+	III
Senecio nemorensis	.	.	+	+	1.3	III
Geranium robertianum	.	+	.	.	+	II
Geum rivale	.	.	+	.	+	II
Parietaria officinalis	.	+	.	.	.	I
Doronicum austriacum	.	.	+	.	.	I
Carduus personata	.	.	.	+	.	I
Myosotis scorpioides	.	.	.	+	.	I
Galio-Urticetea						
Ranunculus repens	+	1.3	+	+	+	V
Urtica dioica	.	1.2	+	+.4	1.3	IV
Anthriscus sylvestris	1.5	+	.	+	.	III
Lapsana communis	+	.	.	+	.	II
Glechoma hederacea	.	+	.	.	+	II
Galium aparine	.	.	+	.	+	II
Silene latifolia	.	.	+	.	+	II

(continued)

Table 6.53 (continued)

Companion						
Dryopteris filix-mas	+	.	+	+	+	IV
Rubus idaeus	.	1.2	+	+	+	IV
Crepis paludosa	1.5	.	+	.	+	III
Taraxacum officinale	+	.	+	.	+	III
Cardamine amara	+	.	+	.	.	II
Caltha palustris	+	.	+	.	.	II
Juncus effusus	+	.	.	+	.	II
Carex remota	+	.	.	.	+	II
Alnus incana	+	.	.	.	+	II
Lunaria rediviva	.	+.3	.	.	+	II
Euphorbia amygdaloides	.	+	.	.	+	II
Hesperis matronalis	.	.	+	+	.	II
Cardamine impatiens	.	.	.	+	+	II
Lycopus europaeus	+	.	.	.	.	I
Cirsium rivulare	+	.	.	.	.	I
Polygonatum verticillatum	.	+	.	.	.	I
Cardamine glanduligera	.	+	.	.	.	I
Mercurialis perennis	.	.	+	.	.	I
Aconitum lycoctonum subsp. *moldavicum*	.	.	.	.	+	I
Angelica archangelica	.	.	.	.	+	I

Relevé: 1. Radeşul Mic (11. 07.1970); 2–3. V. Râul Mare (12.07.1970); 4. V. Lăpuşnicu Mare (15.07.1970); 5. V. Zlătuia (9.08.1985)

number of species characteristic of the *Petasition officinalis* and *Petasitio-Chaerophylletalia* syntaxa (Jarolimek and Šibik 2008; Morariu 1967), as well as species characteristic of the class *Mulgedio-Aconitetea*, supporting its classification in these syntaxa (Kliment et al. 2007).

Class *Carpino-Fagetea* Jakucs ex Passarge 1968

We include in this class the mesophilous, deciduous forests which surround the Retezat massif, extending deep into its central sectors through the western and southern intramontane valleys (Zlătuia, Râul Mare, Lăpuşnicu Mare). Such forest coenoses vegetate on soils that differ in terms of pedogenetic horizons, moisture and trophicity. Among the characteristic species of the class, in the park area are present: *Carpinus betulus, Fagus sylvatica, Abies alba, Acer pseudoplatanus, Acer campestre, Dryopteris filix-mas, Poa nemoralis, Lonicera xylosteum, Euonymus europaeus, Moehringia trinervia, Neottia nidus-avis, Scrophularia nodosa, Symphytum tuberosum, Corylus avellana, Hedera helix, Fraxinus excelsior, Platanthera chlorantha,* and *Viola reichenbachiana* (Boşcaiu 1971; Boublik et al. 2013).

Order *Fagetalia* Pawl. 1928

In this order we include all the mesophilous and mesothermal forest coenoses built up by *Fagus sylvatica, Carpinus betulus, Acer pseudoplatanus, Tilia cordata* and

Alnus incana in the whole montane belt of the Retezat, between 700 and 1250 m altitude. The characteristic herbaceous species designated for this order are: *Actaea spicata, Anemone nemorosa, Adoxa mochatellina, Asarum europaeum, Carex sylvatica, Galium odoratum, Euphorbia amygdaloides, Lamium galeobdolon, Cardamine bulbifera, Lathyrus vernus, Lilium martagon, Sanicula europaea, Veronica urticifolia, Salvia glutinosa, Mycelis muralis, Mercurialis perennis, Pulmonaria officinalis, Impatiens noli-tangere, Paris quadrifolia, Epilobium montanum, Gymnocarpium dryopteris, Luzula luzuloides, Campanula trachelium, Cardamine impatiens,* and *Hordelymus europaeus* (Matuszkiewicz 2008; Oberdorfer 1992). The order includes three alliances.

Alliance *Alnion incanae* Pawl. et al. 1928
Natura 2000: priority habitat type 91EO

In this alliance we include the hygrophilous and mesotrophic forest coenoses of river floodplains located in the intramontane valleys of the Retezat massif, periodically flooded, with moist and nutrient-rich alluvial soils. The characteristic species for the alliance are: *Alnus incana, Stellaria nemorum, Elymus caninus, Chaerophyllum hirsutum, Festuca gigantea, Aegopodium podagraria, Stachys sylvatica, Circaea lutetiana, Geum urbanum, Matteucia struthiopteris, Ajuga reptans, Equisetum hyemale,* and *Equisetum telmateia* (Boublik et al. 2013; Willner and Grabherr 2007).

Ass. *Telekio speciosae-Alnetum incanae* Coldea 1990 (Table 6.54)
– *petasitosum kablikianae* Coldea 1990

We grouped in this unique regional association the coenoses dominated by *Alnus incana* distributed in the area of the Southeastern Carpathians. The Carpathian-Balkan species *Telekia speciosa*, whose geographical distribution area was foreshadowed by Dobolyi (1983), outlines the southeastern European areal distribution of the association (Coldea 2015). The phytocoenoses of the Lăpuşnicul Mare river floodplain show floristic differences from those in Râul Mare, allowing us to distinguish the sub-association *petasitosum kablikianae* Coldea 1990 (rel. 6–10). Some mesohygrophilous and mesotrophic species (*Stellaria nemorum, Elymus caninus, Geum urbanum, Urtica dioica*) present occasionally in this association have high (25–35%) coverage, highlighting the pedoclimatic specificity of the habitats from Retezat where the coenoses of *Alnus incana* occur (Photo 6.9).

Alliance *Lathyro hallersteinii-Carpinion* Boşcaiu 1974
Syn.: *Carpinion dacicum* Soó 1964 (Art. 34)
Type of name: *Lathyro hallersteinii-Carpinetum* Coldea 1975 h.l.

In this regional alliance we group the forest coenoses of *Carpinus betulus*, widespread in Romania and strongly differentiated from those in Central Europe by the regional species *Lathyrus hallersteinii, Lathyrus aureus, L. venetus, L. transsilvanicus, Melampyrum bihariense, Helleborus purpurascens, Helleborus odorus, Cardamine quinquefolia, Carex depressa* subsp. *transsilvanica, Carex brevicollis,* and *Galium kitaibelianum* (Coldea 2015).

Table 6.54 Ass. *Telekio-Alnetum incanae* Coldea 1990

Relevé No.	1	2	3	4	5	6	7	8	9	10	
Altitude (m a.s.l.)	880	720	950	900	920	1060	1070	1120	1200	850	
Aspect	NW	N	N	N	N	W	W	W	W	NW	
Slope (degrees)	3	5	5	5	5	5	5	5	5	5	
Tree height (m)	25	12	14	15	12	15	12	16	18	22	
Tree cover (%)	80	80	70	70	70	60	70	70	60	80	
Herb cover (%)	35	25	40	60	60	60	15	75	75	75	
Litter cover (%)	0	0	30	0	20	0	0	0	0	0	
Sample area (sq. m)	400	400	400	400	400	400	400	400	400	400	K
Char. ass.											
Alnus incana	5.5	5.5	4.5	4.5	5.5	4.5	4.5	4.5	4.5	5.5	V
Telekia speciosa	+	+	+	.	+	.	+	.	.	1.3	III
Diff. subass. petasitetosum											
Petasites kablikianus	.	.	.	.	.	3.5	+.3	4.5	+.5	2.3	III
Aconitum lycoctonum subsp. *moldavicum*	.	.	.	.	.	+	+.3	+	.	+	II
Heracleum palmatum	.	.	.	.	.	+	+	.	+	+	II
Tozzia alpina subsp. *carpathica*	.	.	.	.	.	.	.	+	.	.	I
Alnion incanae											
Stellaria nemorum	1.5	+.5	2.5	3.5	2.5	3.5	1.5	2.5	4.5	+	V
Geranium phaeum	.	+	+.5	.	.	+	+	+	+	+	IV
Chaerophyllum hirsutum	+	+	+	+	+	.	.	.	+	+	IV
Geum urbanum	+	+	+	+	+	+	+	.	.	+	IV
Elymus caninus	+	+	+	+	.	+	1.3	+	+	.	IV
Impatiens noli-tangere	+	.	+.3	+	.	+	+	.	+	+	IV
Cirsium oleraceum	+	+	+	+	+	.	.	+	.	.	III
Aegopodium podagraria	2.5	+	+	+	.	+	.	.	.	+	III
Festuca gigantea	+	+	.	.	.	.	.	.	.	+	II
Ajuga reptans	+	.	.	.	+	+	+	.	.	.	II
Matteucia struthiopteris	+	.	.	.	+	.	.	+	.	.	II
Stachys sylvatica	+	.	.	.	.	.	+	.	.	.	I
Symphyto cordati-Fagion											
Pulmonaria rubra	.	.	+	.	+	.	.	+.3	+	1.2	III
Symphytum cordatum	.	.	+.5	+	.	.	.	+	.	+	II
Campanula abietina	.	.	+	+	.	.	.	.	+	+	II

(continued)

Table 6.54 (continued)

Silene heuffelii	.	.	.	.	.	+	+	.	+	.	II
Cardamine glanduligera	.	.	+	.	+	.	.	.	.	.	I
Fagetalia											
Myosotis sylvatica	.	.	.	+	1.5	+	.	+	+	.	III
Senecio nemorensis	.	.	+	.	+	+	.	.	+	+	III
Milium effusum	.	.	+	+	+	.	.	+	.	+	III
Ulmus glabra	.	+	.	+.3	.	.	.	+	.	1.3	II
Sorbus aucuparia	.	.	.	.	+	.	+.3	.	+	.	II
Anthriscus nitidus	.	.	.	.	.	+	+	+	.	.	II
Mercurialis perennis	.	.	+	+.3	.	.	.	.	.	+	II
Mycelis muralis	.	+	+	+	.	+	.	.	.	.	II
Oxalis acetosella	.	1.5	2.5	.	.	.	+	.	.	+	II
Epilobium montanum	.	+	.	.	+	+	.	+	.	.	II
Asarum europaeum	.	+	+	.	+	.	.	.	.	+	II
Pulmonaria officinalis	.	+	.	.	.	+	1.3	.	+	.	II
Cardamine impatiens	.	+	.	.	.	+	.	.	+	+	II
Fagus sylvatica	.	.	.	.	.	.	.	.	.	+	I
Salvia glutinosa	+	+	.	.	.	.	.	.	.	.	I
Luzula luzuloides	.	+	.	.	+	.	.	.	.	.	I
Carex sylvatica	.	.	+	.	+	.	.	.	.	.	I
Euphorbia amygdaloides	.	.	+	.	.	+	.	.	.	.	I
Carpino-Fagetea											
Glechoma hirsuta	1.3	+.3	+	+	.	+	+	+.3	.	+	IV
Athyrium filix-femina	1.2	+	+	+	+	.	.	.	+	+	IV
Acer pseudoplatanus	.	+	1.3	.	.	.	+	+	.	+	III
Dryopteris filix-mas	1.5	+	+	+	+	.	.	.	.	+	III
Dactylis polygama	.	+	.	.	.	+	.	+	.	.	II
Poa nemoralis	.	+	.	+	.	.	.	.	.	.	I
Symphytum tuberosum	.	.	+	.	+	.	.	.	.	.	I
Epilobietalia											
Rubus idaeus	.	.	+	.	+	+.3	1.3	+	1.5	2.3	IV
Fragaria vesca	.	+	+	+	+	+	+.3	.	+	.	IV
Sambucus racemosa	+	2.5	.	+	+	.	.	.	.	+	III
Salix silesiaca	.	.	.	+.3	+	+	2.3	+	1.3	.	III
Galeopsis tetrahit	.	+	+	+	.	.	+	.	.	+	III

(continued)

Table 6.54 (continued)

Companion											
Urtica dioica	1.5	1.5	1.3	2.3	3.5	+.5	1.4	1.3	1.5	+	V
Geranium robertianum	.	+	+	+	+	+	+	.	.	+	IV
Lamium maculatum	.	+	.	+	+	.	+	+	+	+	IV
Carduus personata	.	+	.	.	.	+	+	+	+	.	III
Valeriana officinalis	.	.	+	+	.	+	+	.	+	.	III
Rumex alpinus	.	+	+	+	.	+	.	+	.	.	III
Ranunculus repens	.	+	2.5	+	.	+	+.3	.	.	+	III
Angelica sylvestris	.	+	.	.	.	+	.	.	+	+	II
Geum rivale	.	.	.	.	.	.	+	+	+	.	II
Picea abies	.	.	1.5	.	.	+	+	.	1.3	.	II
Doronicum austriacum	.	.	+	.	+	.	+	+	.	.	II
Silene latifolia subsp. *alba*	.	+	+	+	.	.	.	+	.	.	II
Deschampsia caespitosa	.	.	.	.	.	+	+	.	+	.	II
Lapsana communis	+	.	.	+	.	.	.	.	.	.	I
Crepis paludosa	.	.	.	+	.	.	+	.	.	.	I
Equisetum hyemale	.	.	.	.	.	+	.	.	+	.	I
Thalictrum aquilegifolium	.	.	.	.	.	.	+	.	.	.	I
Poa trivialis	.	.	.	.	.	.	+	.	.	.	I

Companion species with one occurrence: *Knautia longifolia* 1: +; *Stachys alpina* 1: +; *Betula pendula* 2: +; *Tussilago farfara* 2: +; *Rubus hirtus* 2: +; *Sedum maximum* s.l. 2: +; *Galium mollugo* 2: +; *Peltaria alliacea* 2: +; *Tilia cordata* 2: +; *Circea lutetiana* 2: +; *Vicia sepium* 2: +; *Gymnocarpium dryopteris* 3: +; *Anemone nemorosa* 3: +; *Adoxa moschatellina* 3: +; *Maianthemum bifolium* 3: +; *Trifolium repens* 3: +; *Digilalis grandiflora* 1: +; *Calamagrostis arundinacea* 7: +; *Galium schultesii* 7: +; *Equisetum telmateia* 7: +; *Silene vulgaris* s.l. 7: +; *Ranunculus acris* 9: +; *Rumex acetosa* 9: +; *Scrophularia scopoli* 9: +; *Valeriana sambucifolia* 9: +

Relevé: 1. V. Zlătuia (14.09.1970); 2–5. Râul Mare (2.09, 12.07.1970); 6–9. V. Lăpuşnicul Mare (12–15.07.1970); 10. V. Zlătuia (9.08.1985)

Ass. *Corylo-Tilietum cordatae* Vida 1959 (Table 6.55)

We include in this association the mesothermal phytocoenoses dominated by *Tilia cordata*, *Carpinus betulus*, and *Corylus avellana* populating sunny, steep slopes of the Râul Mare valley. The prevalence of a group of species characteristic of *Carpinus betulus* forests, such as *Lathyrus hallersteinii*, *Galium kitaibelianum*, *Dactylis polygama*, *Glechoma hirsuta*, and *Spiraea chamaedryfolia*, determined us to include it in the alliance *Lathyro hallersteinii-Carpinion* Boşcaiu 1974. The association also comprises characteristic species of the order *Fagetalia* and class *Carpino-Fagetea*. The rocky substrate of the areas populated by these phytocoenoses is revealed by *Cystopteris fragilis*, *Asplenium trichomanes*,

Photo 6.9 *Telekia specioas* (Schreb.) Baumg. (photo M. Ciobanu)

Polypodium vulgare, Polystichum aculeatum, and *Seseli libanotis*. From a floristic-ecologic point of view, this association shows some resemblances with *Seslerio albicantis-Tilietum cordatae* Chytry et Sadlo 1998 from the Czech Republic.

Alliance *Symphyto cordati-Fagion* Vida 1959 em. 1963
Type of name: *Symphyto cordati-Fagetum* Vida 1959
Natura 2000: habitat type 91VO

In this alliance we group the *Fagus sylvatica* forests as well as those mixed with *Acer pseudoplatanus, Abies alba* and *Picea abies* identified in the Retezat National Park. The specific regional Carpathian species of the alliance are: *Cardamine glanduligera, Symphytum cordatum, Hepatica transsilvanica, Silene heuffelii, Ranunculus carpaticus, Aconitum lycoctonum* subsp. *moldavicum, Leucanthemum rotundifolium, Pulmonaria rubra, Hieracium transsilvanicum, Euphorbia carniolica,* and *Festuca drymeja* (Boşcaiu 1971; Doniţă et al. 1990).

Ass. *Symphyto cordati-Fagetum* Vida 1959 (Table 6.56)

The coenoses of this association have a zonal distribution in the northwestern and western part of Retezat (Râul Mare, Zlătuia, Cârligului, Turcului valleys). They

Table 6.55 Ass. *Corylo-Tilietum cordatae* Vida 1959

Relevé No.	1	2	3	4	5	
Altitude (m a.s.l.)	750	770	800	810	820	
Aspect	W	W	W	W	W	
Slope (degrees)	35	45	40	60	45	
Tree height (m)	8	8	10	10	10	
Canopy cover (%)	70	70	90	80	80	
Herb cover (%)	60	30	40	60	60	
Litter cover (%)	80	80	80	50	60	
Sample area (sq. m)	400	400	400	400	400	K
Char. ass.						
Tilia cordata	3.5	1.4	2.5	2.5	2.5	V
Corylus avellana	2.5	3.5	3.5	1.5	2.5	V
Lathyro hallersteini-Carpinion						
Carpinus betulus	+	2.5	+	3.5	3.5	V
Galium kitaibelianum	+.5	+	+.4	+.5	+.3	V
Lathyrus hallersteinii	+	+	+	+	.	IV
Glecoma hirsuta	+	+	+	.	+	IV
Galium schultesii	+	+	+	+	.	IV
Fagetalia						
Veronica urticifolia	+	+	+	+	+	V
Luzula luzuloides	+	+	+	+	+	V
Galium odoratum	+	1.3	2.5	+	.	IV
Epilobium montanum	+	.	+	+	+	IV
Prenanthes purpurea	.	+	+	+	+	IV
Lamium galeobdolon	.	+	+	+	+	IV
Salvia glutinosa	.	+	+	+	+	IV
Euphorbia amygdaloides	.	+	+	+	+	IV
Fagus sylvatica	.	1.4	1.5	.	1.3	III
Pulmonaria officinalis	.	+	+	+	.	III
Mercurialis perennis	.	+	+	+	.	III
Lathyrus vernus	.	+	+	+	.	III
Hepatica nobilis	.	.	1.5	+	+	III
Mycelis muralis	+	+	.	.	.	II
Campanula trachelium	+	.	+	.	.	II
Actaea spicata	.	+	.	+	.	II
Festuca gigantea	.	+	.	+	.	II
Sanicula europaea	.	.	+	+	.	II
Asarum europaeum	.	.	.	+	+	II
Carpino-Fagetea						
Poa nemoralis	1.5	2.5	1.3	3.5	3.5	V
Dactylis polygama	+	+	+	+	1.5	V
Mellitis melissophyllum	+	+	+	+	+	V
Cephalanthera longifolia	+	+	+	+	+	V
Circaea lutetiana	+	+	+	+	.	IV
Hypericum montanum	.	+	+	+	+	IV

(continued)

Table 6.55 (continued)

Digitalis grandiflora	.	+	+	+	+	IV
Fraxinus excelsior	.	.	+	+	+	III
Acer campestre	+	+	.	.	.	II
Scrophularia nodosa	+	+	.	.	.	II
Athyrium filix-femina	.	+	+	.	.	II
Moehringia trinervia	.	.	+	+	.	II
Quercetalia pubescentis						
Campanula persicifolia	+	+	+	+	+.4	V
Sedum maximum s.l.	+	+	.	+	+	IV
Tanacetum corymbosum	+	.	.	+	+	III
Cardaminopsis arenosa	+	.	.	+	.	II
Astragalus glycyphyllus	.	+	+	.	.	II
Stachys officinalis	.	.	+	.	+	II
Valeriana officinalis	.	.	+	.	+	II
Campanula rapunculoides	.	.	+	+	.	II
Euonymus verrucosa	.	.	.	1.3	.	I
Companion						
Solidago virgaurea	+	+	+	+	+	V
Calamagrostis arundinacea	3.5	2.5	+	3.5	1.5	V
Achillea distans	+	+	+	+	1.4	V
Geranium robertianum	+	+	+	+	+	V
Clinopodium vulgare	+	+	+	+	+	V
Asplenium trichomanes	+	+	+	+	+	V
Polypodium vulgare	+	+	+	+	+	V
Fragaria vesca	+	.	+	+	+	IV
Campanula cervicaria	+	+	+	+	.	IV
Stachys alpina	.	+	+	+	+	IV
Cystopteris fragilis	+	+	+	+	.	IV
Aegopodium podagraria	+	+	1.3	+	.	IV
Acer pseudoplatanus	+	+	+	.	.	III
Seseli libanotis	+	+	.	.	+	III
Betula pendula	.	+	.	+	+	III
Galeopsis speciosa	.	+	+	.	+	III
Pteridium aquilinum	+	+	+	.	.	III
Cirsium erisithales	.	.	+	+	+	III
Vicia sepium	+	+	+	.	.	III
Spiraea chamaedryfolia	.	+	.	+.3	.	II
Galium verum	+	+	.	.	.	II
Verbascum nigrum	+	+	.	.	.	II
Urtica dioica	.	+	+	.	.	II

Companion species with one occurrence: *Cardamine flexuosa* 2: +; *Senecio nemorensis* 2: +; *Oxalis acetosella* 2: +; *Polygonatum verticillatum* 2: +; *Gymnocarpium dryopteris* 2: +; *Alnus incana* 2: +; *Sambucus racemosa* 2: +; *Doronicum austriacum* 2: +; *Leucanthemum rotundifolium* 3: +; *Hieracium sabaudum* 3: +; *Genista tinctoria* 3: +; *Polystichum aculeatum* 2: +; *Lunaria rediviva* 2: +; *Cardamine bulbifera* 3: +; *Chaerophyllum aromaticum* 3: +; *Acer platanoides* 3: +; *Symphytum tuberosum* 3: +; *Vicia sylvatica* 3: +; *Brachypodium sylvaticum* 4: +; *Bromus benekeni* 4: +; *Peucedanum oreoselinum* 5: +; *Bilderdykia dumetorum* 5: +

Relevé: 1–5. V. Râul Mare (11.09.1970)

Table 6.56 Ass. *Symphyto cordati-Fagetum* Vida 1959

Relevé No.	1	2	3	4	5	6	7	8	9	10	
Altitude (m a.s.l.)	800	850	950	930	970	990	810	860	870	900	
Aspect	–	–	N	NW	NE	N	W	N	N	N	
Slope (degrees)	0	0	25	20	15	5	10	5	10	10	
Tree height (m)	28	28	27	24	27	27	29	23	24	22	
Canopy cover (%)	90	90	70	80	70	80	90	70	80	90	
Herb cover (%)	30	30	50	40	40	20	45	60	30	20	
Litter cover (%)	60	80	60	60	70	50	75	35	25	75	
Sample area (sq. m)	400	400	400	400	400	400	400	400	400	400	K
Char. ass.											
Fagus sylvatica	5.5	5.5	5.5	5.5	4.5	5.5	5.5	5.5	4.5	5.5	V
Symphytum cordatum	+	+.3	+	+	+	+.3	1.5	+.3	+	+	V
Symphyto-Fagion											
Cardamine glanduligera	+	.	+	+	+	1.3	+	.	.	.	III
Festuca drymeja	.	.	.	.	.	+	+	+	.	+	II
Pulmonaria rubra	+	+.5	.	.	.	.	.	.	.	.	I
Hieracium transsilvanicum	.	.	.	.	+	+	.	.	.	.	I
Fagetalia											
Galium odoratum	1.5	+	+.5	+	+.4	+	2.5	3.5	.	1.5	V
Mercurialis perennis	.	+	+	+	+	+	+	+	+	+	V
Polygonatum verticillatum	+	+	.	+	+	.	+	+	+	+	V
Asarum europaeum	1.3	1.5	+.5	1.5	+.3	.	+	+	+	.	IV
Actaea spicata	+	+	+	+	+	.	+	.	+	+	IV
Gymnocarpium dryopteris	2.5	2.5	.	3.5	3.5	.	.	+	+	+	IV
Paris quadrifolia	+	+	+	+	+	+	+	+	.	.	IV
Lamium galeobdolon	+	.	.	.	+	+	+	+	.	.	III
Senecio nemorensis	.	+	.	+	+	+	.	+	.	+	III
Mycelis muralis	.	+	.	+	.	+	+	+	.	.	III
Salvia glutinosa	.	.	+	+	+	.	+	+	.	.	III
Athyrium filix-femina	+	.	.	.	.	1.5	+	1.5	2.5	+	III
Carex sylvatica	+	+	.	.	.	+	+	+	.	.	III
Daphne mezereum	.	+	+	.	.	+	.	.	.	.	II
Euphorbia amygdaloides	+	.	.	+	.	.	+	.	.	.	II
Adoxa moschatellina	+	+	.	.	.	.	+	+	.	.	II
Anemone nemorosa	.	+	.	.	.	1.3	+	.	.	.	II
Gentiana asclepiadea	.	.	.	+	.	+	+	+	.	.	II
Lilium martagon	.	.	.	+	+	.	+	.	.	.	II
Sanicula europaea	.	.	.	.	+	.	.	.	+	+	II
Pulmonaria officinalis	.	.	.	.	.	.	.	+	+	+	II
Viola reichenbachiana	.	.	.	.	.	.	+	.	+	+	II
Prenanthes purpurea	.	.	+	.	.	.	.	.	.	.	I
Veronica urticifolia	.	+	.	.	.	+	.	.	.	.	I
Cardamine bulbifera	.	.	.	.	.	.	+	.	+	.	I

(continued)

Table 6.56 (continued)

Hordelymus europaeus	.	1.3	.	+	.	.	.	.	.	.	I
Luzula luzuloides	.	.	.	.	.	.	.	+	.	.	I
Impatiens noli-tangere	.	.	.	.	.	.	+	+	.	.	I
Epipactis helleborine	.	+	.	.	.	.	.	.	.	.	I
Carpino-Fagetea											
Glechoma hirsuta	+	+	+.4	+	+	+	.	1.2	+	+	V
Dryopteris filix-mas	+	+	1.3	1.5	1.4	+	+	1.5	+	+	V
Neottia nidus-avis	+	+	+	+	+	.	.	.	+	+	IV
Acer platanoides	.	.	.	+	.	.	+	1.3	.	.	II
Fraxinus excelsior	+	+	.	.	.	.	+	+	.	.	II
Lonicera xylosteum	.	+	+	.	.	.	+	.	.	.	II
Corylus avellana	.	.	.	.	.	+	+	1.3	+	.	II
Symphytum tuberosum	+	+	.	.	.	.	.	.	.	.	I
Scrophularia nodosa	.	+	.	.	.	.	.	+	.	.	I
Brachypodium sylvaticum	.	+	+	.	.	.	.	.	.	.	I
Poa nemoralis	.	.	.	.	.	+	.	.	.	.	I
Vaccinio-Piceetalia											
Picea abies	.	.	.	.	+	.	.	.	1.1	.	I
Sorbus aucuparia	.	.	.	.	.	+	.	.	.	+	I
Calamagrostis arundinacea	.	.	.	.	.	.	.	.	+	.	I
Tilio-Acerion											
Lunaria rediviva	.	.	4.5	+	+	.	+	+	+	.	III
Geranium robertianum	.	+	+	+	.	+	.	+	+	.	III
Acer pseudoplatanus	+	+	.	.	.	.	.	1.3	1.2	.	II
Tilia cordata	+	+	.	.	.	.	+	.	1.2	.	II
Ulmus glabra	.	.	.	.	.	.	.	+	+	.	I
Polystichum aculeatum	.	.	.	.	.	.	+	.	+	.	I
Companion											
Oxalis acetosella	+	+	+	+	+	1.5	3.5	1.5	+	2.5	V
Urtica dioica	.	.	+	.	+		+	+	+	.	III
Ribes uva-crispa	.	.	.	.	.	+	+	+	.	.	II
Aegopodium podagraria	+	+	+	.	.	.	.	+	.	+	II
Petasites hybridus	.	.	.	+	+	.	.	+	+	.	II
Fragaria vesca	.	.	.	.	.	+	.	+	.	.	I
Polypodium vulgare	.	.	.	.	.	+	.	+	.	.	I
Sambucus nigra	.	.	+	.	.	.	+	.	.	.	I
Geum urbanum	.	+	.	.	.	.	.	+	.	.	I
Ajuga reptans	.	.	.	.	.	.	.	+	.	.	I
Abies alba	+	+	.	.	.	.	.	.	.	.	I
Solidago virgaurea	.	.	.	.	.	.	.	+	.	.	I

Companion species with one occurrence: *Equisetum hyemale* 1: +; *Rubus idaeus* 2: +; *Ribes alpinus* 2: +; *Circaea lutetiana* 3: +; *Galeopsis speciosa* 7: +; *Stellaria holostea* 8: +; *Spiraea chamaedryfolia* 8: +

Relevé: 1–2. Râul Mare (12.07.1970); 3–5. V. Cârligului (12.07.1968); 6. V. Turcului (18.06.1969); 7–10. V. Zlătuia (6.09.1969) cover medium deep, slightly acidic, brown eu-mesobasic soils at the base of gentle slopes, saddles and peniplains from the middle montane belt (700–1100 m altitude). *Fagus sylvatica* is the dominant species in the tree layer, achieving on average 85% coverage. *Acer pseudoplatanus, Acer platanoides, Fraxinus excelsior, Tilia cordata,* and *Abies alba* also appear sporadically in these coenoses. The regional Carpathian species *Symphytum cordatum,* designated as characteristic of the association (Vida 1963) has a maximum presence in the herbaceous layer of these beech forests. Other Carpathian species characteristic of the alliance are present too, increasing the phytogeographic specificity of the association. The association also includes mesophilic species characteristic of the order *Fagetalia* and class *Carpino-Fagetea*, like those from the Tatra Mountains (Boublik et al. 2013). Some of these species are important through frequency and coverage, for example: *Galium odoratum, Oxalis acetosella, Asarum europaeum, Actaea spicata, Mercurialis perennis, Gymnocarpium dryopteris, Glechoma hirsuta, Athyrium filix-femina, Dryopteris filix-mas, Paris quadrifolia, Polygonatum verticillatum,* and *Neottia nidus-avis.*

Ass. *Festuco drymejae-Fagetum* Morariu ex Resmeriță 1977 (Table 6.57)

In this association we included the *Fagus sylvatica* coenoses that populate superficial, slightly acidic, less humid, lithic acid brown soils in the upper part of the slopes, sunny ridges and peaks of the montane belt. The characteristic herbaceous species of the association are *Festuca drymeja,* and *Luzula luzuloides*. The areas populated by these phytocoenoses in Retezat (Piciorul Șeselor, Zănoagei, Turcului, Zlătuii) are characterized by lower trophicity and limited soil water provisioning due to reduced soil volume (Doniță et al. 1990). Because of these ecological peculiarities of the above-mentioned areas, few species of the herbaceous layer achieve 5–10% coverage. Some of these are: *Galium odoratum, Oxalis acetosella, Dryopteris filix-mas,* and *Athyrium filix-femina*. The high number of mesophilic nemoral species present in the association fully justifies the association's inclusion in the alliance *Symphyto-Fagion*, order *Fagetalia* and class *Carpino-Fagetea* (Coldea 2015).

Ass. *Phyllitidi-Fagetum* Vida (1959) 1963 (Table 6.58)
Natura 2000: prority habitat type 9180

We grouped in this association the beech forests of steep and rocky slopes between 850 and 1200 m altitude from the western and southern part of Retezat (Râul Mare, Valea Zlătuia, Valea Turcului, Zlata Mountain, Piatra Iorgovanului). The tree layer of the association is dominated by *Fagus sylvatica* and *Acer pseudoplatanus*, achieving on average 75% coverage. In the herbaceous layer of the phytocoenoses, beside *Asplenium scolopendrium*, designated specific to the association, there are also characteristic species of *Tilio-Acerion* and *Symphyto-Fagion*. The higher presence of species of *Symphyto-Fagion* and the beech dominance in such phytocoenoses led the author of the association to name it *Phyllitidi-Fagetum* and regarded it as a geographical synvicariant of the association *Phyllitidi-Aceretum* Moor 1952 in Central Europe (Vida 1963). The characteristic species of *Fagetalia* and *Carpino-Fagetea* are well represented in the structure of the association, similar to those from the Cerna Mountains (Boșcaiu 1971).

Table 6.57 Ass. *Festuco drymejae-Fagetum* Morariu ex Resmeriţă 1977

Relevé No.	1	2	3	4	5	6	7	
Altitude (m a.s.l.)	1000	970	950	800	950	1000	1050	
Aspect	NW	NE	NW	N	SW	SE	SW	
Slope (degrees)	40	25	10	10	30	35	35	
Tree height (m)	25	24	25	23	28	30	25	
Canopy cover (%)	80	80	90	70	80	80	90	
Herb cover (%)	10	20	45	50	40	60	30	
Litter cover (%)	80	85	80	50	90	80	75	
Sample area (sq. m)	400	400	400	400	400	400	400	K
Char. ass.								
Fagus sylvatica	5.5	5.5	5.5	4.5	5.5	5.5	5.5	V
Festuca drymeja	1.3	1.5	3.5	1.5	3.5	3.5	1.5	V
Luzula luzuloides	.	.	+	1.3	+.3	1.5	+	IV
Symphyto-Fagion								
Cardamine glanduligera	+	+	.	.	+	1.2	2.5	IV
Pulmonaria rubra	.	+	.	+	.	.	.	II
Hieracium transsilvanicum	.	.	+	.	.	+	.	II
Hepatica transsilvanica	.	.	+	.	.	.	.	I
Symphytum cordatum	.	.	.	+	.	.	.	I
Fagetalia								
Galium odoratum	+	+	+.3	1.5	+.5	+	+	V
Asarum europaeum	+	+	+	+	+	+	+	V
Senecio nemorensis	+	+	+	+	.	+	+	V
Lamium galeobdolon	+	+	.	+	+	+	+	V
Mercurialis perennis	+	+	+	+	+	+	+	V
Mycelis muralis	+	+	+	+	+	+	+	V
Salvia glutinosa	.	+	+	+	+	+	+	V
Prenanthes purpurea	+	+	+	.	+	+	+	V
Sanicula europaea	+	+	+	+	+.3	+.3	1.5	V
Actaea spicata	+	.	+	.	+	+	+	IV
Polygonatum verticillatum	+	+	+	+	+	+	.	IV
Cardamine bulbifera	+	.	+	.	+	+	1.5	IV
Paris quadrifolia	.	.	.	+	.	+	+	III
Gentiana asclepiadea	.	.	+	+	.	+	.	III
Epilobium montanum	+	.	+	.	.	+	.	III
Viola reichenbachiana	.	+	+	.	+	+	.	III
Gymnocarpium dryopteris	+	+	.	+	.	+	.	III
Euphorbia amygdaloides	.	.	+	.	+.3	+	+	III
Daphne mezereum	.	.	+	.	.	+	.	II
Hordelymus europaeus	.	+.3	+	.	.	.	.	II
Anemone nemorosa	.	.	.	.	.	+	+	II
Veronica urticifolia	.	.	.	.	.	+	.	I
Adoxa moschatelina	.	.	.	+	.	.	.	I

(continued)

Table 6.57 (continued)

Carex sylvatica	.	.	.	+	.	.	.	I
Lilium martagon	.	+	.	.	.	.	.	I
Pulmonaria officinalis	.	.	+	.	.	.	.	I
Carpino-Fagetea								
Glechoma hirsuta	+	+	+	1.4	+	+	.	V
Dryopteris filix-mas	+	.	+	1.5	1.4	+	.	IV
Neottia nidus-avis	+	.	+	.	+	+	+	IV
Acer platanoides	.	+	+	1.3	.	.	.	III
Lonicera xylosteum	.	+	+	.	.	+	.	III
Corylus avellana	.	.	.	1.4	.	+	+	III
Athyrium filix-femina	.	2.5	+	2.5	.	.	.	III
Scrophularia nodosa	.	.	.	+	.	+	+	III
Brachypodium sylvaticum	.	+	+	.	.	.	+	III
Cephalanthera rubra	+	.	.	.	+	+	+	III
Lathyrus vernus	+	.	.	.	+	+	+	III
Symphytum tuberosum	.	.	.	.	.	+	+	II
Poa nemoralis	+	.	.	.	.	.	+	II
Hedera helix	.	.	.	.	+	.	+	II
Melica uniflora	+	.	.	.	.	.	.	I
Fraxinus excelsior	.	.	.	+	.	.	.	I
Vaccinio-Piceetalia								
Oxalis acetosella	+	+	1.5	1.5	+	1.5	1.5	V
Picea abies	+	.	.	.	+	+	+	III
Sorbus aucuparia	.	+	.	.	+	.	+	III
Calamagrostis arundinacea	.	.	+	.	.	2.5	.	II
Monotropa hypopitys	+	+	.	.	.	.	.	II
Tilio-Acerion								
Acer pseudoplatanus	.	.	+	.	.	.	+	II
Lunaria rediviva	.	.	+	+	.	.	.	II
Ribes uva-crispa	.	.	.	+	.	.	.	I
Ulmus glabra	.	+	.	.	.	.	.	I
Companion								
Abies alba	+	.	.	.	+	+	.	III
Geranium robertianum	.	.	+	+	.	.	+	III
Urtica dioica	+	.	.	+	.	.	+	III
Solidago virgaurea	.	.	+	+	.	+	.	III
Ajuga reptans	.	.	.	+	.	+	.	II
Aegopodium podagraria	.	.	+	+	.	.	.	II
Fragaria vesca	.	.	.	+	.	+	.	II
Petasites hybridus	.	.	.	+	.	+	.	II
Polypodium vulgare	+	.	.	+	.	.	.	II
Rubus idaeus	.	+	.	.	.	+	.	II
Plathanthera bifolia	.	.	.	.	.	+	.	I
Geum urbanum	.	.	.	+	.	.	.	I

Relevé location: 1–2. Piciorul Zănoaga (10.07.1968); 3–4. V. Turcului (18.06.1969); 5–6. V. Zlătuia (6.09.1969); 7. Piciorul Şeselor (21.06.1968)

Table 6.58 Ass. *Phyllitidi-Fagetum* Vida (1959) 1963

Relevé No.	1	2	3	4	5	6	7	8	
Altitude (m a.s.l.)	1160	850	1100	1000	1120	1150	1200	1180	
Aspect	SW	N	N	W	W	NW	SE	SE	
Slope (degrees)	45	10	40	50	60	40	50	45	
Tree height (m)	27	25	25	27	20	22	20	21	
Canopy cover (%)	90	80	80	90	80	80	90	90	
Herb cover (%)	60	30	70	25	60	50	25	20	
Litter cover (%)	30	50	30	25	30	40	30	30	
Sample area (sq. m)	400	400	400	400	400	400	400	400	K
Char. ass.									
Fagus sylvatica	2.3	3.5	2.5	4.5	3.5	2.5	5.5	5.5	V
Asplenium scolopendrium	+	.	.	+	+	.	+.2	+.3	IV
Tilio-Acerenion **et** ***Symphyto-Fagion***									
Acer pseudoplatanus	4.5	1.2	3.5	2.5	3.5	3.5	1.5	2.4	V
Pulmonaria rubra	+.3	.	+	+.4	+.3	2.5	+	+	V
Lunaria rediviva	3.5	+	4.5	+	+	1.4	+	.	IV
Polystichum aculeatum	.	+.3	.	+	+.3	+	+	+.2	IV
Symphytum cordatum	+.4	+	.	+	+.5	1.5	.	+	IV
Silene heuffelii	+	.	.	+	.	.	.	.	II
Cardamine glanduligera	.	.	.	+	+	+	.	.	II
Aremonia agrimonoides	.	.	.	.	.	.	+	+.3	II
Festuca drymeja	.	.	.	.	.	.	1.2	+	II
Moehringia muscosa	.	.	.	.	.	.	+	.	I
Geranium macrorrhizum	.	.	.	.	.	.	.	+	I
Aconitum lycoctonum subsp. *moldavicum*	+	.	.	.	+	.	.	.	I
Polystichum setiferum	.	.	.	.	.	.	+	.	I
Fagetalia									
Galium odoratum	+	.	+	1.5	+	1.5	2.3	2.4	V
Dryopteris filix-mas	2.5	+	2.5	+	+	2.5	+	+	V
Ulmus glabra	+	.	+	+	+	.	+	+	IV
Mycelis muralis	+	.	.	+	+	+	.	+	IV
Senecio nemorensis	+	.	+	1.5	+	+	+	.	IV
Mercurialis perennis	1.5	+	1.5	1.5	4.5	.	+	.	IV
Stellaria nemorum	2.5	.	+	+	+	1.5	+	.	IV
Athyrium filix-femina	+	2.5	+	+	.	+	.	.	IV
Hordelymus europaeus	+	.	+	.	.	+	+	+	IV
Actaea spicata	.	+	+	.	+.5	.	+	+	IV
Polygonatum verticillatum	.	+	+	.	+	.	+	+	IV
Euphorbia amygdaloides	.	.	+	.	+	+	+	+.2	IV
Asarum europaeum	+	+	.	+	+	.	.	.	III
Epilobium montanum	+	.	.	+	.	.	+	+	III
Daphne mezereum	+	.	.	.	+	.	+	+	III
Prenanthes purpurea	+	.	.	.	.	+	+	+	III
Carex sylvatica	.	.	.	.	.	.	+	+	II

(continued)

Table 6.58 (continued)

Cardamine bulbifera	.	.	.	.	.	.	+	+	II
Milium effusum	+	.	.	.	.	+	.	+	II
Paris quadrifolia	+	.	.	.	+	.	.	.	II
Lilium martagon	.	.	+	.	+	.	.	.	II
Scrophularia nodosa	+	.	.	.	.	.	+	+	II
Lamium galeobdolon	.	.	.	.	.	.	1.3	+	II
Carpino-Fagetea									
Glechoma hirsuta	+	+	+	+	+	.	+	.	IV
Lonicera nigra	.	.	2.4	.	.	+	+	+	III
Impatiens noli-tangere	+	.	+	+	+	.	.	.	III
Fraxinus excelsior	+	.	.	.	.	.	+	.	II
Tilia cordata	+	2.4	.	.	.	.	.	.	II
Corylus avellana	.	+	.	+	.	.	.	.	II
Brachypodium sylvaticum	+	.	.	.	.	.	.	+	II
Moehringia trinervia	.	.	.	.	+	.	+	.	II
Poa nemoralis	.	.	.	.	.	.	1.3	1.3	II
Vaccinio-Piceetalia									
Picea abies	.	+	1.3	+	+	+	+	+	V
Oxalis acetosella	.	+	+	+	+	+	.	+	IV
Sorbus aucuparia	+	.	.	.	.	1.5	.	.	II
Abies alba	1.3	.	.	.	1.2	+	.	.	II
Calamagrostis arundinacea	+	.	.	.	.	.	+	+	II
Luzula sylvatica	.	.	.	.	.	.	.	+	I
Adenostyletalia									
Adenostyles alliariae	1.5	.	.	+	1.2	2.5	.	+	IV
Streptopus amplexifolius	+	.	.	.	.	+	.	.	II
Doronicum austriacum	+	.	.	.	.	+	.	.	II
Cicerbita alpina	+	.	.	.	.	.	.	.	I
Thalictrum aquilegifolium	+	.	.	.	.	.	.	.	I
Companion									
Urtica dioica	+	+	+	+.5	+	+	.	+	V
Geranium robertianum	+	+	+	.	+	.	+	+	IV
Sambucus nigra	.	+	+	.	.	.	.	.	II
Chrysosplenium alternifolium	.	.	+	.	+	.	.	.	II
Symphytum tuberosum	.	.	.	+	.	.	+	+	II
Saxifraga rotundifolia subsp. *heucherifolia*	.	.	.	.	+	.	+	+	II
Betula pendula	.	1.2	.	.	.	.	.	.	I

Companion species with on or two occurrences: *Telekia speciosa* 1: +; *Ribes alpinus* 1: +; *Pulmonaria officinalis* 2: +; *Gymnocarpium dryopteris* 2: +; *Sanicula europaea* 2: +; *Viola sylvestris* 2: +; *Neottia nidus-avis* 2: +; *Petasites hybridus* 2: +; *Rubus idaeus* 1: +; *Fragaria vesca* 1, 8: +; *Gentiana asclepiadea* 3: +; *Heracleum sphondylium* 3: +; *Galeopsis speciosa* 4: +; *Doronicum columnae* 4: +; *Anthriscus sylvestris* 1: +; *Cystopteris fragilis* 5, 7: +; *Adoxa moscharellina* 5: +; *Chaerophyllum hirsutum* 5: +; *Asplenium trichomanes* 5, 7: +; *Veronica urticifolia* 5: +; *Chaerophyllum aromaticum* 6: +; *Crepis paludosa* 6: +; *Spiraea chamaedrifolia* 6: +

Relevé: 1. V. Râul Mare (12.07.1970); 2. V. Zlătuia (7.09.1969); 3. V. Turcului (9.09.1969); 4–6. Mt. Zlata (11.07.1970); 7–8. Mt. Piciorul Iorgovanului (1.08.1986)

Ass. *Pulmonario rubrae-Fagetum* (Soó 1964) Täuber 1987 (Table 6.59)

The mixed beech-fir forests have an azonal distribution in some valleys of the Retezat Massif (Râul Mare, Lăpuşnicu Mare, Zănoguţa) as sporadic enclaves developed on shaded gentle slopes with moist, acid, brown soils. *Fagus sylvatica* dominates in the tree layer, with small exceptions. In the herbaceous layer, besides the regional Carpathian-Balkan species *Pulmonaria rubra*, the characteristic species of the alliance *Symphyto-Fagion* (Täuber 1987; Vida 1963) are well represented, differentiating them floristically from those described in the Western Carpathians (Boublik et al. 2013). The meso-hygrophilous specificity of these beech and fir forests is revealed by *Milium effusum, Stellaria nemorum, Doronicum austriacum,* and *Adenostyles alliariae*, sporadically present in the association. The species *Lonicera nigra* and *Daphne mezereum* have a higher presence in the shrub layer.

Ass. *Chrysanthemo rotundifolii-Fagetum* (Soó 1964) Täuber 1987 (Table 6.60)

The mixed spruce-beech forests from Retezat have the character of a zonal, montan climax and are mostly located in the central areas of the massif. The *Fagus sylvatica* and *Picea abies* tree species are codominant, their relative abundance varying with altitude (1100–1350 m), soil types (acid brown and feriiluvial brown) and soil trophicity (medium or low). The herbaceous species designated as characteristic of the association are *Leucanthemum rotundifolium* and *Athyrium distentifolium* (Vida 1963). Compared to the beech and spruce forests of the Western Carpathians (Boublik et al. 2013), those from Retezat have in their herbaceous layer a core of species characteristic of the alliance *Symphyto-Fagion*, similarly to other mountain ranges of the Southeast Carpathians (Boşcaiu 1971; Coldea 1990; Mihăilescu 2001; Nechita 2003). The structure of the association includes also a greater number of both mesophilic species characteristic of the order *Fagetalia* and of the class *Carpino-Fagetea* (Coldea 2015), respectively.

Class *Vaccinio-Piceetea* Br.-Bl. in Br.-Bl. et al. 1939

This class groups the coniferous forests of the boreal zone and those of the montane and subalpine belts in the temperate zone of Europe. In the Retezat Mountains, the coniferous forests consist of coenoses dominated by *Picea abies, Pinus cembra, Pinus mugo, Pinus sylvestris,* and *Juniperus communis* subsp. *alpina*. Regarding altitude, these forest/scrub phytocoenoses have a zonal or intrazonal distribution between 1350 and 2200 m (Radeşu Mic, Faţa Retezatului, Judele, Pietrele, Şesele). Cryptopodzols and podzols are the typical soils of such phytocoenoses. Among the species characteristic of the class we mention: *Picea abies, Vaccinium myrtillus, Vaccinium vitis-idaea, Moneses uniflora, Oxalis acetosella, Dryopteris dilatata, Huperzia selago, Orthilia secunda, Deschampsia flexuosa, Dicranum scoparium, Pleurozium schreberi,* and *Hylocomium splendens* (Exner 2007; Matuszkiewicz 2008; Pawlowski and Walas 1949; Pawlowski et al. 1928). The class includes two orders: *Piceetalia* and *Junipero-Pinetalia mugo*.

Table 6.59 Ass. *Pulmonario rubrae-Fagetum* (Soó 1964) Täuber 1987

Relevé No.	1	2	3	4	5	6	7	8	
Altitude (m a.s.l.)	1210	1100	1090	1020	1050	1000	970	950	
Aspect	S	S	SW	W	SW	SW	SW	NE	
Slope (degrees)	10	10	25	5	5	10	15	20	
Tree height (m)	27	28	26	26	26	25	25	30	
Canopy cover (%)	90	90	90	100	90	80	80	90	
Herb cover (%)	40	20	30	40	25	30	30	20	
Litter cover (%)	20	60	80	70	90	80	75	75	
Sample area (sq. m)	400	400	400	400	400	400	400	400	K
Char. ass.									
Fagus sylvatica	1.3	2.5	5.5	5.5	5.5	3.5	4.5	3.5	V
Abies alba	5.5	4.5	1.5	1.5	1.5	2.5	1.5	3.5	V
Pulmonaria rubra	+.5	+	+	+	+.3	+.4	.	+.5	V
Symphyto-Fagion									
Cardamine glanduligera	+	+	+.5	+.5	+.5	+.4	+	+	V
Symphytum cordatum	.	.	+	1.5	+.4	+.5	+	.	IV
Festuca drymeja	.	.	2.5	1.3	1.3	+	.	.	III
Hepatica transsilvanica	.	.	.	.	+	.	+.5	.	II
Silene heuffelii	+	.	.	.	.	.	.	+	II
Campanula abietina	.	+	.	.	.	+	.	.	II
Fagetalia									
Polygonatum verticillatum	+	+	+	+	+	+	+	+	V
Galium odoratum	+	.	+	1.5	+.5	1.3	1.5	1.5	V
Asarum europaeum	+	+	+	.	+.5	+.5	+	.	IV
Milium effusum	1.4	+	+	.	+	+	+	+	IV
Actaea spicata	+	.	+	+	.	+	+	1.4	IV
Senecio nemorensis	.	+	+	+	.	+.3	+	+.5	IV
Paris quadrifolia	+	.	.	+	+	+	+	+	IV
Daphne mezereum	.	.	.	.	+	+	+	+	III
Sanicula europaea	.	.	+	+.3	+	+	1.5	.	III
Dryopteris filix-mas	+	.	+	+	.	+	+	+	III
Athyrium filix-femina	.	.	+	+.3	+	.	+	+	III
Mycelis muralis	+	.	.	.	.	+	+	+	III
Epilobium montanum	+	.	+	.	.	+	+	.	III
Lamium galeobdolon	.	+	+	.	.	+	.	.	II
Ulmus glabra	+	.	.	.	.	.	.	+	II
Prenanthes purpurea	.	+	+	.	.	.	+	+.5	II
Euphorbia amygdaloides	.	.	+	+	+	+	.	.	II
Mercurialis perennis	.	.	.	+	.	+	+	.	II
Stellaria nemorum	.	.	.	.	+	.	.	+	II
Hordelymus europaeus	.	.	.	+.3	.	+	.	.	II
Carpino-Fagetea									
Glechoma hirsuta	+	+	+	+	+	+.5	+	+	V

(continued)

Table 6.59 (continued)

Neottia nidus-avis	+	+	+	.	.	+	+	.	III
Platanthera chlorantha	.	.	+	+	.	+	.	.	II
Lonicera xylosteum	+	+	.	.	.	.	.	.	II
Corylus avellana	.	.	.	.	+	.	.	+	II
Vaccinio-Piceetalia									
Lonicera nigra	.	.	+	+	+	+	+	2.5	IV
Oxalis acetosella	3.5	2.5	.	1.5	2.5	2.5	.	+	IV
Picea abies	1.3	+.3	+	.	.	.	+	.	III
Sorbus aucuparia	+	+	.	+	.	.	.	+	III
Companion									
Calamagrostis arundinacea	+	1.4	.	.	.	1.5	1.5	+	III
Adenostyles alliariae	.	.	.	+	+	+	+	.	III
Doronicum austriacum	+	.	+	+	.	+	.	.	III
Geranium robertianum	+	.	.	.	+	.	.	+	II
Urtica dioica	.	.	.	+	.	.	+	+	II

Companion species with one occurrence: *Spiraea chamaedryfolia* 1: +; *Luzula luzuloides* 1: +; *Cardamine bulbifera* 3: +; *Anemone nemorosa* 3: +; *Gymnocarpium dryopteris* 6: +; *Carex sylvatica* 6: +; *Adoxa moschatellina* 6: +; *Digitalis grandiflora* 6: +; *Tanacetum corymbosum* 6: +; *Gentiana asclepiadea* 7: +; *Rosa pendulina* 8: +; *Acer pseudoplatanus* 8: +; *Cicerbita alpina* 8: +; *Streptopus amplexifolius* 8: +
Relevé*:* 1–2. V. Lăpuşnicului Mare (13.07.1970); 3. V. Râul Mare (12.07.1970); 4–8. Mt. Zlata (9–11.07.1970)

Order *Piceetalia excelsae* Pawł. in Pawł. et al. 1928

This order includes the zonal coniferous forests of Europe dominated by *Picea abies* and *Abies alba*, widely distributed in the upper montane and subalpine belt, and the *Pinus sylvestris* coenoses with azonal distribution in the montane belt. Nutrient-poor soils such as podzolic acid brown and regosols are characteristic of such phytocoenoses, located between 700 and 1700 m altitude in the Retezat Mountains. Subalpine forests in many areas of Retezat have been cleared in the last two centuries for pasture expansion, like in the Ţarcu-Godeanu Mountains (Boşcaiu 1971). The main characteristic species of the order, present in Retezat, are: *Picea abies, Abies alba, Hieracium transsilvanicum, Leucanthemum rotundifolium, Lycopodium annotinum, Orthilia secunda, Campanula abietina,* and *Pleurozium schreberi.*

The order includes three alliances: *Piceion excelsae, Chrysanthemo-Piceion* and *Dicrano-Pinion sylvestris.*

Alliance *Dicrano-Pinion* (Libbert 1933) Matusz. 1962
Natura 2000: habitat type 91T0

We group in this alliance the coenoses of *Pinus sylvestris* from Retezat, which only occur on some granite rocks in the Râului Mare valley, between 650 m and 760 m altitude (Lunca Rotundă). The assignation of this alliance to an order and a class varies from author to author. Oberdorfer (2001) and Matuszkiewicz (2008) assigned it to the order *Cladonio-Vaccinetalia* and the class *Vaccinio-Piceetea*. Eichberger

Table 6.60 Ass. *Chrysanthemo rotundifolii-Fagetum* (Soó 1964) Täuber 1987

Relevé No.	1	2	3	4	5	6	7	8	
Altitude (m a.s.l.)	1100	1150	1100	1130	1000	1124	1340	1160	
Aspect	SW	NW	NW	W	W	SW	SW	–	
Slope (degrees)	30	50	15	15	10	25	15	0	
Tree height (m)	18	30	30	25	30	30	25	27	
Canopy cover (%)	70	80	80	90	90	80	90	90	
Herb cover (%)	30	20	70	40	30	25	30	50	
Litter cover (%)	60	90	30	60	80	70	60	30	
Sample area (sq. m)	400	400	400	400	400	400	400	400	K
Char. ass.									
Fagus sylvatica	3.5	4.5	5.5	5.5	5.5	4.5	2.5	1.5	V
Leucanthemum rotundifolium	+	.	+	.	.	+	.	+	III
Athyrium distentifolium	.	+	.	+	.	+	+	.	III
Symphyto-Fagion									
Festuca drymeja	+	+	4.5	3.5	1.5	1.5	1.5	.	V
Cardamine glanduligera	+	1.5	+	1.3	+	.	.	+	IV
Pulmonaria rubra	+	+	.	.	.	+	+	1.5	IV
Silene heuffelii	.	.	.	.	.	.	+	+	II
Campanula abietina	.	.	.	.	.	.	.	+	I
Tilio-Acerenion									
Acer pseudoplatanus	+	+	.	.	+	.	+	+	IV
Lunaria rediviva	1.5	+	.	.	.	.	.	.	II
Fagetalia									
Milium effusum	+	.	+.3	+	+	+	1.5	2.5	V
Mycelis muralis	+	+	+	+	+	+	+	+	V
Senecio nemorensis	+	+	+	+	+	+	+	+	V
Actaea spicata	+	+	+	+	+	+	+	.	V
Polygonatum verticillatum	+	+	+	+	+	+	+	.	V
Prenanthes purpurea	+	+	+	+	+	+	+.5	.	V
Galium odoratum	+.5	1.5	+	+	+	1.5	1.5	.	V
Asarum europaeum	.	1.5	+	+	+	+	+	+	V
Mercurialis perennis	1.5	1.5	+	+	.	.	+	+	IV
Epilobium montanum	.	+	+	+	.	+	+	.	IV
Lamium galeobdolon	+	.	+	.	.	+	+	+	IV
Euphorbia amygdaloides	+	+	+	+	.	+	+	.	IV
Cardamine bulbifera	+	1.5	+	1.3	+	.	.	+	IV
Paris quadrifolia	+	+	+	+	.	.	+	+	IV
Veronica urticifolia	+	.	.	+	.	+	+	.	III
Daphne mezereum	+	.	+	+	.	.	+	.	III
Luzula luzuloides	.	.	.	.	.	+	+	+	II
Salvia glutinosa	.	+	+	+	.	.	.	.	II
Stellaria nemorum	+	.	.	.	.	.	.	+.3	II

(continued)

Table 6.60 (continued)

Hordelymus europaeus	1.5	.	+.3	.	+.3	.	.	.	II
Gymnocarpium dryopteris	.	+	+	+	.	.	.	.	II
Platanthera chlorantha	.	.	+	+	.	+	.	.	II
Gentiana asclepiadea	.	+	+	.	.	.	.	.	II
Anemone nemorosa	.	+	.	+.3	.	.	.	.	II
Lilium martagon	+	.	.	.	+	.	.	.	II
Carpino-Fagetea									
Glechoma hirsuta	+	+	+	.	+	+	+	+	V
Dryopteris filix-mas	1.5	+	1.5	.	2.5	+	+	+	V
Neottia nidus-avis	+	+	+	+	+	+	+	.	IV
Lathyrus vernus	+	+	+	+	+	.	.	.	IV
Cephalanthera rubra	+	+	+	+	.	.	.	.	III
Lonicera xylosteum	+	.	.	.	.	.	+	1.5	II
Corylus avellana	+	.	.	.	+	.	.	.	II
Brachypodium sylvaticum	+	.	+	.	.	.	.	.	II
Symphytum tuberosum	.	.	.	.	.	+	+	+	II
Viola reichenbachiana	+	.	+	+	.	.	.	.	II
Poa nemoralis	.	.	.	.	.	+	+	.	II
Scrophularia nodosa	.	.	.	.	.	+	+	+	II
Calamagrostis arundinacea	.	.	.	.	+	1.5	1.5	.	II
Euonymus europaeus	.	+	.	.	.	.	.	.	I
Vaccinio-Piceetea									
Picea abies	2.3	1.5	1.5	1.4	1.5	2.5	4.5	3.5	V
Oxalis acetosella	+	+	+	+.5	+	1.5	1.5	3.5	V
Abies alba	.	+	.	+	+	+	1.3	3.5	IV
Sorbus aucuparia	.	.	.	.	+.3	+	+	+	III
Lonicera nigra	+	.	.	.	+.3	+	.	.	II
Dryopteris dilatata	1.5	.	+	.	.	.	.	.	II
Luzula sylvatica	.	+	.	.	.	.	.	+	II
Companion									
Rubus idaeus	+	.	.	+	+	+	.	+	IV
Geranium robertianum	+	+	.	.	+	+	+	.	IV
Sambucus racemosa	.	.	.	.	+	+	+	+	III
Aegopodium podagraria	+	.	+	.	.	.	.	+	II
Doronicum austriacum	.	.	.	.	.	+	+	+	II
Cirsium oleraceum	.	.	.	.	.	+	+	.	II
Ranunculus platanifolius	.	.	.	.	.	.	+	+	II
Lamium maculatum	.	+	+	+	.	.	.	.	II
Urtica dioica	+	.	+	.	.	.	.	.	II
Fragaria vesca	.	.	.	.	.	+	+	+	II

Companion species with one occurrence: *Ulmus glabra* 2: +; *Cicerbita alpina* 1: +; *Cirsium erisithales* 1: +; *Vicia sylvatica* 1: +; *Heracleum sphondylium* 1: +; *Scrophularia alata* 3: +; *Aconitum lycoctonum* subsp. *moldavicum* 4: +; *Petasites hybridus* 4: +; *Ribes uva-crispa* 5: +; *Lapsana communis* 6: +; *Adenostyles alliariae* 7: +; *Myosotis sylvatica* 8: +; *Solidago virgaurea* 8: +; *Streptopus ampexifolius* 8: +; *Epipactis latifolia* 7: +

Relevé*:* 1–2. V. Turcului (18.06.1968); 3, 5. V. Zlătuia (7.09.1969); 4. Şesele (10.07.1968); 6–7. Mt. Zlata (8.07.1970); 8. V. Lăpuşnicul Mare (9.07.1970)

et al. (2007) included it into the class *Erico-Pinetea* and the order *Vaccinio-Pinetalia sylvestris*. Rodwell et al. (2002) and Pedrotti (2004) classified it in the order *Piceetalia* and the class *Vaccinio-Piceetea*. In recent years, Chytry et al. (2013) and Mucina (2016) placed this alliance in the order *Pinetalia sylvestris* Oberd. 1957 and in the class *Vaccinio-Piceetea*. Following the floristic analysis of these phytocoenoses, we consider more appropriate the inclusion of the alliance in the order *Piceetalia*. Among the species characteristic of the alliance, *Pinus sylvestris, Dicranum scoparium,* and *Cladonia sylvatica* are present in the area investigated.

Ass. *Vaccinio myrtilli-Pinetum sylvestris* Juraszck 1928 (Table 6.61)
– *jovibarbetosum heuffelii* subass. nova hoc loco
Holotype: Table 6.61, rel. 4 hoc loco

In this association we group the relict coenoses of *Pinus sylvestris* which occur on superficial, poorly developed, nutrient-depleted, slightly acidic lithosols of steep granite rocks of the Râul Mare valley. The dominant tree species are *Pinus sylvestris, Quercus petraea* and *Betula pendula*. In the herbaceous layer some acidophilic species typical of spruce forests (*Vaccinium myrtillus, Deschampsia flexuosa, Luzula luzuloides, Calamagrostis arundinacea, Dicranum scoparium*) together with several neutrophilic Carpathian-Balkan species, such as *Jovibarba heuffelii, Moehringia pendula, Silene nutans* subsp. *dubia, Dianthus tenuifolium, Symphyandra wanneri* have a high presence and ocassionally 5–15% coverage. Based on these regional species, we describe a new regional sub-association—*jovibarbetosum heuffelii*, specific to the Southeastern Carpathians. The pre-boreal age of these rocky pine forests in Retezat is attested by the spore-pollen diagram from Clopotiva (Boşcaiu 1971). The persistence of these phytocoenoses was ensured by the extreme conditions of the steep rocky substrate and the lithosols that prevented the installation of more competitive, late-seral cenoses of *Fagus sylvatica*.

Alliance *Piceion excelsae* Pawl. in Pawl. et al. 1928
Natura 2000: habitat type 9410

We include in this alliance the *Picea abies* and *Abies alba* forest coenoses vegetating mainly in the upper montane belt (1300–1700 m) of the Retezat Mountains. The eco-pedological conditions in the habitats where such phytocoenoses occur are, in general, similar, differing in terms of soil moisture and trophicity. Among the characteristic species of the alliance, we mention: *Luzula sylvatica, Homogyne alpina, Blechnum spicant, Dryopteris dilatata, Soldanella major, Lonicera nigra, Plagiothecium undulatum,* and *Sphagnum girgensohnii* (Pawlowski et al. 1928; Szafer et al. 1923).

Ass. *Hieracio transsilvanici-Piceetum* Pawl. et Br.-Bl. 1939 (Table 6.62)

The spruce forests that we grouped in this association are frequently distributed in the upper montane and subalpine belts of the Retezat, populating moderately steep slopes, with acid brown and cryptospodic acid soils, rich in moder type humus but with low moisture. We observed representative coenoses for the association on Zlata Mountain, Piciorul Şeselor, Radeşul Mare, Judele, and Pietrele. *Picea abies* is

Table 6.61 Ass. *Vaccinio myrtilli-Pinetum sylvestris* Juraszek 1928 *jovibarbetosum heuffelii* subass. nova hoc loco

Relevé No.	1	2	3	4	5	6	7	
Altitude (m a.s.l.)	650	670	700	750	760	765	700	
Aspect	SW	SW	SW	SW	SW	SW	SW	
Slope (degrees)	20	25	30	35	35	50	40	
Tree height (m)	14	14	15	16	15	16	16	
Canopy cover (%)	60	70	70	70	70	60	70	
Herb cover (%)	30	40	20	25	20	25	45	
Litter cover (%)	10	20	20	30	20	10	10	
Sample area (sq. m)	200	200	250	200	200	400	400	K
Char. ass.								
Pinus sylvestris	2.4	3.4	4.5	2.5	3.5	4.5	3.5	V
Spiraea chamaedryfolia	.	.	+	+	.	+	+	III
Diff. subass. nova								
Jovibarba heuffelii	+	+	1.5	1.5	+	+.3	+	V
Silene nutans subsp. *dubia*	+	+	+.2	+.3	+	1.5	1.3	V
Moehringia pendula	+	2.4	1.3	1.3	+	+	+	V
Dianthus tenuifolius	+	+	+	+	+	+.3	+.2	V
Symphyandra wanneri	+	+.3	.	.	+	.	.	III
Galium kitaibelianum	+	1.3	.	.	.	+	+	III
Veronica bachofenii	.	+	.	+	.	.	.	II
Dicrano-Pinion et ***Piceetalia***								
Betula pendula	1.3	1.3	+	2.5	2.5	1.2	.	V
Vaccinium myrtillus	+	+	3.5	+.3	2.4	2.5	2.3	V
Deschampsia flexuosa	1.3	2.5	1.4	2.5	2.5	1.5	1.5	V
Luzula luzuloides	+	+	1.3	1.3	1.5	+.3	2.5	V
Calamagrostis arundinacea	1.4	+	+	+	.	+	+	V
Dicranum scoparium	1.3	+	3.4	2.4	2.5	1.3	1.5	V
Cytisus hirsutus	+	.	+	+	+	+	+	IV
Sorbus aucuparia	.	.	+	+	+	+	.	III
Picea abies (juv.)	.	.	.	.	.	+	+	II
Veronica officinalis	+	+	.	.	+			III
Prenanthes purpurea	.	.	.	+	+	.	.	II
Melampyrum pratense	+	.	.	.	.	.	+	II
Hylocomium splendens	.	.	2.3	.	+	.	.	II
Carpino-Fagetea s.l.								
Quercus petraea	3.5	2.4	+	2.5	1.4	1.2	2.3	V
Poa nemoralis	2.4	2.3	.	+	+	+	+	V
Fraxinus excelsior	+	+	+	.	.	.	.	III
Fagus sylvatica	.	.	+	.	+	+	2.2	III
Corylus avellana	+	.	+	.	+	.	.	III
Tilia cordata	+	.	+	.	.	.	.	II

(continued)

Table 6.61 (continued)

Companion								
Sedum telephium subsp. *maximum*	+	+	+	+	+	+	+	V
Solidago virgaurea	1.3	1.3	+	+.5	.	+	+	V
Polypodium vulgare	+	+	+	.	.	+	+	IV
Hypnum cupressiforme	.	.	1.3	1.3	+	1.3	1.4	IV
Hypericum perforatum	+	+	.	+	+	.	.	III
Asplenium trichomanes	+	.	+	.	.	.	+	III
Hieracium umbellatum	+	+	+	.	+	.	.	II
Polytrichum juniperinum	.	.	+	.	.	+	.	II
Sedum acre	+	.	.	+	.	.	.	II
Cladonia sylvatica	.	.	+	.	.	+	.	II
Achillea distans	+	+	.	.	.	.	.	II
Genista tinctoria	.	.	.	.	.	+	.	II
Campanula persicifolia	.	+	.	.	.	.	+	II
Seseli libanotis	+	.	.	.	.	.	.	I
Trifolium medium s.l.	+	.	.	.	.	.	.	I
Cardaminopsis arenosa	.	.	.	.	+	.	.	I

Relevé: 1–5. Lunca Rotundă-Râul Mare (14.09.1970); 6–7. Lunca Rotundă-Râul Mare (15.07.2020)

monodominant in the tree layer, achieving on average 75% coverage. *Abies alba, Acer pseudoplatanus,* and *Sorbus aucuparia* are also present sporadically. The characteristic species of the association, *Hieracium transsilvanicum*, has a high presence in the herbaceous layer. Along with it, some species characteristic of alliance, order and class, such as *Vaccinium myrtillus, Vaccinium vitis-idaea, Homogyne alpina, Soldanella major, Campanula abietina, Oxalis acetosella, Dryopteris dilatata* are also present and in some places have a high cover. The bryophyte layer is dominated by *Dicranum scoparium, Hylocomium splendens,* and *Pleurozium schreberi* (Photo 6.10).

Alliance *Chrysanthemo rotundifolii-Piceion* (Krajina 1933) Brezina et Hadač in Hadač 1962

We group in this Carpathian specific alliance the meso-hygrophilous *Picea abies* coenoses around the intramontane valleys and springs in the upper montane and subalpine belts of the Retezat, with moist, histic, cryptospodic acid brown soils, rich in nutrients. Some transgressive hygrophilous species of *Adenostyletalia* are often present in these spruce forests. The main characteristic species of the alliance are: *Leucanthemum rotundifolium, Senecio subalpinus, Phegopteris connectilis, Aconitum variegatum* subsp. *paniculatum, Aconitum napellus* subsp. *tauricum, Cicerbita alpina,* and *Gentiana asclepiadea* (Jarolimek and Šibik 2008).

Ass. *Chrysanthemo rotundifolii-Piceetum* Krajina 1933 (Table 6.63)

Table 6.62 Ass. *Hieracio transsilvanici-Piceetum* Pawł. et Br.-Bl. 1939

Relevé No.	1	2	3	4	5	6	7	8	9	10	
Altitude (m a.s.l.)	1650	1680	1740	1750	1760	1780	1730	1490	1600	1590	
Aspect	W	W	W	SW	E	NW	W	S	W	NW	
Slope (degrees)	25	10	20	20	40	35	15	40	5	5	
Tree height (m)	15	25	30	35	20	16	21	22	20	22	
Canopy cover (%)	70	80	70	70	80	80	70	80	80	80	
Herb cover (%)	60	70	70	70	60	30	20	40	40	40	
Litter cover (%)	80	80	60	70	80	65	40	50	40	45	
Sample area (sq. m)	400	400	400	400	400	400	400	400	400	400	K
Char. ass.											
Picea abies	5.5	5.5	4.5	4.5	4.5	5.5	4.5	4.5	5.5	5.5	V
Hieracium transsilvanicum	+	2.5	.	.	+	+	.	.	+.3	+	III
Laserpitium krapfii	.	+	.	+	.	+	.	+.2	.	.	II
Piceion abietis											
Homogyne alpina	+.3	+.3	+.3	1.5	+.5	1.3	1.3	.	1.5	1.5	V
Soldanella major	1.4	+	1.5	1.5	2.5	+	1.3	.	2.5	1.5	V
Calamagrostis villosa	+.4	3.5	1.3	1.3	+	1.3	1.3	3.3	+	2.3	V
Campanula abietina	+	+	+	+	+	+	+	+	.	.	IV
Lycopodium annotinum	.	+	2.4	+	.	.	.	.	+	+	III
Lonicera nigra	.	+	.	+	.	.	.	.	+	.	II
Piceetalia											
Luzula sylvatica	2 5	+.5	1.4	1.3	+.5	+.2	+.3	+	1.5	+.3	V
Sorbus aucuparia	+	+	+	.	+	1.3	+	+	+	+	V
Melampyrum sylvaticum	.	.	+	.	+	.	+	1.3	.	.	II
Corallorrhiza trifida	.	.	+	.	.	.	+	+	.	.	II
Juniperus communis subsp. *alpina*	.	.	.	.	.	+	+	+	.	.	II

(continued)

Table 6.62 (continued)

Vaccinio-Piceetea											
Vaccinium myrtillus	+.3	3.4	+.4	+.3	1.3	.	4.5	+	+.3	2.5	V
Oxalis acetosella	4.5	2.5	4.5	4.5	2.5	+	+	.	2.5	1.5	V
Pleurozium schreberi	1.3	2.5	2.5	3.5	2.4	+	2.5	1.3	1.3	1.5	V
Dryopteris dilatata	1.3	+	+	+	+	.	.	.	+	+	IV
Dicranum scoparium	1.3	.	2.5	2.5	1.3	1.5	+	1.5	2.3	.	IV
Polytrichum formosum	.	1.3	1.3	+	1.5	.	.	+	2.4	1.3	IV
Athyrium distentifolium	1.3	+	+	.	+	+	+	.	.	.	III
Deschampsia flexuosa	+	+	+	+	.	.	.	.	+	+	III
Rhytidiadelphus triquetrus	.	1.3	1.3	.	.	1.3	.	.	+	+	III
Vaccinium vitis-idaea	.	+	+	.	.	.	+	+	.	.	II
Moneses uniflora	.	.	.	.	+	.	.	+	.	.	I
Huperzia selago	.	.	.	.	.	.	.	.	+	+	I
Fagetalia s.l.											
Prenanthes purpurea	+	+	+	+	+	+	+	.	+.3	+	V
Gentiana asclepiadea	.	+	.	+	.	+	+	.	+	+	III
Valeriana tripteris	.	.	+	.	+	+	.	.	+	+	III
Polygonatum verticillatum	.	+	+	+	.	+.2	.	.	+	+.2	III
Stellaria nemorum	+	.	.	.	.	.	+	.	.	.	II
Dryopteris filix-mas	.	.	.	+	+	.	+	.	.	.	II
Lamium galeobdolon	.	.	.	+	.	.	+	+	.	.	II
Luzula luzuloides	.	+	+	+	.	.	.	+	.	.	II
Veronica urticifolia	.	.	.	+	.	.	.	+	.	.	I
Poa nemoralis	.	.	.	.	.	.	+	.	.	.	I
Pulmonaria rubra	.	.	.	+	.	.	.	.	.	.	I
Lilium martagon	.	.	+	.	.	+	.	.	.	.	I

Adenostyletalia											
Adenostyles alliariae	+	+	+	2.4	.	+	+	.	+.3	+	IV
Doronicum austriacum	+	.	+.3	+.3	+	1.3	+	.	1.3	+.3	IV
Senecio fuchsii	+	+	+	+	.	+	+.3	.	+	.	IV
Veratrum album	+	+	+	+	.	+	+	.	+	+	IV
Cicerbita alpina	.	+	+	+	.	+	+	.	+	.	III
Ranunculus platanifolius	.	.	+	.	.	+	.	.	+	.	II
Companion											
Hylocomium splendens	1.3	4.5	4.5	.	.	+	1.5	+	+	2.5	IV
Hieracium pseudobifidum	.	+	+.4	+	.	+	+	+	.	.	III
Mnium cuspidatum	.	+	1.4	+	.	+	.	+	.	.	III
Solidago virgaurea subsp. *minuta*	+	.	.	.	.	+	+	+	.	.	II
Plagiothecium undulatum	.	.	+	+	.	.	.	.	1.3	1.5	II
Mnium punctatum	.	+	.	+	.	.	.	.	.	.	I
Isothecium myurum	.	+	+	+	.	.	.	.	.	.	II
Eurynchium striatum	.	+	.	+	.	.	.	.	.	.	I

Companion species with one occurrence: *Hieracium sabaudum* 1: +; *Lamium maculatum* 1: +; *Sambucus racemosa* 7: +; *Rubus idaeus* 6: +; *Crocus vernus* 4: +; *Brachythecium vellutinum* 4: +; *Ptilium crista-castrensis* 4: +; *Tetraphis pellucida* 3: +; *Mnium rostratum* 3: +; *Hypnum cupressiforme* 2: +; *Polytrichum juniperinum* 1: +; *Potentilla ternata* 7: +; *Phyteuma vagneri* 7: +

Relevé: 1. Mt. Zlata (8.07.1970); 2. V. Zlătuia (10.07.1969); 3–4. V. Zlătuia-Scoaba Retezatului (10.07.1969); 5. V. Judele (14.07.1970); 6. V. Rea (7.08.1970); 7–8. Cioaca Radeșului (8.07.1970). 9–10. Cascada-Gemenea (10.06.1981)

Photo 6.10 *Soldanella major* (Neilr.) Vierh. (photo M. Ciobanu)

The forest coenoses grouped in this Carpathian association are distributed sporadically within the intramontane valleys of Retezat (Zlătuia, Zănoaga, Judele, Radeşul, Slăveiul), frequently reaching the subalpine belt (1500–1800 m.s.m.). They vegetate on the gentle slopes of the valleys with brown, moist, cryptospodic acid soils, rich in humus of moder type, that provide abundant edaphic and atmospheric humidity. *Picea abies* is monodominant in the tree layer, achieving 85% coverage. In addition to the characteristic species *Leucanthemum rotundifolium,* several meso-hygrophilous species indicated as specific to the alliance are present in the herbaceous layer, including *Cicerbita alpina, Senecio subalpinus, Gentiana asclepiadea,* and *Aconitum napellus* subsp. *tauricum*. In the pedo-climatic environment of these subalpine phytocoenoses, with numerous meso-hygrophilous species typical of the order *Adenostyletalia*, some regional Carpathian-Balkan species, characteristic of the beech forests, such as *Pulmonaria rubra, Silene heuffelii,* and *Festuca drymeja* survived the last glaciation.

Order *Junipero-Pinetalia mugo* Boşcaiu 1971

The authors intended to bring together within this order, in a distinct unit, the *Pinus mugo* and *Juniperus communis* subsp. *alpina* scrubs, which have a distinct physiognomy and occupy a well individualized subalpine areal distribution in the mountains of southeastern Europe. Although such coenoses survived the last glaciation in the Retezat area, their maximum extension seems to have coincided with the Alleröd interstade and Dryas III (Boşcaiu 1971). This syntaxon has recently been accepted by Rodwell et al. (2002), Willner and Grabherr (2007), Mucina (2016) for classifying these shrub communities separately from the boreal coniferous forests. The right

Table 6.63 Ass. *Chrysanthemo rotundifolii-Piceetum* Krajina 1933

Relevé No.	1	2	3	4	5	6	
Altitude (m a.s.l.)	1810	1660	1490	1675	1580	1800	
Aspect	SW	E	NE	W	W	W	
Slope (degrees)	10	10	25	20	15	5	
Tree height (m)	30	25	20	25	30	22	
Canopy cover (%)	70	80	80	80	70	80	
Herb cover (%)	40	30	40	40	30	45	
Litter cover (%)	60	70	80	80	70	50	
Sample area (sq. m)	400	400	400	400	400	400	K
Char. ass.							
Picea abies	4.5	5.5	5.5	5.5	5.5	5.5	V
Leucanthemum rotundifolium	+	1.5	+	+	+	+.3	V
Chrysanthemo-Piceion **et** ***Piceetalia***							
Luzula sylvatica	1.4	+.5	2.5	1.5	1.4	+.5	V
Soldanella major	1.5	1.5	1.5	2.5	.	1.5	V
Homogyne alpina	+.5	1.5	+	+	.	1.5	V
Sorbus aucuparia	.	+	1.3	+	+	.	IV
Campanula abietina	+	+	+	.	+	.	IV
Calamagrostis villosa	.	+	2.5	1.5	.	+	III
Hieracium transsilvanicum	.	.	1.5	.	+	+	III
Senecio subalpinus	+	.	.	.	.	+	II
Phegopteris connectilis	+	.	.	.	.	+	II
Lonicera nigra	.	.	+	.	.	.	I
Melampyrum sylvaticum	.	.	+	.	.	.	I
Corallorrhiza trifida	.	.	+	.	.	.	I
Streptopus amplexifolius	.	.	+	.	.	.	I
Vaccinio-Piceetea							
Vaccinium myrtillus	+.4	+	.	3.4	+	1.5	V
Oxalis acetosella	+.3	2.5	2.5	+.5	1.4	+5	V
Athyrium distentifolium	+	1.5	1.5	2.5	1.3	.	V
Dryopteris dilatata	.	+	+	.	.	+	III
Pinus mugo	+	.	.	.	.	I	II
Pinus cembra	+	.	.	.	.	+	II
Juniperus communis susbp. *alpina*	+	.	.	.	.	+	II
Vaccinium vitis-idaea	+	.	.	.	.	.	I
Moneses uniflora	.	.	+	.	.	.	I
Adenostyletalia **s.l.**							
Adenostyles alliariae var. *kerneri*	1.4	2.5	+	1.4	1.3	3.4	V
Doronicum austriacum	+	1.5	+	+.3	+	+	V
Senecio fuchsii	+	+	+	+	+.4	+	V
Ranunculus platanifolius	+	+	.	+	.	+.3	IV
Veratrum album subsp. *lobelianum*	+	.	.	+	.	+	III
Geranium sylvaticum	+	+	.	.	.	+	III

(continued)

Table 6.63 (continued)

Cicerbita alpina	.	+	+	.	.	+	III
Angelica archangelica	+	+	.	.	.	+	III
Rumex alpestris	.	+	.	.	+	.	II
Aconitum napellus subsp. *tauricum*	.	+	.	.	.	+	II
Chaerophyllum hirsutum	+	+	.	.	.	.	II
Senecio subalpinus	+	.	.	.	.	+	II
Heracleum palmatum	.	+	.	.	.	.	I
Viola biflora	.	+	.	.	.	.	I
Salix silesiaca	.	.	+	.	.	.	I
Fagetalia s.l.							
Prenanthes purpurea	.	.	+	+	+	.	III
Gentiana asclepiadea	+	.	+	.	.	+	III
Stellaria nemorum	.	+	.	.	1.4	.	II
Lamium galeobdolon	.	.	+	.	+	.	II
Veronica urticifolia	.	+	+	.	.	.	II
Pulmonaria rubra	.	+	+	.	.	.	II
Dryopteris filix-mas	.	+	.	.	+	.	II
Mycelis muralis	.	.	.	.	+	.	I
Polygonatum verticillatum	.	.	+	.	.	.	I
Companion							
Polytrichum commune	2.5	1.4	+	2.5	.	+	V
Rubus idaeus	+	+	.	+	.	+	IV
Veleriana tripteris	.	+	+	+	.	.	III
Rhytidiadelphus triquetrus	.	.	1.3	1.3	.	1.3	III
Hieracium pseudobifidum	.	+	+.3	+	.	.	III

Companion species with one or two occurrences: *Crepis paludosa* 1: +; *Coeloglossum viride* 1: +; *Epilobium angustifolium* 1: +; *Hypericum maculatum* 1, 2: +; *Geum montanum* 1, 6: +; *Caltha laeta* 1, 6: +; *Sambucus racemosa* 2, 3: +; *Myosotis sylvatica* 2: +; *Saxifraga rotundifolia* subsp. *heucherifolia* 2: +; *Solidago virgaurea* subsp. *minuta* 2, 6: +; *Festuca drymeja* 2: +; *Deschampsia flexuosa* 2, 4: +; *Cirsium erysithales* 2: +; *Pyrola minor* 2: +; *Thalictrum aquilegifolium* 2: +; *Polystichum lonchitis* 2: +; *Huperzia selago* 3: +; *Silene heuffelii* 5: +; *Cladonia pyxidata* 1: +; *Rhitidiadelphus squarrosus* 4: +; *Hylocomium splendens* 3: +; *Pleurozium schreberi* 2, 4: 3.5; *Rhacomitrium sudeticum* 1: +; *Sphagnum girgensohnii* 1, 6: +; *Plagiothecium roeseanum* 1: +; *Mnium cuspidatum* 1: +; *Mnium punctatum* 1, 6: +

Relevé: 1. Casa Gemenele (7.07.1969); 2. V. Zănoaga-V. Judele (14.07.1970); 3. Plaiul Radeșului (12.07.1970); 4. Plaiul Slăveiului (14.07.1970); 5. Stâna Slăveiului (14.07.1970); 6. Casa Gemenea (7.06.1981)

syntaxonomic position of the order is in the class *Vaccinio-Piceetea* (Boşcaiu 1971; Roussakova 2000; Willner and Grabherr 2007) and not in *Erico-Pinetea* (Mucina 2016). The order includes two alliances.

Alliance *Pinion mugo* Pawl. 1928

This syntax has been proposed to group in a single alliance the *Pinus mugo* coenoses within the Carpathians, which constitute the climax vegetation in the subalpine belt and are located between the upper limit of spruce and alpine grasslands (1500–2100 m.s.m.). Subsequently, the alliance was accepted and used to syntaxonomically assign several plant community types in the Carpathians (Boşcaiu 1971; Chytrý 2013; Coldea 1985, 1993; Malynovski and Kricsfalusy 2002). Its areal distribution was later expanded both in the Balkans (Horvat et al. 1974) and Central Europe (Karner 2007; Theurillat et al. 1995). The main characteristic species of the alliance are: *Pinus mugo, Pinus cembra, Ribes petraeum, Sorbus aucuparia* subsp. *glabrata, Calamagrostis villosa, Solidago virgaurea* subsp. *minuta,* and *Salix silesiaca*.

Ass. *Rhododendro myrtifolii-Pinetum cembrae* (Borza 1934) Coldea 2015 (Table 6.64)
Natura 2000: habitat type 9420

As stated by Alexandru Borza in 1934, representative coenoses for this association are located in the Gemenele, Judele and Pietrele glacial cirques, over 1750 m altitude. The *Pinus cembra* populations are becoming more compact and, in some places, are replaced by timberline spruce woodlands. In the shrub layer of these phytocoenoses, *Pinus mugo* and *Juniperus communis* subsp. *alpina* become increasingly present. The soils under these phytocoenoses are superficial, often skeletal, acidic, and sometimes represented by moist podzols and prepodzols. In the herbaceous layer are present, beside *Rhododendron myrtifolium,* numerous species characteristic of the higher syntaxa to which we assigned the association (Täuber 1984). Floristically, this woodland association is similar to the *Vaccinio-Pinetum cembra* Oberd. 1962 described from the Eastern Alps, but differs from it through the absence of *Larix decidua, Lonicera caerulea, Rhododendron ferrugineum,* and *Peucedanum ostruthium* (Karner 2007).

Ass. *Rhododendro myrtifolii-Pinetum mugo* Coldea 1985 (Table 6.65)
Natura 2000: prority habitat type 4070

We grouped in this regional association the *Pinus mugo* coenoses from the Romanian Carpathians, well individualized phytogeographically from the syntaxa described in the Czech Republic (Chytrý 2013) and Ukraine (Malynovski and Kricsfalusy 2002) by the species *Rhododendron myrtifolium, Soldanella major, Campanula abietina,* and *Laserpium krapfii*. Such coenoses occupy large areas both in the glacial cirques (Gemenele, Judele) and on slopes of the Zlata, Judele and Şesele peaks. They vegetate on superficial, skeleton-rich, lithic and ferric-illuvial podzols, with acid reaction (pH = 4.1–4.6) and poor in bases (11–15%). The ecological optimum of the juniper scrubs in the Retezat is reached on the plateaus and slight slopes covered abundantly by snow in winter and exposed to the northwestern humid winds (Coldea 1985). The *Pinus mugo* scrubs on the steep slopes (Slăvei, Judele) have an exceptional protective role in preventing the erosion processes. In their floristic structure are present, beside mesophilous species characteristic of the alliance and order, also a group of meso-hygrophilous species typical

Table 6.64 Ass. *Rhododendro myrtifolii-Pinetum cembrae* (Borza 1934) Coldea 2015

Relevé No.	1	2	3	4	5	6	7	8	9	10	
Altitude (m a.s.l.)	1750	1780	1790	1750	1750	1770	1790	1790	1800	1810	
Aspect	N	N	N	W	W	N	N	N	N	N	
Slope (degrees)	45	50	20	30	30	45	20	25	40	45	
Tree cover (%)	60	70	60	50	60	60	50	50	60	50	
Sample area (sq. m)	200	200	200	200	300	200	200	300	200	200	K
Char. ass.											
Pinus cembra	2.5	2.5	2.5	3.5	1.5	2.5	2.5	2.5	1.5	2.5	V
Pinus cembra (juv.)	1.5	+	.	.	+.5	+	+	.	+	1.5	IV
Rhododendron myrtifolium	+	1.1	+	+	.	1.3	+	+	.	3.5	IV
Pinion mugo et ***Junipero-Pinetalia***											
Calamagrostis villosa	3.5	3.5	2.5	4.5	2.5	2.5	3.5	3.5	1.2	1.3	V
Deschampsia flexuosa	1.5	2.5	2.5	1.3	+	2.5	+	1.5	1.3	1.2	V
Pinus mugo	.	.	.	2.5	1.5	.	2.5	3.5	+	.	III
Polytrichum alpinum	+	.	.	+	+	+	+	.	.	.	III
Laserpitium krapfii	.	+	.	.	+	.	+	.	+	.	II
Solidago virgaurea subsp. *minuta*	+	.	.	+	+	.	.	.	.	.	II
Sphagnum girgensohnii	.	.	.	.	.	.	.	.	2.5	3.5	I
Sphagnum nemoreum	.	.	.	.	.	.	.	.	2.5	1.4	I
Polytrichum juniperinum	.	.	.	.	.	.	.	.	+	2.5	I
Vaccinio-Piceetea											
Picea abies	3.5	2.5	2.5	2.5	3.5	2.5	2.5	2.5	3.4	2.5	V
Picea abies (juv.)	2.5	+	+	.	1.5	+.5	+	.	+	2.5	V
Vaccinium myrtillus	2.5	2.5	+	+	+	1.3	2.5	2.5	+	2.5	V
Luzula sylvatica	1.5	2.5	+	2.5	1.5	1.3	+	2.5	+	.	V
Juniperus communis subsp. *alpina*	+	+	+	1.5	1.5	+	2.5	2.5	.	.	IV

Soldanella major	2.5	3.5	+	2.5	+	2.5	2.3	1.5	.	.	IV
Homogyne alpina	2.5	+	+	+	+	2.5	+	2.5	.	.	IV
Sorbus aucuparia subsp. *glabrata*	.	+	+	.	.	.	+	+	.	+	III
Vaccinium vitis-idaea	+	.	+	.	+	.	.	.	2.5	2.5	III
Gentiana asclepiadea	+	+	.	+	+	+	.	.	.	.	III
Orthilia secunda	.	.	.	+	+	.	.	+	.	.	II
Athyrium distentifolium	.	.	.	+	.	+	+	+	.	.	II
Hylocomium splendens	.	.	.	.	+	.	.	.	3.5	4.5	II
Dicranum scoparium	.	.	.	.	+	.	.	.	+	1.5	II
Campanula abietina	.	.	+	.	.	+	.	.	.	.	I
Dryopteris dilatata	.	+	.	+	.	.	.	.	.	.	I
Huperzia selago	.	.	.	+	.	.	.	.	.	.	I
Hieracium transsilvanicum	.	.	.	.	.	+	.	.	.	.	I
Melampyrum sylvaticum	.	.	+	.	.	+	.	.	.	.	I
Leucanthemum rotundifolium	.	.	.	+	.	.	.	+	.	.	I
Oxalis acetosella	.	.	.	+	.	.	.	.	+	.	I
Pleurozium schreberi	.	.	.	.	.	.	.	.	+	1.5	I
Rhytidiadelphus triquetrus	.	.	.	.	.	.	.	.	+	2.5	I
Lycopodium annotinum	.	.	.	.	.	.	+	.	.	.	I
Adenostyletalia											
Adenostyles alliariae var. *kerneri*	1.3	+	+	+	.	+	+	+	.	.	IV
Senecio nemorensis	+	+	+	.	+	+	+.5	.	.	.	III
Doronicum austriacum	+	+	+	.	.	+	.	+	.	.	III
Ranunculus platanifolius	.	.	.	+	.	+	+	+	+	.	III
Aconitum napellus subsp. *tauricum*	.	.	.	.	.	.	+	+	.	.	I
Cicerbita alpina	+	.	+	.	.	.	.	.	.	.	I
Doronicum columnae	.	.	+	.	+	.	.	.	.	.	I

(continued)

Table 6.64 (continued)

Veratrum album	.	.	.	.	.	+	.	+	.	.	I
Alnus viridis	.	.	.	.	.	.	.	.	+	.	I
Rumex alpestris	.	.	.	+	.	.	+	.	.	.	I
Companion											
Rubus idaeus	.	.	.	+	.	+	.	+	+	+	III
Dryopteris filix-mas	+	.	.	+	+	+	.	+	.	.	III
Cystopteris fragilis	+	+	.	.	+	.	.	.	.	.	II
Sambucus racemosa	.	.	.	.	+	.	.	.	.	.	I
Gentiana punctata	.	.	.	+	.	.	+	.	.	.	I

Companion species with one or two occurrences: *Geum montanum* 6, 8: +; *Potentilla ternata* 6, 8: +; *Hypericum maculatum* 6. 8: +; *Knautia longifolia* 5: +; *Epilobium montanum* 7, 8: +

Relevé: 1–5. Faţa Retezatului (8.07.1969); 6–8. Căldarea Gemenele (7.07.1969); 9–10. Căldarea Judele (7.07.1982)

Table 6.65 Ass. *Rhododendro myrtifolii-Pinetum mugo* (Borza 1959) Coldea 1990

Relevé No.	1	2	3	4	5	6	7	8	9	10	
Altitude (m a.s.l.)	1890	2010	1760	1940	1980	1900	2010	1920	1830	1910	
Aspect	S	SW	SW	W	E	NW	SW	SE	E	SE	
Slope (degrees)	5	15	20	25	15	15	25	30	25	10	
Herb cover (%)	100	100	100	100	100	100	100	100	100	100	
Sample area (sq. m)	100	100	100	100	100	100	100	100	100	100	K
Char. ass.											
Pinus mugo	5.5	.5.	5.5	4.5	5.5	5.5	5.5	5.5	5.5	5.5	V
Rhododendron myrtifolium	+	1.2	.	+	+	+	1.3	+	+	.	IV
Pinion mugo											
Juniperus communis subsp. *alpina*	1.3	+	+	1.4	+	2.5	+	.	.	.	IV
Pinus cembra	.	+	.	+	+	+	.	+	.	.	III
Sorbus aucuparia subsp. *glabrata*	.	.	.	.	+	.	.	.	+	.	I
Laserpitium krapfii	.	.	.	.	.	+	.	.	.	.	I
Junipero-Pinetalia et ***Vaccinio-Piceetea***											
Vaccinium myrtillus	4.5	5.5	+	4.5	.	4.5	3.5	+	3.5	4.5	V
Soldanella major	+	+.3	+.5	+	+	+.5	+.5	+.3	1.5	1.5	V
Homogyne alpina	+	.	1.5	+.5	+	+.5	2.5	+.4	+	+	V
Calamagrostis villosa	+	1.5	4.5	2.5	2.4	3.5	2.5	3.5	2.5	+	V
Luzula sylvatica	+.4	.	+	.	+	+	.	+	+.4	+	IV
Hylocomium splendens	1.3	1.3	4.5	1.4	4.5	1.4	1.3	.	.	1.3	IV
Pleurozium schreberi	2.4	3.5	2.5	2.5	2.5	2.5	4.5	.	1.3	.	IV
Deschampsia flexuosa	+	+	+	+.5	+	+	.	.	.	.	III
Dryopteris dilatata	.	.	+	+	+	.	.	+	+	.	III
Vaccinium vitis-idaea	+	.	.	.	.	+	1.5	+	.	+	II
Picea abies (juv.)	.	+	.	+	.	+	.	+	.	.	II
Oxalis acetosella	.	.	+.4	+	+.5	.	.	.	.	.	II

(continued)

Table 6.65 (continued)

Huperzia selago	.	.	.	.	+	.	+	.	.	+	II
Campanula abietina	+.3	.	+.4	.	.	.	+	+	.	.	II
Sphagnum girgensohnii	.	.	.	.	.	.	.	3.5	3.5	4.5	II
Leucanthemum rotundifolium	.	.	.	.	+	.	.	.	+.4	.	I
Melampyrum sylvaticum	.	.	.	.	.	+	.	.	.	+	I
Adenostyletalia s.l.											
Veratrum album subsp. *lobelianum*	.	+	.	+	+	+	+	1.3	+	.	IV
Adenostyles alliariae	+	.	+	2.5	2.4	.	.	+	+	.	III
Senecio fuchsii	+	.	.	+	+	+	.	+	+	.	III
Aconitum napellus subsp. *tauricum*	+	.	+	.	+	+	.	.	+	.	III
Ranunculus platanifolius	.	.	.	+.3	+	.	.	+	+	.	II
Cicerbita alpina	+	.	.	+	+	.	.	+	.	.	II
Rumex alpestris	.	.	+	+	+	.	.	+	.	.	II
Doronicum austriacum	.	.	+	+	+	.	.	+.3	+	.	II
Athyrium distentifolium	+	.	.	.	+	.	.	+	+	.	II
Geranium sylvaticum	.	.	+	.	.	.	.	.	+	.	I
Alnus viridis	.	.	.	.	+	.	.	.	.	.	I
Companion											
Dicranum scoparium	+	1.3	1.5	1.3	1.5	1.3	1.3	+	+	+	V
Luzula luzuloides	+	.	+	+.5	.	+	+	+	.	.	III
Ligusticum mutellina	+	.	+	.	+	.	+	+	+	.	III
Gentiana punctata	.	.	.	.	+	.	+	+	+	.	II
Cetraria islandica	.	+	+	.	+	.	.	.	.	.	II

Companion species with one or two occurrences: *Salix silesiaca* 4: +; *Milium effusum* 5: +; *Rumex alpinus* 8: +; *Geum montanum* 1, 9: +; *Rubus idaeus* 1: +; *Solidago virgaurea* subsp. *minuta* 4, 5: +; *Polygonatum verticillatum* 4: +; *Silene vulgaris* 5: +; *Stellaria nemorum* 5: +; *Allium victorialis* 8: 1.4, 10: +; *Crepis paludosa* 9: +; *Senecio subalpinus* 9: +; *Viola biflora* 9: 1.3; *Valeriana tripteris* 9: +; *Eriophorum vaginatum* 10: +; *Polytrichum formosum* 3, 6: +; *Polytrichum commune* 5: +, 6: 1.4; *Brachythecium rutabulum* 5: +; *Diplophyllum albicans* 5, 6: +; *Ptillium crista-castrensis* 6: 4.5; *Rhytidiadelphus triquetrus* 6: 1.4; *Polytrichum juniperinum* 6, 7: +; *Polytrichum piliferum* 7: +; *Aulacomium palustre* 7: +
Relevé: 1. Vf. Zlata (8.07.1970); 2. Fața Retezatului (9.07.1969); 3. Piciorul Colțului (9.07.1969); 4. Slăveiul (14.07.1970); 5. Tăul Negru (8.07.1969); 6. Șesele (8.07.1969); 7. Tăul Secat (8.07.1969); 8. Lacul Gemenele (16.07.1970); 9. V. Judele (12.07.1970); 10. Lacul Lia (6.07.1969)

of the order *Adenostyletalia*, revealing the high water content of the soil in some locations.

Alliance *Junipero alpinae-Bruckenthalion* Boşcaiu 1971
Natura 2000: habitat type 4060

In this regional Carpathian-Balkan alliance are grouped the acidophilic subalpine associations built up by *Juniperus communis* subsp. *alpina* in the Southeastern Carpathians, Pyrenees, and Dinarics. The geological substrate in these areas is crystalline rock on which podzolic soils developed. The characteristic species of the alliance are *Bruckenthalia spiculifolia, Campanula abietina, Campanula serrata, Potentilla ternata,* and *Thymus praecox* subsp. *polytrichus* (Boşcaiu 1971). We point out that this Southeast European alliance is a geographical synvicariant of the *Juniperion nanae* Br.-Bl. et al. 1939, circumscribed to the western and central Alps (Rodwell et al. 2002).

Ass. *Bruckenthalio-Juniperetum alpinae* Horvat (1938) 1974 nomen inversum (Table 6.66)

We grouped in this Balkan-Carpathian association the juniper scrubs from the subalpine belt of the Retezat on sunny, heavy slopes, with superficial, acidic and nutrient-poor soils (Faţa Retezatului, Piciorul Colţului, Peleaga, Zlata Mountain). The dominant species in this association, hachieving on average 40% coverage, are *Juniperus communis* subsp. *alpina* and *Bruckenthalia spiculifolia*. Along with them are present species characteristic of both the alliance *Junipero-Bruckenthalion* and the order and class. Since such shrub coenoses in the Retezat mountains come often in contact with the acidophilic grasslands of the alliance *Potentillo ternatae-Nardion*, in some places in the herbaceous layer there are present even with a coverage of 5–15% *Festuca nigrescens, Potentilla ternata, Nardus stricta* and *Geum montanum*. Some species of the alliance *Caricion curvulae*, such as *Festuca supina, Agrostis rupestris, Phyteuma confusum* and *Veronica bellidioides* also penetrate in the herbaceous layer of these coenoses at higher altitudes.

Ass. *Campanulo abietinae-Juniperetum alpinae* Simon 1966 (Table 6.67)

The *Juniperus communis* subsp. *alpina* coenoses grouped in this association have their ecological optimum in the upper montane and subalpine belts of the southern and central part of the Retezat, on sunny, gentle slopes (Zlata Mountain, Zănoguţa, Şesele, Slăveiu). Sometimes they are also found in the glades of timberline spruce forests. They vegetate on superficial podzols with acidic reaction and relatively low moisture. The herbaceous layer includes, beside the regional Carpathian-Balkan

Table 6.66 Ass. *Bruckenthalio-Juniperetum alpinae* Horvat (1938) 1974 nom. invers.

Relevé No.	1	2	3	4	5	6	7	8	
Altitude (m a.s.l.)	1930	1960	1980	1980	1760	1750	1680	1750	
Aspect	SW	SW	SW	SW	SW	S	S	SW	
Slope (degrees)	45	45	25	40	30	20	30	50	
Herb cover (%)	100	100	90	100	100	70	90	80	
Sample area (sq. m)	100	100	100	100	100	100	100	100	K
Char. Ass.									
Juniperus communis subsp. *alpina*	3.5	2.3	2.3	3.5	4.5	2.3	2.5	3.4	V
Bruckenthalia spiculifolia	1.3	3.5	2.4	1.3	3.5	1.3	1.3	+.2	V
Junipero-Bruckenthalion									
Campanula abietina	+.4	+.3	+	+	+.3	+	+.3	+.3	V
Thymus pulegioides subsp. *polytrichus*	+.3	+.4	1.5	+	+	+	1.3	+.4	V
Centaurea nervosa	+	+	.	+	.	+	+	+	IV
Laserpitium krapfii	+	+	.	+.3	.	+	.	+	IV
Phyteuma vagneri	+	.	.	+	.	+	.	+	III
Campanula serrata	.	+.4	.	.	.	.	+	1.5	II
Junipero-Pinetalia mugo (incl. ***Vaccinio-Piceetea***)									
Vaccinium myrtillus	3.5	3.5	.	3.5	2.5	1.5	2.5	3.5	V
Pinus mugo	.	+	1.3	+	+	.	1.3	.	IV
Vaccinium vitis-idaea	1.5	1.5	2.5	+.5	+	1.5	.	.	IV
Calamagrostis villosa	3.5	1.5	.	2.5	.	+	.	2.5	IV
Deschampsia flexuosa	+.5	+.5	+.3	+	.	1.5	+	+	V
Homogyne alpina	+.3	.	.	.	+	+	+	+	III
Picea abies	.	.	.	.	+	1.3	.	+	II
Rhododendron myrtifolium	.	.	.	+	.	.	+	+	II
Soldanella major	.	.	.	.	.	+	.	+	II
Melampyrum sylvaticum	.	.	.	.	.	+	.	.	I
Potentillo-Nardion									
Potentilla ternata	+	.	+.4	+	1.5	+	+.5	+	V
Antennaria dioica	+.4	+.3	+.4	+.3	+.3	+	+	.	V
Geum montanum	+	.	+	+	1.5	.	+	+	IV
Scorzonera rosea	+	+	.	+	+.3	+	+	.	IV
Festuca nigrescens	.	.	.	3.5	.	3.5	2.5	.	II
Luzula multiflora	.	.	.	.	+	+	+	.	II
Gentiana acaulis	.	.	.	+	+	+	.	.	II
Nardus stricta	.	.	.	.	2.5	.	2.5	.	II
Caricion curvulae									
Festuca supina	.	.	+.4	.	3.5	.	.	1.5	II
Avenula versicolor	.	.	+.5	+	+	.	.	.	II
Phyteuma confusum	.	+	1.5	+	.	.	.	.	II

(continued)

Table 6.66 (continued)

Veronica bellidioides	.	.	+.4	+	+	.	.	.	II
Campanula alpina	.	.	.	+	.	.	.	.	I
Agrostis rupestris	.	.	.	.	3.5	.	.	.	I
Companion									
Anthoxanthum alpinum	+	+.5	+.5	+	+	.	+	+	V
Achillea stricta	+	.	.	+	+	+	+	.	IV
Hypericum maculatum	+	.	.	+	+	2.4	.	+	IV
Luzula luzuloides	.	1.5	+.5	2.5	2.5	+	.	1.5	III
Avenula praeusta	+	.	.	+	+	+	.	.	III
Carlina acaulis	+	+	.	+	+	.	.	.	III
Gymnadenia conopsea	+	+	.	+	.	+	.	.	III
Hypochaeris uniflora	+	1.5	.	+	.	+	.	.	III
Polytrichum piliferum	2.5	3.5	2.5	3.5	2.5	.	.	.	III
Polytrichum juniperinum	+	.	+	+	3.5	.	+	.	III
Ceratodon purpureum	1.4	+.3	+	1.4	3.5	.	.	.	III
Salix caprea	.	.	+	+	.	1.3	.	.	II
Rhinanthus alpinus	.	+	.	+	.	.	+	.	II
Pseudorchis albida	.	.	+	+	.	+	.	.	II
Pulsatilla alba	.	+	+.3	+	.	.	.	.	II
Cetraria islandica	.	1.3	1.5	.	+	.	.	.	II
Cladonia pyxidata	+	+	+	.	.	.	.	.	II
Cladonia sylvatica	+	1.3	1.4	.	.	.	.	.	II

Companion species with one or two occurrences: *Viola dacica* 5, 7: +; *Poa alpina* 4: +; *Thlaspi dacicum* 4: +; *Hieracium aurantiacum* 6, 7: +; *Aconitum napellus* subsp. *tauricum* 1: +; *Athyrium distentifolium* 1: +; *Thesium alpinum* 4, 6: +; *Betula pubescens* subsp. *carpatica* 6: +; *Epilobium angustifolium* 6: +; *Pogonatum urnigerum* 1: +; *Tortella tortuosa* 1: +; *Pleurozium schreberi* 4: +; *Crocus vernus* 7: +; *Senecio fuchsii* 6: +; *Platanthera bifolia* 6: +; *Lotus corniculatus* 6: +; *Trollius europaeus* s.l. 6: +; *Achillea lingulata* 6, 7: +; *Genista tinctoria* subsp. *oligosperma* 7: +.3; *Poa media* 7: +; *Gnaphalium sylvaticum* 7: +; *Solidago virgaurea* subsp. *minuta* 8: +; *Euphrasia tatrae* 8: +; *Gentianella praecox* 8: +

Relevé*:* 1–4. Faţa Retezatului (6.07.1969); 5. Piciorul Colţului (9.07.1969); 6. Mt. Zlata (8.07.1970); 7. V. Pelegii (7.08.1970); 8. V. Bucuri (7.08.1970)

species, *Campanula abietina*, also characteristic of the alliance *Junipero-Bruckenthalion* (Simon 1966) and the order *Junipero-Pinetalia mugo* (Boşcaiu 1971). From a floristic point of view, this association has major similarities with the *Juniperus communis* subsp. *alpina* coenoses from the Ukrainian Carpathians (Malynovski and Kricsfalusy 2002), which, in our opinion, are also attributable to this association, and not to *Juniperetum nanae* described in the Alps.

Table 6.67 Ass. *Campanulo-Juniperetum nanae* Simon 1966

Relevé No.	1	2	3	4	5	
Altitude (m a.s.l.)	1850	1880	1820	1710	1800	
Aspect	W	W	S	W	W	
Slope (degrees)	5	20	20	5	5	
Herb cover (%)	100	100	100	70	100	
Sample area (sq. m)	100	100	100	100	100	K
Char. ass.						
Juniperus communis subp. *alpina*	5.5	4.5	4.5	3.5	5.5	V
Campanula abietina	+.3	+.5	+	1.5	1.3	V
Junipero-Bruckenthalion						
Campanula serrata	.	+	+	+	.	III
Bruckenthalia spiculifolia	+	.	+	.	.	II
Centaurea nervosa	.	.	+	.	+	II
Laserpitium krapfii	.	+	+	.	.	II
Phyteuma vagneri	.	.	+	.	+	II
Thymus pulegioides subsp. *polytrichus*	.	.	+	+	.	II
Solidago virgaurea subsp. *minuta*	+	.	.	+	.	II
Junipero-Pinetalia mugo (incl. ***Vaccinio-Piceetea***)						
Vaccinium myrtillus	4.5	4.5	4.5	.	3.5	IV
Vaccinium vitis-idaea	+	+	+	.	+	IV
Soldanella major	+	1.5	+	.	1.2	IV
Homogyne alpina	+	+	+.5	.	+.3	IV
Pinus mugo	+	.	1.3	.	+	III
Picea abies	+	.	.	.	+	II
Deschampsia flexuosa	2.5	+	.	.	.	II
Calamagrostis villosa	.	2.5	.	.	2.5	II
Hylocomium splendens	.	2.4	3.5	.	.	II
Potentillo-Nardion						
Geum montanum	+	+	+.3	.	+	IV
Potentilla ternata	.	.	+.5	1.5	+	III
Festuca nigrescens	.	+	1.5	2.5	.	III
Antennaria dioica	.	+	+	+	.	III
Luzula multiflora	.	.	+	+	.	II
Caricion curvulae						
Poa media	+	.	+	.	+	III
Festuca supina	.	+	.	.	.	I
Avenula versicolor	.	.	+	.	.	I
Vaccinium gaultherioides	.	.	+	.	.	I
Phyteuma confusum	.	.	+	.	.	I
Companion						
Senecio fuchsii	+	+	+	+	+	IV
Pleurozium schreberi	+	2.4	1.3	.	+	IV
Luzula luzuloides	+	1.4	+	.	.	III

(continued)

Table 6.67 (continued)

Ligusticum mutellina	.	+	+	.	+	III
Veratrum album subsp. *lobelianum*	.	+	+	.	+	III
Hypericum maculatum	.	+	.	.	+	II
Anthoxanthum odoratum			+		+	II

Companion species with one or two occurrences: *Hieracium aurantiacum* 5: +; *Veronica bellidioides* 4, 6: +; *Betula pendula* 1: +; *Rubus idaeus* 2: +; 5: +.4; *Epilobium angustifolium* 2, 3: +; *Adenostyles alliariae* var. *kerneri* 2, 6: +; *Sedum annuum* 2: +; *Cerastium fontanum* 2, 4: +; *Luzula sylvatica* 2, 3: +; *Aconitum toxicum* 2: +; *Angelica archangelica* 3: +; *Crocus vernus* 4: +.5; *Achillea stricta* 4, 6: +; *Pulsatilla alba* 4, 6: +; *Gentiana acaulis* 4: +; *Cladonia sylvatica* 4: +; *Cetraria islandica* 2: 2.5, 4: +; *Rhytidiadelphus squarrosus* 3: 1.5; *Dicranum scoparium* 3: +; *Mnium cuspidatum* 3: +; *Brachythecium rutabulum* 3: +; *Brachythecium salebrosum* 3: +; *Brachythecium reflexum* 3: +; *Polytrichum juniperinum* 4: +

Relevé: 1. Mt. Zlata (8.07.1970); 2. Şesele (8.07.1969); 3. Tăul Zănoguţii (10.07.1970); 4. Plaiul Zănoguţa-Slăvei (14.07.1970); 5. Mt. Zlata (10.07.1970)

Chapter 7
Considerations on the Flora, Vegetation and Conservation of the Plant Gene Pool in the Retezat National Park

The 1152 plant species and 104 subspecies of the phyla *Pteridophyta* and *Spermatophyta* identified in the Retezat National Park area have high affinities with the flora of other Central European mountain systems, particularly the alpine and boreal species, but are also connected through a smaller group of southern species with the flora of the mountains from southeastern Europe. The European element dominates in the areal-geographical spectrum of the Retezat flora (24.5%), followed by the Eurasian (23.2%), European alpine (13.3%) and circumpolar (12.2%) ones. This massif shelters many Carpathian and Dacian endemics (12.2%), among which we mention *Draba dorneri, Draba simonkaiana, Cerastium transsilvanicum, Gypsophila petraea, Dianthus henteri, Thymus comosus, Hepatica transsilvanica, Galium kitaibelianum, Achillea oxyloba* subsp. *schurii, Carduus kerneri* subsp. *lobulatiformis, Centaurea phrygia* subsp. *retezatensis, Dianthus glacialis* subsp. *gelidus, Festuca stricta* subsp. *rumelica* and *Poa granitica* subsp. *disparilis.* Along with them, many Dacian-Balkan species (12.2%) are present such as *Bruckenthalia spiculifolia, Campanula abietina, Campanula transsilvanica, Asperula capitata, Potentilla ternata, Centaurea atropurpurea, Dianthus compactus, Lathyrus hallersteinii, Silene heuffelii, Plantago gentianoides, Phyteuma confusum, Rhododendron myrtifolium, Symphyandra wanneri, Veronica bachofenii, Veronica baumgartenii*, giving a distinct southern specificity to the plant associations in which they occur. Both the European and Eurasian species, as well as the Carpathian and Dacian-Balkan endemic species, have different altitudinal distributions in the Retezat, depending on their biological requirements towards the environmental factors.

Out of the total of 1256 plant species and subspecies, only 360 are considered alpine, of which only 110 are truly alpine, while 250 are subalpine species distributed between 1800 and 2200 m altitude. Approximately 460 plant species are present in the coniferous forests and the natural and quasi-natural grasslands of the upper montane belt (1350–1800 m), and the remaining 432 taxa are distributed in

G. Coldea, V. Cristea, *The Vascular Plant Communities of the Retezat National Park (Southern Carpathians)*, Geobotany Studies,
https://doi.org/10.1007/978-3-031-05618-5_7

deciduous forests and secondary mesophilic grasslands of the submontane and middle montane belts of the Retezat (700–1300 m).

The communities of natural, woody and herbaceous plants, formed and established in different phases of the postglacial period (Boşcaiu 1971), functioned as true biological filters and retained in their floristic structure those plant species adapted to their coenotic environment. The endemic and rare species present in these plant communities (associations) can persist as long as the pedo-climatic conditions remain unchanged. In the first half of the last century (1936), Alexandru Borza predicted that the most efficient way to protect rare plant species is the one that protects the ecosystems (associations) in which they live.

The Habitats Directive 92/43/EEC of 21st of May 1992 (Evans 2006) supports the same idea of protecting the native flora. In order to reveal the presence and conservation status of the rare and endemic species of the 67 plant associations described in the Retezat, we classified them in the habitat types distinguished in Romania (Doniţă et al. 2005; Gafta and Mountford 2008; Schneider and Drăgulescu 2005), and tracked the location of each habitat in the park area (over 38,000 ha) (Munteanu et al. 2003; Sârbu 2007). Currently, the park has a single strictly protected area, the Gemenele Scientific Reserve, located in the western part of the massif. In the perimeter of this reserve, which covers about 2000 ha and whose geological substrate consists exclusively of crystalline schists, 52 plant associations were distinguished and grouped in the following 16 types of Natura 2000 habitats: 3220, 4060, 4070*, 6150, 6230*, 6430, 6520, 7110*, 8110, 8220, 9180*, 9410, 9420, 91E0*, 91T0, 91V0, of which five (marked with *) are priority habitats. Many rare and endemic species for the Southeastern Carpathians live in the coenotic assemblages of these habitats, such as *Alopecurus laguriformis, Aconitum toxicum, Barbarea vulgaris* subsp. *lepuznica, Hepatica transsilvanica, Heracleum palmatum, Chrysosplenium alpinum, Campanula transsilvanica, Phyteuma vagneri*, and some are of European-wide interest: *Campanula abietina, Campanula serrata, Draba dorneri, Poa granitica* subsp. *disparilis, Plantago atrata* subsp. *carpatica, Sempervivum montanum* subsp. *carpaticum, Achillea oxyloba* subsp. *schurii, Tozzia alpina* subsp. *carpathica*. Through the existence of this reserve, the protection and conservation of these species and the plant associations in which they live is ensured.

The Calcareous Retezat, with a predominantly calcareous geological substrate, is located in the southern and southeastern part of the park. In this area we found and described 11 plant associations grouped in the following three types of Natura 2000 habitats: 6170, 8120 and 8220. The floristic structure of these associations includes many calcophile, Carpathian and Carpathian-Balkan species, some rare or vulnerable and protected by law in Romania (*Leontopodium alpinum, Nigritella rubra, Nigritella nigra, Gentiana lutea, Carduus kernerii* subsp. *lobulatiformis, Eritrichium nanum* subsp. *jankae, Onobrychis montana* subsp. *transsilvanica, Papaver alpinum* subsp. *corona sancti-stephani*). Others, such as *Artemisia eriantha, Cypripedium calceolus, Lilium jankae, Bupleurum falcatum* subsp. *dilatatum, Adenophora liliifolia,* and *Pulsatilla vulgaris* subsp. *grandis*, are of European interest.

Although many basophilic species, rare in Romania, and phytogeographically important plant associations with prominent southern specificity occur in the

Calcareous Retezat, there is no strictly protected area in this part of the park. The transhumance in the Calcareous Retezat occurs in summer against the will of the park administration. To stop the loss of rare and vulnerable species in this area (e.g., *Draba simonkaiana, Oxytropis pyrenaica, Pedicularis baumgartenii, Scutellaria alpina, Salix alpina, Sorbus chamaemespilus*) and to safeguard the protected plant species in Romania and those of European interest mentioned above, it is imperative to establish two botanical reserves with strict protection within the integral protection areas, occupying 19,784 ha, where grazing should not be allowed. The first reserve, proposed to cover an area of ca. 3500 ha, should include the Albele, Scorota and Piatra Iorgovanului peaks, whereas the second one, with an area of 2500 ha, should protect the Piule Mountain. Grazing and ecotourism should be banned in these reserves.

The establishment of these two reserves would protect the rare and endangered basophilic species in the Calcareous Retezat area for the future and would conserve the biodiversity of plant communities within the park.

References

Aichinger E (1933) Vegetationskunde der Karawanken Pflanzensoziologie, 2. Jena

Alexiu V (1998) Vegetația masivului Iezer-Păpușa. Studiu fitosociologic. Cultura, Pitești

Baumgarten C (1816) Enumeration stirpium Magno Trassilvanie principatui, I–III. Vindobonae

Beldie A (1967) Flora și vegetația Munților Bucegi. Acad. Române, București

Borhidi A (1996) Anannotated checklist of the Hungarian plant communities. I. The non-forest vegetation. In: Borhidi A (ed) Critical revision of the Hungarian plant communities. Pécs, pp 43–94

Borza A (1934) Studii fitosociologice în Muntii Retezatului. Bul Grăd Bot Cluj 14(1):21–84

Borza A (1959) Flora și vegetația văii Sebeșului. Acad. Române, București

Borza A (1963) Pflanzengesellschaften der rumänischen Karpathen. Biologia, XVIII 11:856–864

Borza A, Boşcaiu N (1965) Introducere în studiul covorului vegetal. Acad. Române, București

Boşcaiu N (1971) Flora și vegetația Munților Țarcu, Godeanu și Cerna. Acad. Române, București

Boşcaiu N, Plămadă E, Péterfi L (1972) Studii fitocenologice asupra complexului de mlaștini alpine și subalpine din Valea Judele - Parcul Național Retezat. Ocrotirea Naturii, București 15(2): 175–185

Boşcaiu N, Täuber F, Coldea G (1978) Asociații vegetale rupicole și petrofile din Munții Retezat. St și Comun de Ocrot Nat Suceava 4:253–264

Boşcaiu N, Boşcaiu V, Coldea G, Täuber F (1982) Sintaxonomia făgetelor carpatine. In: Boşcaiu N (ed) Făgetele carpatine. Semnificația lor bioistorică şi ecoprotectivă. Cluj-Napoca, pp 228–303

Boublik K, Douda J, Hedl R, Chytrý M (2013) Mesic and wet decidous broad-leaved forest. In: Chytrý M (ed) Vegetation of the Czech Republic, 4, Forest and shrub Vegetation. Academia, Praha, pp 193–292

Braun-Blanquet J (1930) Zentralalpen und Tatra eine planzensoziologische Parallele. Veröff Geobot Inst Rübel 6:81–123

Braun-Blanquet J (1948) La végétation apline des Pyrénées orientales. Monogr Estacion Est Piren, Commun Sigma, 98, 306 p

Braun-Blanquet J (1964) Pflanzensoziologie, 3 Aufl. Wien-New York

Braun-Blanquet J, Jenny H (1926) Vegetationsentwicklung und Bodenbildung in der alpinen Stufe der Zentralalpen. Denkschr Schweiz Naturf Ges 6:183–349

Buia A, Păun M, Pavel C (1962) Studiul geobotanic al pajiștilor. In: Buia A (ed) Pajiștile din Masivul Parîng și îmbunătățirea lor. Agro-Silvică, Bucuresti, pp 143–274

Chiriță DC, Păunescu C, Teaci D (1967) Solurile României. Agro-Silvică, București

Chytrý M (ed) (2007) Vegetation of Czech Republic 1. Grassland and Heathland Vegetation. Academia, Praha

G. Coldea, V. Cristea, *The Vascular Plant Communities of the Retezat National Park (Southern Carpathians)*, Geobotany Studies,
https://doi.org/10.1007/978-3-031-05618-5

Chytrý M (ed) (2011) Vegetation of Czech Republic 3. Aquatic and Wetland Vegetation. Academia, Praha, pp 31–42

Chytrý M (ed) (2013) Vegetation of Czech Republic 4. Forest and Shrub Vegetation. Academia, Praha, pp 157–163

Ciurchea M, Crişan F (1993) Vegetaţia lichenologică saxicolă din rezervaţia ştiinţifică a Parcului Naţional Retezat. In: Popovici I (ed) Parcul Naţional Retezat. Studii ecologice. West Side, Braşov, pp 58–78

Codarcea A, Pavelescu L (1963) Considération sur la génése des roches granitoides de l'autochtone Danubien des Carpathes Méridionales. Congr. V. Carpato-Balcanic, II, Bucureşti

Coldea G (1982) Principalele sigmaasociaţii din rezervaţia ştiinţifică Retezat. Şt şi Cerc Biol Ser Biol Veget 34(2):112–117

Coldea G (1983) Precizări sintaxonomice asupra cenozelor cu *Deschampsia caespitosa*. St şi Cerc Biol Veget, Bucureşti 35(2):96–100

Coldea G (1984) L'étude des pâturages basiphiles sousalpines des Monts Rodnei (Les Carpathes Orientales). Contrib Bot, Cluj-Napoca 24:137–154

Coldea G (1985) Phytozönologisches Studium der Krummholzgebüschen in der Südostkarpaten. Feddes Repert 96(5–6):397–405

Coldea G (1990) Munţii Rodnei. Studiu geobotanic. Acad. Române, Bucureşti

Coldea G (1991) Prodrome des associations végétales des Carpates du sud-est (Carpates roumaines). Doc Phytosociol, Camerino, N. S. 13:317–539

Coldea G (1993) Cormofite. Sintaxonomia şi descrierea asociaţiilor vegetale. In: Popovici I (ed) Parcul Naţional Retezat. Studii ecologice. West Side, Braşov, pp 31–48

Coldea G (ed) (2012) Les association végétales de Roumanie, 2: Les associations anthropogènes. Presa Univ, Clujeană, Cluj-Napoca

Coldea G (ed) (2015) Les association végétales de Roumanie, 3: Les associations forestières et arbustives. Presa Univ, Clujeană, Cluj-Napoca

Coldea G, Täuber F, Lupşa V (1972) Specii noi şi rare pentru flora munţilor Retezat. Studia Univ, vol 2. Babeş-Bolyai, Cluj, pp 23–25

Cristea V, Gafta D, Pedrotti F (2015) Fitosociologia. TEMI, Trento

Csató L (1868) Retyezát helyviszonyai és természetrajzi tekintetben. Erd Muzeum-Egylet evkönyvei II:72

Csűrös S, Csűrös-Kaptalan M, Papp S (1956a) Contribuţii la studiul vegetaţiei zonei de calcar din vecinătatea sudică a Munţilor Retezat. St şi Cerc Biol, Cluj, VII 1–4:33–56

Csűrös S, Kaptalan M, Nagy F (1956b) Contribuţii la studiul vegetaţiei alpine a regiunii centrale a Muntilor Retezat. St şi Cerc Biol, Cluj, VII 1–4:57–76

Csűrös S, Gergely I, Pap S (1962) Material pentru flora munţilor Piule-Piatra Iorgovanului. Contrib Bot, Cluj 3:131–150

Csűrös S, Kovács A, Moldovan I (1964) Cercetări de vegetaţie în rezervaţia ştiinţifică a Parcului Naţional Retezat. Contrib Bot, Cluj 4:167–188

Dierschke H (1981) Syntaxonomische Gliederung der Bergwiesen Mitteleuropas (*Polygono-Trisetion*). In: Dierschke H (ed) Syntaxonomie, Vaduz, pp 311–341

Dierssen K (1977) Zur Synökologie von *Betula nana* in Mitteleuropa. Phytocoenologia 4(2): 180–205

Dierssen K (1978) Some aspects of the classification of oligotrophic and mesotrophic mire communities in Europe. Coll Phytosoc Sols tourbeux, Lille 7:399–423

Dihoru G (1975) Învelişul vegetal din muntele Siriu. Acad. Române, Bucureşti

Dihoru G, Negrean G (2009) Cartea roşie a plantelor vasculare din România. Acad. Române, Bucureşti

Dobolyi ZK (1983) Die Zönologischen Verhältnisse von *Telekia speciosa* (Schreb.) Baumg. in den Karpaten und an der Balkanhalbinsel. Ann Hist Nat Mus Nat Hung 75:71–79

Doniţă N, Chiriţă C, Stănescu V (eds) (1990) Tipurile de ecosisteme forestiere din România. ICAS, Seria II. Tehnică Agricolă, Bucureşti

Doniță N, Popescu A, Paucă-Comănescu M, Mihăilescu S, Biriş IA (2005) Habitatele din România. Tehnică Silvică, Bucureşti

Dúbracová Z, Jarolimek I (2007) *Caricetea curvulae*. In: Kliment J, Valachovič M (eds) Rastlinné spoločenstvá Slovesnka, 4. Vysokohorska mokradi, Veda, Bratislava, pp 321–339

Eggler J (1952) Übersicht der höheren Vegetationseinheiten der Ostalepen. Mitt Naturwiss Ver Steiermark, Graz 81(82):28–41

Eichberger C, Heiselmayer P, Grabner S (2007) *Dicrano-Pinion*. In: Willner W, Grabher G (eds) Die Wälder und Gebüsche Österreichs. Spektrum Akademischer Verlag, München, pp 177–182

Ellenberg H (1979) Zeigerwerte der Gefässpflanzen Mitteleuropas. 2. Aufl. Scripta Geobotanica 9: 1–122

Ellenberg H (1996) Vegetation Mitteleuropas mit den Alpen. 5. Aufl. Ulmer Verlag, Stuttgart

Evans D (2006) The habitats of the European Union Habitats Directive, Biology and Environment. Proceedings of the Royal Irish Academy 106B(3):167–173

Exner A (2007) *Picetalia* Pawl. 1928. In: Willner W, Grabherr G (eds) Die Wälder und Gebüsche Österreichs. Elsevier Spektrum Akademischer Verlag, München, pp 183–208

Fărcaş I, Sorocovschi V (1993) Clima munților Retezat. In: Popovici I (ed) Parcul Național Retezat, Studii ecologice. West Side, Braşov, pp 13–20

Fărcaş S, De Beaulieu J-L, Reille M, Coldea G, Diaconeasa B, Goeury C, Goslar T, Jull T (1999) First 14C datings of Lake Glacial and Holocene pollen sequences from Romanian Carpathes. CR Acad Sci Paris, Sciences de la vie 322:799–807

Florea N, Munteanu I (2003) Sistemul român de taxonomie a solurilor (SRTS). Estfalia, Bucureşti

Gafta D, Mountford O (eds) (2008) Manual de interpretare a habitatelor Natura 2000 din România. Risoprint, Cluj-Napoca

Gafta D, Muncaciu S (2016) Large habitat range but low floristic variation: the case of *Festuco rubrae-Agrostietum capillaris* grasslands. In: Box EO (ed) Vegetation structure and function at multiple spatial, temporal and conceptual scales. Geobotany studies. Springer, Berlin, pp 331–355

Gafta D, Roman A, Puşcaş M, Şuteu A (2012) Floristic heterogeneity of *Festuca airoides*-dominated grasslands in the Romanian Carpathians. Plant Biosystems 146:603–613

Geanana G M (1975) Limita superioară a pădurii în Munții Retezat. Teza de doctorat. Universitatea Bucureşti

Ghișa E (1940) Contribuțiuni la studiul fitosociologic al munților Făgăraş. Bul Grăd Bot, Cluj 20(3–4):127–141

Grabherr G (1993) *Loiseleurio-Vaccinietea*. In: Grabherr G, Mucina L (eds) Die Pflanzengesellschaften Österreichs. Teil II. Natürliche waldfreie Vegetation. Gustav Fischer Verlag, Jena, pp 447–467

Hadač E (1956) Rostlinná společenstva Temnosmrečinoké doliny ve Vysokých Tatrách. Biol. práce SAV, Bratislava 2/1:1–88

Hájek Z (2007) *Anthoxantho odorati-Agrostietum tenuis*. In: Chytrý M (ed) Vegetation of the Czech Republic, 1. Grassland and heathland vegetation. Academia, Praha, pp 200–203

Hájková P, Hájek M (2011) *Montio-Cardaminetea*. In: Chytrý M (ed) Vegetation of the Czech Republic, 3. Aquatic and wetland vegetation. Academia, Praha, pp 580–609

Hájková P, Navrátilová J, Hájek M (2011) *Oxycocco-Sphagnetea*. In: Chytrý M (ed) Vegetation of the Czech Republic, 3. Aquatic and wetland vegetation. Academia, Praha, pp 705–733

Haret M (1922) Flore du Retezat et des montagnes du Haut-Banat. Lucr Inst de Geogr Univ Cluj I:301–307

Hegedüsová Vantarová K (2014) *Poion aplinae*. In: Hegedűsová Vantarová K, Škodova I (eds) Rastlinné spoločenstvá Slovenska, 5. Travinno-bylinna vegetacia, Vied, Bratislava, pp 268–271

Hegedüsová K, Korzeniak J, Májeková J, Stoica A, Coldea G, Kuzemko A, Budzhak V, Tokaryuk A, Chorney I, Skodová I (2020) Syntaxonomical revision of the *Triseto flavescentis-Polygonion bistortae* alliance in the Carpathians. Plant Biosyst 155(1):16–41

Heuffel J (1858) Enumeratio plantarum in Banatu Temesiensi Sponte crescentium et frequentius cultarum, Viennae

Hinterlang D (1992) Vegetationsökologie der Weichwasserquellgesellschaften zentraleuropaischer Mittllgebierge. Crunoeciots Soligen 1:5–117
Horvat I, Glavač V, Ellenberg H (1974) Vegetation Südosteuropas. Gustav Fischer Verlag, Stuttgart
Jarolimek J, Šibik J (eds) (2008) Diagnostic, constant and dominant species of the higher vegetation units of Slovakia. Veda Publishing House of the Slovak Acadamy of Sciences, Bratislava
Jávorka SA (1911) Retyezát flórának újabb érdekességei. Bot Közl X:27–32
Karner P (2007) *Pinion mugo* Pawł. 1928. In: Willner W, Grabherr G (eds) Die Wälder und Gebüsche Österreichs. Elsevier, München, pp 209–218
Kliment J (2007) *Nardetea strictae*. In: Kliment J, Valachovič M (eds) Rastlinné spoločenstvá Slovesnka, 4. Vysokohorska mokradi, Veda, Bratislava, pp 343–353
Kliment J, Jarolimek I, Šibik J (2007) *Mulgedio - Aconitetea*. In: Kliment J, Valachovič M (eds) Rastlinné spoločenstvá Slovesnka, 4. Vysokohorska mokradi, Veda, Bratislava, pp 23–129
Koči M (2007a) *Elyno-Seslerietea*. In: Chytrý M (ed) Vegetation of the Czech Republic, 1. Grassland and heathland vegetation. Academia, Praha, pp 84–88
Koči M (2007b) *Mulgedio-Aconitetea*. In: Chytrý M (ed) Vegetation of the Czech Republic, 1. Grassland and heathland vegetation. Academia, Praha, pp 91–131
Koči M, Chytrý M (2007) *Loiseleurio-Vaccinietea*. In: Chytrý M (ed) Vegetation of the Czech Republic, 1. Grassland and heathland vegetation. Academia, Praha, pp 65–69
Krajina V (1933) Die Pflanzengesellschaften des Mlynica-Tales in den vysoké Tatry (hohe Tatra), I-II. Beih Bot Cbl 50–51:774–957; 1–224
Malynovski K, Kricsfalusy V (2002) Plant communities of the Ukrainian Carpathien highland. Carpathian Foundation Uzhgorod
Matuszkiewicz W (2008) Przewodnik do oznaczania zbiorowisk roślinnych Polski. Wydawnictwo Naukowe PWN, Warszawa
Mihăilescu V (1963) Carpații Sud-Estici. Științifică, București
Mihăilescu S (2001) Flora şi vegetaţia Masivului Piatra Craiului. Virgiliu, Bucureşti
Morariu I (1967) Clasificarea vegetației nitrofile din România. Contrib Bot, Cluj 7:233–246
Mucina L (ed) (2016) Vegetation of Europe: hierarchical floristic classification system of vascular plant communities. Applied Vegetation Science (Suppl 1):3–223
Mucina L, Grabherr G, Wallnöfer S (1993) Die Pflanzengesellschaften Österreichs. Teil III. Wälder und Gebüsche. Gustav Fischer Verlag, Jena
Munteanu D, Mihăilescu S, Coldea G (2003) Parcurile naționale, naturale și rezervații ale biosferei din România. MAPAM, Bucureşti
Murgoci MG (1912) The geological synthesis of the south Carpathians. CR Congr Geol Intern XI, Stockholm
Nechita N (2003) Flora şi vegetaţia cormofitelor din masivul Hăşmaş, Cheile Bicazului şi Lacul Roşu. Constantin Matasă, Piatra-Neamţ
Nyárady EI (1958) Flora și vegetația munților Retezat. Acad. Române, București
Nyárady EI (ed) (1965) Flora R.P. Române, X. Acad. Române, București
Oberdorfer E (ed.) (1977) Süddeutsche Pflanzengesellschaften. Teil I., 2. Aufl. Jena
Oberdorfer E (ed.) (1978) Süddeutsche Pflanzengesellschaften. Teil II., 2. Aufl., Jena
Oberdorfer E (ed) (1983) Süddeutsche Pflanzengesellschaften. Teil III. Wirtschaftswiesen und Unkrantgesellschaften, Jena
Oberdorfer E (ed) (1992) Süddeutsche Pflanzengesellschaften. Teil IV. Wälder und Gebüsche. Gustav Fischer Verlag, Jena
Oberdorfer E (2001) Pflanzensoziologische Excursionsflora. 8. Auflage, Stuttgart
Oprea A, Sîrbu C (2009) Munţii Stânişoarei. Studiu fitosociologic. Univ. "Al. I. Cuza", Iaşi
Oroianu S (1998) Flora și vegetația defileului Mureșului între Toplița și Deda. Casa de Editură Mureș, Târgu Mureș
Pavelescu L (1958) Etude géologique et petrographique de la partie centrale et de SE de Monts Retezat. Ann Com Géol, București: 14–15
Pawlowski B (1939) Notulae floristicae ad Carpatos Austro-Orientales pertinentes. Bul Grăd Bot, Cluj XIX(1–2):1–20

Pawlowski B, Walas J (1949) Les associations des plantes vasculaires des Monts de Czywczyn. Bull Acad Polon Scien et Lett, Ser B, Sc Nat 1948(1):117–180

Pawlowski B, Solokowski M, Wallisch K (1928) Die Pflanzenassoziationen des Tatra-Gebirge. VII Teil. Die Pflanzenassozionen und die Flora des Morskie Oko-Tales. Bull Acad Scien et Lett, Ser B, Sc Nat 205–272

Pax F (1898, 1909) Grundzuge der Pflanzenverbreitung in der Karpathen. I-II, Leipzig

Pax F (1919) Pflanzengeographie von Rumänien. Nova Acta Leop 35

Pedrotti F (2004) Ricerche geobotaniche al "Laghestel di Piné" (1967-2001). Braun-Blanquetia, Camerino 35:1–55

Péterfi LŞ (1993) Flora şi vegetaţia algală a mlaştinilor, lacurilor glaciare şi a apelor curgătoare. In: Popovici I (ed) Parcul Naţional Retezat, Studii ecologice. West Side, Braşov, pp 78–93

Pişota I (1971) Lacurile glaciare din Carpaţii Meridionali. Academiei Române, Bucureşti

Plămadă E (1993) Bryophyta. In: Popovici I (ed) Parcul Naţional Retezat, Studii ecologice. West Side, Braşov, pp 49–57

Pop E (1960) Mlaştinile de turbă din R.P. Română. Acad. Române, Bucuresti

Pop I (1968) Conspectul asociatiilor ierboase de pe masivele calcaroase din cuprinsul Carpatilor româneşti. Contrib Bot, Cluj 8:267–275

Pop E, Lupşa V, Boşcaiu N, Soran V (1970) Analyse pollinique du sédiment de l'étang du Tăul Zănoguţii (Massif du Retezat, Roumanie). Fragm Fl Geobot 16(1):160–166

Pott R (1995) Die Pflanzengesellschaften Deutschlands. E. Ulmer Verlag, Stuttgart

Pott R (2000) Paleoclimate and vegetation-long-term vegetation dynamics in Central Europe with particular reference to beech. Phytocoenologia 30:285–333

Prodan I (1930) Centaureele României. Bul Acad Agron Mem I, Cluj

Prodan I (1931) Achileele României. Bul Acad Agron Mem II, Cluj

Puşcaru D, Puşcaru-Soroceanu E, Pauca A, Şerbănescu I, Beldie A, Ştefureac T, Cernescu N, Saghin F, Creţu V, Lupan L, Tascenco V (1956) Păşunile alpine din Munţii Bucegi. Acad. Române, Bucureşti

Puşcaru-Soroceanu E, Csűrös S, Puşcaru D, Popova-Cucu A (1981) Die Vegetation der Wisen und Weiden des Făgăraş-Gebirge in den Südkarpaten. Phytocoenologia 9(3):257–309

Puşcaş M, Gafta D, Cristea V (2005) L'analyse éco-coenotique des prairies édifiées par *Carex curvula* All. des Carpates roumaines. Acta Botanica Gallica 152:497–505

Resmeriţă I (1977) Cl. *Carpino-Fagetea* (Br.-Bl. et Vlieger 1927) Jakucs ex Passarge 1968 din Munţii Maramureş. Contrib Bot, Cluj-Napoca 17:91–110

Resmeriţă I (1987) Vegetaţia alpină de pe muntele Drăgşanu din munţii Retezat. Şt şi Cerc Biol Ser Biol Veget 39(1):34–39

Richard JL (1972) La végétation des Crêtes rocheuses du Jura. Ber Schweiz Bot Ges 82(1):68–112

Rochel A (1828) Plantae Banatus rariores iconibus et descriptionibus illustratae. Pestini

Rodwell JS, Schaminée JHJ, Mucina L, Pignatti S, Dring J, Moss D (2002) The diversity of European vegetation, Wageningen

Roussakova V (2000) Végétation alpine et sous alpine supérieure de la montagne de Rilla (Bulgarie). Braun-Blanquetia, Camerino 25:1–132

Sămărghiţan M (2005) Flora şi vegetaţia văii Gurghiului. University Press, Târgu Mureş

Sârbu A (ed) (2007) Arii speciale pentru protecţia şi conservarea plantelor în România. Victor B Victor, Bucureşti

Sârbu I, Ştefan N, Oprea A (2013) Plante vasculare din România. Victor B Victor, Bucureşti

Schneider E, Drăgulescu C (2005) Habitate şi situri de inters comunitar. Univ. Lucian Blaga, Sibiu

Schreiber W, Sorocovschi V (1993) Munţii Retezat. Condiţiile fitogeografice. In: Popovici I (ed) Parcul Naţional Retezat, Studii ecologice. West Side, Braşov, pp 8–12

Schur F (1866) Enumeratio plantarum Transsilvaniae. Vindobonae

Seibert P (1992) *Vaccinio-Piccetea*. In: Oberdorfer E (ed) Süddeutsche Pflanzengesellschaften, Teil IV, Wälder und Gebüsche. Gustav Fischer Verlag, Jena, pp 53–90

Šibik J, Petrik A, Valachovič M, Dubracová Z (2007a) *Carici rupestris-Kobresietea bellardii*. In: Kliment J, Valachovič M (eds) Rastlinné spoločenstvá Slovesnka, 4. Vysokohorska mokradi. Veda, Bratislava, pp 321–339

Šibik J, Kliment J, Jarolimek I, Dubracová Z (2007b) *Loiseleurion-Vaccinietea*. In: Kliment J, Valachovič M (eds) Rastlinné spoločenstvá Slovesnka, 4. Vysokohorska mokradi. Veda, Bratislava, pp 285–317

Simon T (1958) Uber die alpinen Pflanzengesellschaften des Pirin-Gebirges. Acta Bot Acad Sci Hung 4(1–2):159–189

Simon T (1962) Vergleichende Torfmoorstudien in den Karpaten. Act Bot Acad Sci Hung 8:187–203

Simon T (1966) Beiträge zur Kenntnis der Vegetation des Bihar (Bihor)-Gebirges. Ann Univ Scient Budapest, Sect Biol 8:253–273

Simonkiai L (1886) Enumeratio Florae Transsilvanicae vasculosae critica. Budapest

Soltés R, Hájek M, Valahovič M (2001) *Oxycocco-Sphagnetea*. In: Valachovič M (ed) Rastlinné spoločenstvá Slovesnka, 3, Vegetácia mokradi. Veda, Bratislava, pp 277–296

Steiner GM (1993) *Scheuzerio-Caricetea fuscae* und *Oxycocco-Sphagnetea*. In: Grabherr G, Mucina L (eds) Die Pflanzengesellschaften Österreichs, Teil II. Fischer Verlag, Jena, pp 131–181

Szafer W, Pawlowski B, Kulczynski S (1923) Die Pflanzenassozionen des Tatra-Gebirges I. Teil Die Pflanzenassozionen des Chocholawska-Tales. Bull Acad Polon Sci et Lett, Ber B, Sci Nat 1–66

Täuber F (1984) Beitrage zur zönologie der Zirbelfeifer (*Pinus cembra* L.) in Retezat National park. Recherches écologiques dans le Parc National de Retezat, Cluj-Napoca, pp 83–90

Täuber F (1987) Contribuţii la sintaxonomia făgetelor carpato-dacice (*Symphyto-Fagenalia* subordo novum). Contrib Bot, Cluj-Napoca 27:179–191

Theurillat JP, Aeschimann D, Küpfer P, Spichiger R (1995) The higher vegetation units of Alps. Coll Phytosoc 23:189–239

Uhliarova E, Janisova M, Ujhazy K, Skodova I, Hajek M (2014) *Arrhenatherion elatioris*. In: Hegedüsová Vantarová K, Škodova I (eds) Rastlinné spoločenstvá Slovenska, 5. Travinno-bylinna vegetacia. Veda, Bratislava, pp 202–239

Ujvari I (1972) Geografia apelor Române. Ştiinţifică, Bucureşti

Unar J, Unarová M, Šmarda J (1985) Vegetacni pomery Tomanovy doliny a zlebu spod Diery V Západnich Tatrách 1 Fytocenologické tabulky. Folia Fac Sci Nat Univ Purkynianae Brun, Ser Biol 25(10):5–101

Valachovič M (ed) (1995) Rastlinné spoločenstvá Slovenska 1. Pionierskaia Vegetacia. Veda, Bratislava

Valachovič M (ed) (2001) Rastlinné spoločenstvá Slovenska. 3. Vegetácia mokradí. Veda, Bratislava

Vida G (1963) Die zonalen Buchenwälder des Ostkarpatischen Florenbezirches (Transsilvanicum) auf Grund Untersuchungen in Parâng-Gebirge. Acta Acad Sci Hung 9(1–2):177–196

Voik W, Schneider-Binder E (1978) Cercetări asupra asociaţiilor de grohotişuri (*Thlaspietea rotundifolii* Br.-Bl. 1926) din etajul alpin al munţilor Făgăraş. St şi Comun Şt Nat, Sibiu 22: 189–202

Willner W, Grabherr G (eds) (2007) Die Wälder und Gebüsche Österreichs. Ein Bestimmungswerk mit Tabellen. Elsevier, München

Zahn HK (1928) Hieracia transsilvanica. Bul Grăd Bot Cluj 8:33–86

Zechmeister H (1993) *Montio-Cardaminetea*. In: Grabherr G, Mucina L (eds) Die Pflanzengesellschaften Österreichs, Tell II, Naturliche waldfreie Vegetation. Fischer Verlag, Jena, pp 213–240

Index

G. Coldea, V. Cristea, *The Vascular Plant Communities of the Retezat National Park (Southern Carpathians)*, Geobotany Studies,
https://doi.org/10.1007/978-3-031-05618-5

MIX
Papier aus verantwortungsvollen Quellen
Paper from responsible sources
FSC® C105338

If you have any concerns about our products,
you can contact us on
ProductSafety@springernature.com

In case Publisher is established outside the EU,
the EU authorized representative is:
Springer Nature Customer Service Center GmbH
Europaplatz 3, 69115 Heidelberg, Germany

Printed by Libri Plureos GmbH
in Hamburg, Germany